Essentials of Biological Security

Essentials of Biological Security

A Global Perspective

Edited by

Lijun Shang
Biological Security Research Centre
School of Human Sciences
London Metropolitan University
London, UK

Weiwen Zhang
Center for Biosafety Research and Strategy
Tianjin University
Tianjin, China

Malcolm Dando
School of Social Sciences
University of Bradford
Bradford, UK

This edition first published 2024

Registered Office(s)
John Wiley & Sons, Inc., 111 River Street, Hoboken, NJ 07030, USA
John Wiley & Sons Ltd, The Atrium, Southern Gate, Chichester, West Sussex, PO19 8SQ, UK

For details of our global editorial offices, customer services, and more information about Wiley products visit us at www.wiley.com.

Wiley also publishes its books in a variety of electronic formats and by print-on-demand. Some content that appears in standard print versions of this book may not be available in other formats.

Library of Congress Cataloging-in-Publication Data

Names: Shang, Lijun, editor. | Zhang, Weiwen (Professor of microbiology and biochemical engineering), editor. | Dando, Malcolm, editor.
Title: Essentials of biological security : a global perspective / edited by Lijun Shang, Weiwen Zhang, Malcolm Dando.
Description: Hoboken, NJ : Wiley, 2024. | Includes index.
Identifiers: LCCN 2023053440 (print) | LCCN 2023053441 (ebook) | ISBN 9781394189014 (cloth) | ISBN 9781394189021 (adobe pdf) | ISBN 9781394189038 (epub)
Subjects: LCSH: Biosecurity.
Classification: LCC JZ5865.B56 E88 2024 (print) | LCC JZ5865.B56 (ebook) | DDC 363.325/3–dc23/eng/20231201
LC record available at https://lccn.loc.gov/2023053440
LC ebook record available at https://lccn.loc.gov/2023053441

Cover Design: Wiley
Cover Image: © Image Source/Getty Images

Set in 9.5/12.5pt STIXTwoText by Straive, Pondicherry, India
Printed and bound by CPI Group (UK) Ltd, Croydon, CR0 4YY

C9781394189014_210224

Contents

Endorsement

Impressive dedication displayed by prominent experts, led by Lijun Shang, Weiwen Zhang and Malcolm Dando, has culminated in the creation of this remarkable book, a testament to their commitment. It delves into the pressing issue of biosecurity with a specific focus on preventing intentional disease outbreaks in humans, animals, and plants. It underscores the intricate connections between natural, accidental, and deliberate disease prevention, with a particular emphasis on preventing acts of malevolence.

The realm of biosecurity is extensive, extending far beyond the boundaries of life sciences and the Biological and Toxin Weapons Convention. This publication is in harmony with the WHO's Global Framework, placing significant emphasis on the collaborative role of various stakeholders in bolstering the management of biohazards and focusing on the crucial areas of biosecurity education and code of conduct.

The book comprehensively covers the threat posed by biosecurity concerns, the international responses to these challenges, the pivotal role played by scientists and the prospective future of biosecurity. Notably, it underscores the pressing need for the establishment of an International Biosecurity Education Network and the responsible engagement of scientists.

Prof. Aamer Ikram, HI(M)
Ex-CEO National Institute of Health Pakistan
Chair Board International Federation of Biosafety Associations
Chair Advisory Board TEPHINET
Member Executive Board IANPHI, Gavi; The Vaccine Alliance, ISTR

The global biosecurity situation has become increasingly severe in recent years due to emerging threats from natural epidemic outbreaks, the misuse and abuse of biotechnology and bioterrorism, among others. Biosecurity has become an important part of overall national security in many nations. Such challenges highlight the needs to reform the existing biosecurity education system. I am thus very delighted to see the book by a group of world-prominent experts led by Profs. Lijun Shang, Weiwen Zhang and Malcolm Dando specifically address the needs and create an updated biosecurity resource book that could easily be utilised by practicing life scientists in an educational programme. Under the framework of the Tianjin Biosecurity Guidelines of the UN Biological Weapon Convention, the book presents an excellent overview of the entire landscape of key fundamentals essential for biological security education for future life scientists. The broad topics covered by the book include most of the issues related to biosafety and biosecurity and therefore will attract experts and students, which will eventually help build suitable biosecurity education systems across the world.

Prof. Ying-Jin Yuan
Professor of Synthetic Biology
Academician of Chinese Academy of Sciences
Tianjin University, P.R. China

The new book edited by Prof. Lijun Shang, Dr. Weiwen Zhang and Em. Prof. Malcolm Dando on different aspects of biological security will drive the readers on a journey that goes from framing the problem by first understanding the past events in the biological weapons arena; to then putting in context the current threats and possible responses and, finally, thinking of future scenarios, particularly in the current fast-evolving scientific and technological world.

Following the logic presented in the WHO Global framework, when it comes to the importance of understanding the roles and needs of different actors, the book includes chapters from experts from different countries, representing universities, international organisations, NGOs and government and private companies, collecting, in this way, the opinions of the major biological security stakeholders and, consequently, building a broad and multidisciplinary perspective on biorisks management.

One of the outstanding conclusions is the importance of increasing education on biosecurity and biorisks management as a way to improve biological security in a sustainable way.

Dr. Maria Espona
ArgIQ codirector, professor at UCA, Catholic University Argentina

List of Figures

List of Tables

List of Contributors

Mayra Ameneiros
Centre for Science and Security Studies
King's College London
London, UK

Lela Bakanidze
EU CBRN Centers of Excellence Regional Secretariat for Central Asia
Tashkent, Uzbekistan

Brian Balmer
Department of Science and Technology Studies
University College London
London, UK

Yuhan Bao
iGEM Foundation
Cambridge, MA, USA
and
Tsinghua University, Beijing, China

Gemma Bowsher
School of Security Studies
King's College London
London, UK

Nancy Connell
Rutgers New Jersey Medical School
Newark, NJ, USA

Michael Crowley
Bradford University
Bradford, UK

Malcolm Dando
School of Social Sciences
University of Bradford
Bradford, UK

Brett Edwards
Department of Politics, Languages and International Studies
University of Bath
Bath, UK

Alonso Flores
iGEM Foundation, Cambridge
MA, USA

Gigi Gronvall
Center for Health Security
Johns Hopkins University Bloomberg School of Public Health
Baltimore, MD, USA

Jaroslav Krasny
Weapons of Mass Destruction Programme, UNIDIR
Geneva, Switzerland

Jez Littlewood
Independent Consultant, Edmonton
AB, Canada

Louison Mazeaud
Weapons of Mass Destruction Programme, UNIDIR
Geneva, Switzerland

Kathryn Millett
Biosecure Ltd
Cheltenham, UK

Kathryn Nixdorff
Department of Microbiology and Genetics
Technical University of Darmstadt
Darmstadt, Germany

Tatyana Novossiolova
Law Program-Center for the Study
of Democracy
Sofia, Bulgaria

Dana Perkins
Former member of the Group of Experts
supporting the United Nations Security
Council 1540 Committee on Weapons of
Mass Destruction Non-Proliferation
New York, NY, USA

James Revill
Weapons of Mass Destruction Programme
and Space Security Programmes, UNIDIR
Geneva, Switzerland

Lijun Shang
Biological Security Research Centre,
School of Human Sciences
London Metropolitan University
London, UK

Nariyoshi Shinomiya
National Defense Medical College
Tokorozawa, Japan

Xinyu Song
Center for Biosafety Research and Strategy
Tianjin University
Tianjin, China

Ralf Trapp
Independent Consultant
Chessenaz, France

Leifan Wang
Center for Biosafety Research and Strategy
Tianjin University Law School
Tianjin, China

Yang Xue
Center for Biosafety Research and Strategy
Tianjin University
Tianjin, China

Jean Pascal Zanders
Independent disarmament researcher at
The Trench

Weiwen Zhang
Center for Biosafety Research and Strategy
Tianjin University
Tianjin, China

Vivienne Zhang
Weapons of Mass Destruction Programme
and Space Security Programmes, UNIDIR
Geneva, Switzerland

Foreword

Studying biosecurity after the great COVID-19 pandemic has two different levels of importance. The first is a relative value, represented by the dynamics of the pandemic crisis itself. It has affected the entire world's society in a historically novel framework because it is marked by disruptive advances in the life sciences. This scientific aspect is apparently secondary to the growth of the world's population and the consequent mobility of large masses of people, which, together with the concentration of even larger masses of people, has outlined an unprecedented scenario for the viral spread and thus for the pandemic. Indeed, the growth of the world's population has also been made possible by advances in the life sciences and access to medicines for ever-larger sectors of the world's population. An advance in applied knowledge has allowed a technological shift in some biomedical research laboratories, transforming them from low-security sites into fully fledged critical infrastructures. The effects of a security/safety breach can now go far beyond the loss of materials and equipment and constitute a serious risk whose extent is still being defined. The very solutions offered to the pandemic crisis can be placed within this framework of advancement in the life sciences, in particular for mRNA vaccines, but also for virus detection. Possible technical solutions are intertwined with possible safety solutions in their application, shaping the options on which political decisions have been exercised.

The second value of biosecurity studies is an absolute one, which invests in institution-building and the work of the institutions themselves, starting with research institutions and moving on to international institutions via national ones. Working on biosafety means embracing a range of issues that affect the economy and society, as well as the policymaking that drives both. But it also means interacting with other applied research and security paradigms such as cybersecurity and in some cases radionuclear security. Therefore, biosecurity research must and can, as the chapters in this book testify, bring security applied to other technologies under a new approach. This leads us to reflect on the specifics of biosecurity in relation to the other securities that make up CBRN+Cy (Chemical, Biological, Radiological and Nuclear plus Cyber). If we think about the possibility of controlling and tracking the material, biosecurity is at the other end of the controllability scale compared to nuclear security. While nuclear material is undoubtedly the most controlled and controllable for mankind, the materials that biosecurity has to control are the most uncontrollable. This material fact has implications for security institutions and their activities.

Material difficulty, even before the secrecy of information, has informed the development and spread of nuclear technology. A material difficulty that was and is composed of various elements, such as the costs of nuclear fuel cycle management, nuclear material procurement and reprocessing equipment. The difficult technical and economic viability constituted a first line of 'passive' limitation to the uncontrolled spread of nuclear materials, effectively restricting the possibilities of the use of nuclear technology by another State other than the 'club' of nuclear powers. This point, which is politically fundamental, is useful in comparative terms with respect to the biological. For the first period in the history of nuclear weapons, the US monopoly subsumed control of nuclear materials, which continued after the Soviet test in a shared and normed control. Added to the impossibility for the majority of States in the international community to bear the costs of uranium enrichment was the model of international cooperation agreement that the US offered after the launch of the 'Atoms for Peace' programme. An agreement that conditioned the transfer of materials and technology on the US verifying their use by the receiving state. This model was gradually extended to other nuclear technology and materials providers through the technical role assigned to the IAEA.

On these assumptions, the safeguards system was built, which from the point of view of international instruments differed profoundly from an arms prohibition agreement. Indeed, it was not through prohibition that the superpowers exercised control over the spread of nuclear weapons, but through cooperation and the promotion of nuclear technology itself. Hypothetically, cooperation could provide the means for a nuclear weapons programme, but instead it implemented the exact opposite, subjecting the recipient state to controls that prevented the realisation of a weapons programme. This form of control implied the distinction between dual-use and non-dual-use technologies and materials and equally implied the creation of institutions to control dual-use. Why this choice of controlled openness instead of maintaining the previous closure and secrecy? By the end of the 1950s, the US leadership had realised that the secrecy of information and the cost and technical difficulties of enriching uranium and separating plutonium could not forever prevent many States from acquiring nuclear weapons. It was in some respects what Thomas P. Hughes called 'technological momentum'. With 'Atoms for Peace', a phase of building a technodiplomatic system was opened that led up to the Nuclear Non-Proliferation Treaty (NPT), which remains in force to this day. The creation of this system was made possible by a substantial convergence between the superpowers, a détente that benefited the entire international system.

On the other hand, along with the control of weapons programmes, the only other problem that affected the community of States when the NPT came into force was that of the safety of nuclear facilities, hence of the personnel working in them, and subsequently of the environment. Safety covered the possibilities of accidents or unintentional mishandling of nuclear materials and radiological sources. There was therefore no question of malicious use, hence security. Everything was attributable to State action, and therefore non-State actors such as criminals and terrorist organisations could only make malicious use of nuclear materials and radiological sources in James Bond films. But the continuation of the aforementioned technological dynamic led not only to a vast number of States being able to approach a nuclear weapons programme, but also to non-State actors appropriating nuclear materials and technology. The nuclear black market obviously saw other States as its first customers, but it was conducted by criminal organisations. This made it possible not only for North Korea to have a nuclear weapons programme but also for the risk of nuclear terrorism.

Thus, a distinction had to be made between safety and security, in languages where this distinction is possible. Obviously, the linguistic distinction followed a conceptual distinction, which developed a consequent institutional creation in the IAEA system. This distinction does not develop in parallel because security is an additional element of safety. If a plant is not safe, it will also inevitably have a security risk. If a plant is only safe, it cannot respond to a security risk. So, safety is a necessary condition for developing security. What has been said may logically seem a banality, but it is precisely on the common factors of safeguards, safety and security that the three 'S' approach has been promoted in Japan, which rationalises costs and optimises systems. An optimisation that, it must be said, is by no means the majority both at the level of States and in international organisations, for different reasons between the two plans (national and international). In fact, security belongs primarily to States, while safeguards and part of safety belong to international organisations, mainly the IAEA.

This summary of things that are already known may seem too long and distracting, but these are considerations arising from the coherent structure of the chapters in this book and from the intentions stated by the editors. Precisely because the biological is as far removed from the radionuclear, it is possible and necessary to organise the three 'S's differently. In the half century from the Geneva Protocol of 1925 to the Biological and Toxin Weapons Convention of 1975, it has not been possible to establish a control regime similar to that of nuclear safeguards. Following the Convention, however, it was possible to establish a system of confidence-building measures, which is similar to what the IAEA is promoting in nuclear security. Or, if one looks at safety, it is immediately apparent from reading the chapters of the book how much more intertwined it is with security than in the nuclear realm. There is a natural tendency, due to the very materiality of the biological, to move without hiatus from the interdiction of bacteriological and toxin weapons to biosafety and biosecurity, as the book makes clear.

Hence, biosecurity, rather than being an additional element of safety, conforms to a possible synthesis between safety and confidence-building measures, suited to the current international risk scenario. The value of the book, however, is not only contained in the quality of the individual chapters but also in the technopolitical proposal to create an international network for education in biosecurity, with the dramaturgical acronym IBSEN. Experience would be all in favour of such an attempt, promoted by the WHO just as the IAEA has promoted the international network for nuclear security education. With the significant difference of being able to develop and promote transnational education programmes in a booming technoscientific field. And in a field in which biosecurity must not be limited by a strict separation from safety, which is bureaucratic before being conceptual, as is the case with nuclear power. As Harvard's Belfer Center also reports, the life sciences are fundamental to driving scientific progress, benefiting everything from public health to agriculture and environmental preservation. Yet, it is vital to recognise and mitigate potential risks, especially in the realms of biosafety and biosecurity. Training and education are needed to do this because the critical element in safety/security in life sciences is the human element, much more so than in other CBRN. Anyone wishing to develop an education programme in biosecurity will find in the following pages not only insightful researches but also a relevant contribution to many courses of study.

Professor Matteo Gerlini
Chair of the International Nuclear Security Education Network (INSEN)
University of Siena, Department of Political and International Sciences
Siena
Italy

Acknowledgements

We would like to thank all of the authors for submitting their chapter outlines, drafts and final versions according to the very tight agreed timetable, and some of the authors for engaging with us in discussions of their chapters. The chapter authors, nevertheless, have sole responsibility for their own chapters. We would also like to thank the members of the Tianjin University Faculty who kindly assisted us in checking the format of all of the chapters and references.

The Editors

Acronyms

A

ABEO	Advisory Board for Education and Outreach (of OPCW/SAB)
ACE2	Human receptor type
Ag-RDTs	Antigen detecting rapid diagnostic tests
AI	Artificial intelligence
AM	Additive manufacturing
APHIS	Animal and Plant Health Inspection Service (USA)
APP3	Action Package Prevent-3, Biosafety and Biosecurity (of the GHSA)
ASAP	Artificial starch anabolic pathway
ASPR	Administration for Strategic Preparedness and Response (USA)

B

BACAC	Biosafety Association for Central Asia and Caucasus
BMBL	Biosafety in Microbiological and Biomedical Laboratories (Guidance)
BSAT	Biological select agents and toxins
BSE	Bovine spongiform encephalitis
BSL	Biosafety level
BTRP	Biological Threat Reduction Program
BTWC	Biological and Toxin Weapons Convention
BW	Biological weapon
BWC	Biological Weapons Convention (short form of BTWC)

C

CAR T	Cell therapy
CAT	Chloramphenicol acetyl transferase (Enzyme)
CB	Chemical and biological
CBP	Customs and Border Protection (USA)
CBRN	Chemical, biological, radiological and nuclear
CBRS	Center for Biosafety Research and Strategy (Tianjin University)

CBW	Chemical and biological weapons
CCD	Conference of the Committee on Disarmament
CD	Conference of Disarmament
CDC	Centers for Disease Control and Prevention (USA)
CIA	Central Intelligence Agency (USA)
CIDTP	Convention against Torture and Other Cruel, Inhuman or Degrading Treatment or Punishment
CND	Campaign for nuclear disarmament
CO_2	Carbon dioxide
CoE	Centres of Excellence (EU CBRN Risk Mitigation)
COVID-19	COVID-19 pandemic (virus)
Cpf1	CRISPR effector
CPT	Committee for the Prevention of Torture
CRISPR/Cas	Genome editing technique
CTBT	Comprehensive Test Ban Treaty
CTR	Cooperative threat reduction
CW	Chemical weapon
CWC	Chemical Weapons Convention
CWS	Chemical Warfare Service (USA)

D

3D	Three dimensional (protein folding)
DHB	District Health Board (Waikato, New Zealand)
DNA	Deoxyribonucleic acid
DOD	Department of Defense (USA)
DTRA	Defense Threat Reduction Agency (USA)
DURC	Dual Use Research of Concern

E

E&O	Education and outreach
EAR	Export Administration Regulations (USA)
EDP	Extremely dangerous pathogens
EEAS	European External Action Service
ELBI	Emerging Leaders in Biosecurity Fellowship
ENMOD	Convention on the Prohibition of Military or Any Other Hostile Use of Environmental Modification Techniques
ENDC	Eighteen Nation Disarmament Committee
ePPPs	Enhanced potential pandemic pathogens
EU	European Union
EU CBRN CoE	EU CBRN Risk Mitigation Centres of Excellence Initiative
EWARS	Early Warning, Alert, and Response System (of the WHO)

F

FAO	Food and Agriculture Organization
FBI	Federal Bureau of Investigation
FDA	Food and Drug Administration (USA)
FESAP	Federal Experts Security Advisory Panel (USA)
FIRES	Documentary Video Project (OPCW)
FSAP	Federal Select Agent Program (USA)

G

GASR7	Gene
GeBSA	Georgian Biosafety Association
GHSA	Global Health Security Agenda
GHSI	Global Health Security Initiative
GMO	Genetically Modified Organism
GOARN	Global Outbreak Alert and Response Network
GOF	Gain-of-Function (experiment)
GPC	General purpose criterion
gRNA	Guide ribonucleic acid

H

H1N1	Novel influenza virus
H5N1	Avian influenza virus
H7N5	Avian influenza virus
HIV	Human immunodeficiency virus
HHS	Department of Health and Human Services (USA)

I

IAEA	International Atomic Energy Agency
IAP	InterAcademy Panel
IAP	International Association of Prosecutors
IBSEN	International Biosecurity Security Education Network
ICCA	International Council of Chemical Associations
ICCPR	International Covenant on Civil and Political Rights
ICPO-INTERPOL	International Criminal Police Organization
ICRC	International Committee of the Red Cross
ICST	International Collaborations in Science and Technology
ICT	Information computer technology
ICTA	International Chemical Trade Association
IFBA	International Federation of Biosafety Associations

iGEM	International Genetic Engineering Machine Competition
IHL	International Humanitarian Law
IHR	International Health Regulations
IHRL	International Human Rights Law
IL-4	Interleukin-4
INB	International Network on Biotechnology
INSEN	International Nuclear Security Education Network
INTERPOL	International Criminal Police Organization
IO	International Organisation
IRB	Institutional Review Board
ISO	International Organisation for Standardisation
ISTC	International Science and Technology Center
ISU	Implementation Support Unit (of the BTWC)
ITA	International Traffic in Arms Regulations (USA)
INTERPOL	International Criminal Police Organisation
IUPAC	International Union of Pure and Applied Chemistry
IWG	International Working Group

J

JEE	Joint External Evaluation Tool (of the IHR)

L

LMO	Living modified organisms

M

MERS	Virus
MIT	Massachusetts Institute of Technology
ML	Machine learning
MMUST	Masinde Muliro University of Science and Technology (Kenya)
MoJ	Ministry of Justice (Georgia)
mRNA	Messenger ribonucleic acid
MSP	Meeting of States Parties (BTWC)
MX	Meeting of Experts (BTWC)
MYBPC3	Gene

N

NASEM	National Academies of Science, Engineering and Medicine (USA)
NCDC	National Center for Disease Control and Public Health (Georgia)
NGO	Non-government organisation
NPT	Nuclear Non-Proliferation Treaty

NRC	National Research Council (USA)
NSABB	National Science Advisory Board for Biosecurity (USA)
NSC	White House National Security Council (USA)
NUSEC	Nuclear Security Information Portal

O

OPCW	Organisation for the Prohibition of Chemical Weapons
OSH Act	Occupational Safety and Health Act (USA)
OSHA	Occupational Health and Safety Administration (USA)
OSTP	White House Office of Science and Technology Policy (USA)

P

P3CO	Potential Pandemic Pathogen Care and Oversight (USA)
PAM	Protospacer adjacent motif
PDCA	Plan–do–check–act (cycle)
POC	Point of care
PPE	Personal protective equipment

R

R&D	Research and development
RCAs	Riot control agents
RISE	Group of American teenage criminals

S

S&T	Science and technology
SAB	Scientific Advisory Board (of the OPCW)
SARS	Virus
SARS-CoV-2	Virus
SEB	Staphylococcal enterotoxin B
SGTEB	Sous Groupe de Travail et d'Etudes Biologiques
SIRUS	Superfluous injury or unnecessary suffering (prohibition)
STCU	Science and Technology Center in Ukraine

T

TALENs	Traditional genome editing tools
TCR	Genetically engineered T cells
TJU-CBRS	Tianjin University Center for Biosafety Research and Strategy
TPNW	Treaty on the Prohibition of Nuclear Weapons

U

UK	United Kingdom of Great Britain and Northern Ireland
UN	United Nations
UNBP	UN Basic Principles on the Use of Force and Firearms by Law Enforcement Officials
UNCoC	UN Code of Conduct for Law Enforcement Officials
UNEP	United Nations Environment Program
UNESCO	United Nations Educational, Scientific and Cultural Organization
UNICRI	United Nations Interregional Crime and Justice Research Institute
UNIDIR	United Nations Institute for Disarmament Research
UNODA	United Nations Office of Disarmament Affairs
UNSC	United Nations Security Council
UNSCR 1540	United Nations Security Council 1540
UNSGM	United Nations Secretary General's Mechanism
US$	United States dollars
US(A)	United States of America
USDA	US Department of Agriculture
USG	US Government
USSR	Union of Soviet Socialist Republics

V

VEREX	Verification Experts Group (of the BTWC)
VX	Nerve agent

W

WAHIS	World Animal Health Information System
WHO	World Health Organization
WILPF	Women's International League for Peace and Freedom
WINSI	Women in Nuclear Security Initiative
WMD	Weapons of Mass Destruction
WOAH	World Organisation for Animal Health
WWI	World War One
WWII	World War Two

Z

ZFNs	Traditional genome editing tools

1

Biological Security After the Pandemic

Lijun Shang[1], Weiwen Zhang[2], and Malcolm Dando[3]

[1] *Biological Security Research Centre, School of Human Sciences, London Metropolitan University, London, UK*
[2] *Center for Biosafety Research and Strategy, Tianjin University, Tianjin, China*
[3] *School of Social Sciences, University of Bradford, Bradford, UK*

Key Points

1) Biosecurity concerns the prevention of natural, accidental and deliberate disease in humans, animals and plants. All three aspects of preventing natural, accidental and deliberate disease are interrelated and improvements in each can support the others, but the focus here is on preventing deliberate disease.
2) While the focus is on the life sciences and the Biological and Toxin Weapons Convention (BTWC), we take a broad view of the threat and the regulatory regime, particularly to include mid-spectrum agents such as toxins and bioregulators and the Chemical Weapons Convention (CWC).
3) We follow the World Health Organisation's Global Framework in seeing multiple stakeholders having a range of tools and mechanisms that can be used to improve biorisk management. The focus here is on the need for biosecurity education, particularly for life and associated scientists in support of the Tianjin Guidelines.
4) The book is divided into 20 chapters in five sections: Introduction and Overview (1 chapter); The Threat (7 chapters); The International Response (4 chapters); The Role of Scientists (6 chapters); and The Future (2 chapters).
5) In the longer term, we see the need for an International Biosecurity Education Network (IBSEN) similar to the International Nuclear Security Education Network (INSEN) run by the International Atomic Energy Agency (IAEA).

Essentials of Biological Security: A Global Perspective, First Edition. Edited by Lijun Shang, Weiwen Zhang, and Malcolm Dando.

Summary

The chapter begins by stressing the importance of improving biosecurity after the pandemic and defines biosecurity as the prevention of natural, accidental and deliberate disease in humans, animals and plants. While stressing the all three aspects are interrelated and critical to each other, the chapter makes it clear that the focus of this book is on the prevention of deliberate disease. Moreover, while the focus is on the life sciences and the BTWC, a broad view is taken of the threat and the regulatory regime so as to include mid-spectrum agents such as toxins and bioregulators and particularly the CWC. It is argued that the WHO Global Framework is the best approach to biosecurity, with multiple stakeholders being seen as having a range of tools and mechanisms available to help improve biorisk management and that the essential component that is of concern here is biosecurity education in support of the Tianjin Guidelines. The sections and chapters of the book are then outlined before it is suggested that in the longer term an International Biosecurity Education Network (IBSEN) similar to The International Nuclear Security Education Network (INSEN) run by the IAEA will be needed to effectively improve biosecurity.

1.1 The Objective of the Book

In 2019, a large group of government and non-government experts were drawn together in an exercise to produce a list of critical questions for future biosecurity in the United Kingdom. The exercise had been organised by the Biosecurity Research Initiative at St. Catherine's College, University of Cambridge, and consisted of a three-phase process in which firstly a panel identified 59 experts from a range of disciplines, then secondly these experts were asked to draw on their own contacts to propose lists of tractable but unanswered biosecurity questions and finally a subset of 32 experts voted anonymously to select the top 10% of the 450 questions (which had been divided into 6 categories). And then 35 of the experts met at St. Catherine's to discuss, vote and rank the questions to provide a final list of 80 questions. The categories and the number of questions in each category are set out in Table 1.1.

It is clear from this exercise that ensuring future biological security was considered a complex task even before the pandemic struck in late 2019, and the pandemic has obviously made taking effective action to improve biological security for every nation much more urgent today.[1]

For this exercise, biological security was defined broadly to encompass the prevention of natural, accidental and deliberately caused diseases to humans, animals and plants. It suggested that:[2]

> ... Consistently emerging themes included: the nature of current and potential biological security threats, the efficacy of existing management actions, and the most appropriate future options ...

Table 1.1 Topics and numbers of biosecurity questions.

Topic
1) Bioengineering technologies Questions 1–6
2) Communication and behavioural change Questions 7–17
3) Disease threats Questions 18–35
4) Governance and policy Questions 36–52
5) Invasive alien (non-native) species Questions 53–67
6) Securing against misuse Questions 68–80

Source: Adapted from reference 2.

And the authors added that the resulting questions provided an ‘agenda for biological security’ in the future. They also noted that:

> Many emerging biosecurity dilemmas, such as the malicious use of synthetic biology, require new approaches to biosecurity, including the engagement of social scientists and policy-makers in forecasting ...

In this book, we take a similarly broad science and social science approach, as we concentrate on the specific aspect of the potential deliberate disease in future biological security and the role that scientists can play in preventing the hostile misuse of their benignly intended work, but the much wider context of biological security must also be kept in mind.

Our understanding of deliberate biological threats has evolved rapidly in recent years. In the final decades of the last century, the threat was widely considered to be from biological warfare conducted by States, but early in this century, the rising concern about terrorism expanded that concern to include non-State actors and sometimes indeed eclipsed the concerns about States. Then in the last two decades, concerns about experiments in the life and associated sciences that seemed to perhaps enable dangerous malicious activities by those with hostile intentions led to increasing concerns also about the direction and governance of the life and associated sciences. A concept that proved to be useful in considering how to deal with this expanded range of threats in a coherent manner started out as a ‘web of deterrence’ against State actions but has evolved into the idea of a ‘web of prevention’ linking a wide range of policies that together can help to minimise the possibility of the deliberate misuse of the life sciences.

As the authors of one study of this concept noted:[3]

> ... Biological threats are complex and multifaceted and hence, their effective prevention and countering require multiple lines of collaborative action and sustained cross-sectorial coordination ...

Therefore, they argued, actions required for biosafety and actions required for biosecurity, including the problem of the malign dual use of benignly intended work, should be integrated and seen as an:

> ... *integrated and comprehensive web of prevention* in which the efforts aimed at preventing the accidental release of biological agents or toxins, including naturally occurring disease and the efforts aimed to prevent the deliberate release of biological agents and toxins and the misuse of life sciences are complementary and reinforce each other ... (Emphasis added)

We follow this reasoning here with biosafety, biosecurity both within the laboratory and outside of the laboratory being seen as complementary to other means of minimising the possibility of deliberate misuse of the life sciences such as export controls and codes of conduct for scientists, and regulatory measures seen as covering both pathogens and toxins (defined to include natural bioregulatory chemical agents such as neurotransmitters if used in unusual amounts or by unusual means).

As has been pointed out by many authors, one of the major difficulties anticipated for the future of assuring biosecurity is the very rapid rate of the advances being made in the life and associated sciences. The precise topics of concern will obviously vary given the different backgrounds and interests of various groups of authors, but this theme emerges strongly in all attempts to assess future developments over the next decades. For example, the World Health Organisation (WHO) published a horizon scan in 2021 titled *Emerging technologies and dual-use concerns: a horizon scan for global public health*. This study was organised by the Science Division of the organisation and involved numerous outside experts. For the purpose of this study, dual-use research of concern (DURC) was defined as:[4]

> ... life science research that is intended for benefit but which might be misapplied to do harm ...

The study was organised to assess how such research that could have high impact might evolve over three time periods, up to 5 years, from 5 to 10 years and after 10 years. As in the UK study of biosecurity questions, the answers were derived from a systematic process involving topic identification, scoring, expert discussion and refinement and re-scoring of the refined list. Table 1.2 shows the topics that were identified in the five-year timescale.

Table 1.2 Priority DURC issues identified in the five-year timescale.

Bioregulators
Cloud laboratories
De novo synthesis of variola virus
Research on SARS-Cov-2
Synthetic genomics platforms for virus reconstruction

Source: Adapted from reference 4.

Now, most of these topics would not be of great surprise, few would dissent from the finding that:

> ... The expert group expected that there will be significant research into the determinants of the infectivity, severity and host specificity of SARS-CoV-2 within the next 5 years, as well as of immune evasion strategies ...

However, it has to be stressed that the expert group saw potential dual-use problems well beyond the frequent concentration on pathogens and genomics. In the five-year timeframe, for example, it suggested that:

> Bioregulators are biochemical compounds, such as peptides, that affect cellular processes. Research has identified a number of bioregulators and synthetic analogues that can modify life processes, including cognition, reproduction and development .... They can ... be misused, and ... have profound effects within minutes of exposure ...

It should also be noted that in the longer beyond 10-year period the 'Hostile Exploitation of Neurobiology' was identified as an additional field of dual-use concern.

In 2022, the WHO published its up-to-date definitive *Global guidance framework for the responsible use of the life sciences*. This document was again the result of the work of numerous experts from diverse parts of the work that had been organised by the WHO, and it clearly identified the multiple stakeholders (Table 1.3) who can play a critical role in ensuring future biological security, and the many tools and mechanisms that are available to those stakeholders.

Important to note here is the clear identification of scientists and their institutions as having a key role to play in ensuring future biological security and that codes of ethics are one of the tools and mechanisms that can be utilised. As the guidance notes:[5]

> ... Codes of ethics can be a useful tool to raise awareness of the need for biorisk management and provide norm setting standards .... There have ... been initiatives to outline high-level principles that can serve as references in developing or amending codes of conduct .... The most recent is the Tianjin Biosecurity Guidelines for Codes of Conduct ...

Table 1.3 Stakeholders identified by the WHO.

1) National governments
2) Scientists
3) Research institutions
4) Funding bodies
5) Publishers and editors
6) Standard-setting institutions
7) Educators
8) International organisations
9) Civil society networks and publics
10) The private sector

Source: Adapted from reference 5.

However, the guidance stresses a major problem in its rationale for the global guidance framework stating that:

> ... A chronic and fundamental challenge is a widespread lack of awareness that work in this area – which is predominantly undertaken to advance knowledge and tools to improve health, economies and societies – could be conducted or misused in ways that result in health and society risks to the public. Also, incentives to identify and mitigate such risks are lacking.

Our objective in producing this book is to provide a one-stop-shop, where any interested scientist or other stakeholder can quickly grasp the main issues involved in dealing with the problem of dual use and ensuring biological security more generally. We additionally hope that it can easily be used by educators to add material about biological security to their teaching of life and associated science courses at multiple levels.

The difficulties in adding biosecurity to the education and culture of the life and associated scientist should not be under estimated in view of the vast numbers of such scientists around the world, the disparate nature of the fields within which they work and the rate of the advances being made in many of these fields of research. A further factor that needs to be taken into account is that scientists will increasingly have to take part in discussion with governments about how dual-use dangers are to be regulated. An example of what may increasingly become matters of concern to scientists occurred in January 2023 when two Working Groups of the United States National Science Advisory Board for Biosecurity (NSABB) put forward new recommendations for the regulation of such experiments within the United States.[6] The proposals related first to 'Research with enhanced potential pandemic pathogens (ePPPs)' and second to 'Dual Use Research of Concern (DURC)'. The proposals were based on consideration of the efforts within the United States, particularly in the last decade, to minimise the dangers from particular types of experiments that could be of concern. The technical details of the recommendations need not be dealt with in this introduction and overview, except to note (Table 1.4) that should the proposals be accepted by government extra responsibilities will certainly fall upon practicing scientists.

Table 1.4 Some recommendations made by the NSABB.

Recommendation 3.1. Amend the US Government (USG) potential pandemic pathogen care and oversight (P3CO) framework to include and articulate specific roles, responsibilities and expectations for investigators and institutions in the identification, review and evaluation of research for potential involvement of ePPPs, taking into account existing review and oversight processes.
Recommendation 3.2. Local, institutional compliance procedures must be better harmonised, strengthened where needed and adequate technical and financial assistance provided.
Recommendation 8.2. Any updates to USG DURC policies, particularly updates regarding the scope of research subject to review and/or the relevant entities to which the policies apply, must involve relevant stakeholders and be accompanied by robust USG outreach and education and an adequate implementation period.

Source: Adapted from reference 6.

Table 1.5 The sections of the book.

1) Introduction and overview
Chapter 1
2) The threat
Chapters 2–8
3) The international response
Chapters 9–12
4) The role of scientists
Chapters 13–18
5) The future
Chapters 19–20

It also seems most unlikely that other governments will not, in the not-too-distant future, also be moving along similar lines of requiring much more involvement of practicing life and associated scientists in the regulation of their benignly intended work.

1.2 The Structure of the Book

In order to meet our objective of providing a comprehensive, but also easy to use, source of information on biological security after the pandemic, we have divided the subject into 19 short chapters that are grouped into four sections following this introduction and overview section (Table 1.5). References have been kept to a small number but chosen so that they can provide a quick route into the more detailed literature for those who need or are interested in following up the issues discussed (however, some chapters have larger numbers of references, as the authors considered the material, they were covering, would be less familiar to readers than most of the book). The main sections of the WHO global guidance framework document, for example, have 154 references, and there are additional references in the annexes to the report.

The next part of this chapter briefly introduces the chapters and themes of the book.

1.3 Overview of the Chapters

Section 2 on the threat begins with an extended account, in Chapter 2 by Jean-Pascal Zanders, of the way in which our understanding of poisons and infections has developed over the last 200 years as the sciences of first chemistry and then biology became established, and of how the international community has attempted to prevent the hostile use of these new sciences since the nineteenth century. Then in Chapter 3, Gemma Bowsher reviews how the many different actors with many different purposes have sought to use biological weapons in the past and also how the possible use of biological weapons is now being utilised in disinformation campaigns and infodemics. Brett Edwards describes the way in which the context and state of scientific knowledge influenced the potential to use biological weapons in antiquity through to 1946 (Chapter 4), and Brian Balmer reviews the offensive biological weapons programmes of States during the latter half of the twentieth century (Chapter 5). In Chapter 6, Kathryn Nixdorff investigates the developing concerns about dual-use raised by a series of experiments in the early years of this century, and Xinyu Song and Weiwen Zang describe some of the key cutting-edge technologies of concern today focusing particularly on synthetic biology and genome editing (Chapter 7). The section on the threat is rounded off by Chapter 8 in which Ralf Trapp discusses the convergence of other technologies, such as artificial intelligence (AI) and machine learning (ML) with biotechnology, that is causing increased concern about dual-use applications.

Section 3 of the book begins with an account of the origin and development to date of the idea of the web of prevention by Tatyana Novossiolova in Chapter 9, and this is followed by a review of the structure and functions of the 1925 Geneva Protocol and the BTWC by Jez Littlewood (Chapter 10) and other relevant international agreements such as the Chemical Weapons Convention (CWC) by Michael Crowley (Chapter 11). The section ends with a review of the role of relevant international organisations such as the WHO and the International Committee of the Red Cross by Louison Mazeaud, James Revill, Jaroslav Krasny and Vivienne Zhang (Chapter 12).

Section 4 of the book turns to the key role that life and associated scientists can play in improving biosecurity and begins with a review of the elements of biorisk management by Mayra Ameneiros (Chapter 13), and this is followed by a description of two national regulatory systems one in the United States and the other in Georgia by Dana Perkins and Lela Bakanidze (Chapter 14). The lessons that can be derived from our recent experiences of ePPP research and the COVID-19 pandemic are then investigated by Nariyoshi Shinomiya in Chapter 15. The Hague Ethical Guidelines and the Tianjin Biosecurity Guidelines are reviewed by Yang Xue (Chapter 16), and then the problem of engaging scientists in biorisk management is described by Yahan Bao and Alonso Flores in Chapter 17. Underpinning all the chapters lies the question of appropriate ethics and how this is to be applied, and this issue is discussed in the final Chapter 18 of the Section by Leifan Wang.

The book ends with two chapters in Section 5 which look towards the future. In Chapter 19, Nancy Connell and Gigi Gronwall examine the multi-layered system of different components that are becoming interwoven in the efforts to effectively prevent the misuse of the life and associated sciences, and finally, in Chapter 20, Kathryn Millett and Lijun Shang stress the need for an IBSEN to be established quickly to support the effective biosecurity education of life and associate scientists in support of the Tianjin Biosecurity Guidelines.

Given the current shortage in resource books for teaching biosecurity, we hope that the book makes a useful contribution in helping to assist lecturers and teachers in universities and colleges to add elements of this subject to their courses, and we would like to particularly thank members of the faculty at Tianjin University who helped with the task of finalising the editing of this book.

Author Biography

Dr. Lijun Shang, BSc, MSc, PhD, FPhysoc., London Metropolitan University (LMU), United Kingdom. Lijun Shang is a professor of Biomedical Sciences at School of Human Sciences in London Metropolitan University (LMU). He is the founding director of

Biological Security Research Centre at LMU. His research focuses mainly on ion channels in Health and Disease. Since 2015, he expanded his research interest into biochemical weapons and science convergence, as he wished to incorporate studies of the social impact of the advances in the life sciences. Since 2020, Professor Shang has been leading a series of projects in the effort to provide a civil society input into the broad Biological and Toxin Weapons Convention (BTWC).

Dr. Weiwen Zhang, Baiyang Chair Professor of Tianjin University; Director for Laboratory of Synthetic Microbiology, and Center for Biosafety Research and Strategy (CBRS) at Tianjin University of China. Dr. Zhang has broad research experience in microbial synthetic biology and has authored more than 250 peer-reviewed scientific papers. Dr. Zhang is currently chief scientist for the National Key R&D Research Program of China – Synthetic Biology program and chief investigator for the Key Strategic Project of the Chinese Association for Science and Technology on dual-use biotechnology governance. Dr. Zhang is also the founding director of Center for Biosafety Research and Strategy (CBRS) and has served on a number of scientific advisory boards on biosecurity, biosafety, food science and technology and so on for multiple ministries and agencies in China.

Dr. Malcolm Dando, BSc, PhD, DSc, Emeritus Professor, Peace Studies and International Development, University of Bradford, United Kingdom. Prof. Dando is a Fellow of the UK Royal Society of Biology. His research combines two themes: the advances in science that could be of concern in relation to the Chemical Weapons Convention (CWC) and the Biological and Toxin Weapons Convention (BTWC) and what might be done to mitigate these risks through awareness-raising and education of life and associated scientists. He has authored, co-authored and edited 20 books on these subjects, most recently *The Chemical and Biological Nonproliferation Regime after the COVID-19 Pandemic: Dealing with the Scientific Revolution in the Life Sciences*. Palgrave Springer/Nature, Switzerland (March 2023).

References

In each chapter of the book, a small number of key references that would be most useful for the reader to follow up are marked with a star*.

***1** Lane Warmbrod, K. *et al* (2023) *8. Biosecurity, Biosafety, and Dual Use: Will Humanity Minimise Potential Harms in the Age of Biotechnology?* Open Book Publishers. https://doi.org/10.11647/OBP.0336.08.

2 Kemp, L. *et al* (2021) 80 questions for UK biological security. *PLoS One*, 16 (1). e0241190, 2–3. https://doi.org/10.1371/journal.pone.0241190.

3 Novossiolova, T. *et al* (2021) The vital importance of a web of prevention for effective biosafety and biosecurity in the twenty-first century. *One Health Outlook*, 3, 17. https://doi.org/10.1186/s42522-021-00949-4.

***4** World Health Organization. (2021) *Emerging Technologies and Dual-Use Concerns: A Horizon Scan for Global Public Health*, World Health Organisation, Geneva, 1–6. ISBN: 978-92-4-003616-1. (electronic version).

***5** World Health Organization. (2022) *Global Guidance Framework for the Responsible Use of the Life Sciences: Mitigating Biorisks and Governing Dual-Use Research*, World Health Organization, Geneva, pp. 61–62 and 6. ISBN: 978-92-4-005610-7. (electronic version).

***6** National Science Advisory Board for Biosecurity. (2023) *Proposed Biosecurity Oversight for the Future of Science: Draft Recommendations of Two National Science Advisory Board Working Groups*, NSABB, Washington, DC, 3–4.

2

Falling Between the Cracks and by the Sides: Can Disarmament Treaties Respond to Scientific and Technological Developments?

Jean Pascal Zanders

Independent disarmament researcher at The Trench

Key Points

1) Science and technology (S&T) drive armaments and support disarmament.
2) S&T together with industrialisation in the nineteenth century laid the basis for chemical and biological weapons (CBW) but also contributed to the early efforts to constrain their use in war.
3) Diplomatic processes can be slow to address advances in S&T but in the context of the Biological and Chemical Weapons Conventions, assorted assistance and cooperation programmes mobilise national and international stakeholder communities to prevent the re-emergence of CBW.
4) Nevertheless, States parties to the Conventions must continue to evaluate S&T advances to ensure the long-term viability of the prohibitions and guide priorities for the assistance and cooperation programmes.

Summary

Science and technology (S&T) contribute to armaments and support disarmament. S&T advances assimilate quickly into military hardware and risk making international weapon control treaties obsolete fast. However, the connections between S&T, armament and disarmament are multifarious and intricate. In the area of chemical and biological weapons (CBW), the underlying sciences – chemistry, toxicology, biology, medicine, etc. – contributed greatly to societal and economic development during the nineteenth century. At the same time, they also prepared the ground for modern CB warfare. The rising sciences and new technologies also fed into the industrial revolutions that first took off in Western Europe and North America.

In 1899, the first Hague Peace Conference codified the laws of war and adopted a declaration banning the use of shells disseminating asphyxiating and deleterious

Essentials of Biological Security: A Global Perspective, First Edition. Edited by Lijun Shang, Weiwen Zhang, and Malcolm Dando.

gases. It was separate from the customary rule prohibiting poison use. The semantic bifurcation between 'poison' and 'asphyxiating and deleterious gases' reflected the impact of S&T and industrialisation on the armament dynamic, on the one hand, and the impact of the new toxic chemicals on public health, on the other hand. The declaration thus constrained the use of a weapon not yet in the military arsenals.

The Hague Declaration was the forerunner of the 1925 Geneva Protocol, 1972 Biological and Toxin Weapons Convention (BTWC) and the 1993 Chemical Weapons Convention (CWC). However, in the early 1920s, diplomats faced the problem that many of the war gases used in World War I also had widespread industrial and commercial applications. Once addressed, they solved the legal problems associated with the dual-use dilemma and thus opened the path to disarmament.

States party to the BTWC and CWC must address today's S&T challenges to maintain the treaties' relevancy. Both Conventions are of unlimited duration, which means that after completion of weapon destruction, work must continue to prevent the re-emergence of the proscribed weaponry. The five-yearly review conferences always have S&T review on the agenda but rising global geopolitical tensions have complicated the process. With the rise of new CBW threats, State Parties have set up diverse cooperation and assistance programmes, which in many ways help national and regional stakeholder communities to keep up with the latest S&T developments and prevent their misuse.

2.1 Introduction

Disarmament seeks to eliminate a discrete category of weaponry and once completed, prevent its re-emergence. Science and technology (S&T) drive innovation and sustain economic development, international cooperation and societal progress. S&T are crucial to successfully implementing disarmament policies, but they drive progress on a much broader front. Even without any conscious policy to violate or undermine disarmament, product and process innovation may challenge the long-term viability of any international disarmament agreement. Questions may arise about whether the treaty prohibition covers the latest advances or newest concepts, whether the reporting requirements and verification provisions still apply to the newest processes, and whether parties to a treaty can update its provisions and common understandings.

Meanwhile, S&T also propels weapon development and production. This is not a detached pursuit: any society-wide advance is screened for potential military exploitation. Other lines of research and development receive exclusive or significant grants from defence establishments or industries. S&T are also central to the defence, protection and mitigation of the effects of certain types of weaponry, which implies understanding their offensive properties. Exploiting the dual-use potential of many research and development activities continues to pose a significant challenge to preventing the re-emergence of internationally proscribed weaponry.

This chapter explores the multifarious and complex relationships between disarmament and S&T concerning chemical and biological weapons (CBW). It first describes the rise of chemistry and medicine, the attendant industrial revolutions through the nineteenth century, and how they transformed the social order so scientists and industrialists could advance their interests. By laying the foundations for organic chemistry and advocating the germ theory, they provoked semantic shifts for the 'poison' concept. By the end of the century, disease was no longer part of the concept, and 'asphyxiating and other deleterious gases' semantically bifurcated from the poisons from the mineral, vegetable and animal kingdoms.

The second part traces the codification of the customary ban on poison use in war and the emergence of asphyxiating and other deleterious gases as a novel weapon category. After World War I, the latter category proved beyond regulation because diplomats faced the dual-use dilemma. As described in the third part, the negotiations for the Geneva Protocol in 1925 set the stage for a breakthrough because diplomats approached the problem from a technical rather than a humanitarian viewpoint. It paved the way for CBW disarmament. Nevertheless, competing aims of weapon control – disarmament, arms control, non-proliferation, humanitarian arms limitation – would lead to different outcomes for different arms categories. Shifting security policies, fresh threat constructions and new S&T breakthroughs also affected existing weapon control treaties. The latter aspects are the subject of parts four and five. The final section draws conclusions about the relationships between CBW disarmament and S&T and reflects on particular approaches to manage and nurture the disarmament norm adequately.

2.2 Concepts of Disease and Toxicants in Relationship to CBW

The terms 'chemical weapon' (CW) or 'biological weapon' (BW) are relatively recent. They acquired general usage after World War II and reflected a long evolution of the concepts of poison and disease behind them. In 1817, Mathieu Orfila, the Spanish-born founder of toxicology, defined poison as 'any substance, which, taken inwardly, in a very small dose, or applied in any kind of manner to a living body, impairs health or destroys life'.[1] As captured in the first edition of the *Encyclopaedia Britannica* (1771), the mineral, animal and vegetable kingdoms each have their peculiar poisons.[2] Discussed under the lemma 'Medicine', mineral poisons comprised 'arsenicals and mercurials'. The other two categories represented toxins, i.e. poisons from living organisms. (Another group of toxins was unknown back then because science had not yet characterised the microbial life forms that excrete them.) Whilst perhaps unusual to the present-day mind, the generic notion of 'poison' also referred to illness. For instance, one conception of disease held that telluric emissions or other sources of corruption 'poisoned' the air, thus sickening exposed people and animals. Another theory posited that illness and toxic substances upset the balances (the humours) or the flow of liquids and solids, thereby bringing about 'dis-ease' in the patient. In contrast, 'poison' did not cover all diseases since contemporaries also attributed major plagues to divine wrath and associated certain afflictions with social status.

Prevailing customs of war typically forbade using poison or poisoned weapons on the battlefield. Contemporary legal treatises confirmed the generic understanding of 'poison'. But at the end of the eighteenth century, the semantic content of 'poison' began shifting under the influence of the rising sciences, technological innovation, the nascent industrial revolution and the social and cultural transformations they necessitated.

2.2.1 The Impact of Germ Theory on the 'Poison' Concept

Several early microbiologists in the seventeenth century posited that 'animalcules' caused infection by invading the body. Their arguments justified quarantines, promoted early hygienic measures and contributed to the acceptance of inoculation as a countermeasure. They also prompted colonialists to transfer smallpox-infected blankets to Native Americans in 1763. From the 1830s on, microbial research coalesced into a broader germ theory. The idea of invasive particles causing epidemics also stimulated diplomatic activity supporting sanitary control (e.g. harmonisation of national quarantine policies to avoid hampering the burgeoning international trade) and the drafting of treaties on international infectious control after 1850.[3] Yet, it still took more than half a century before science accepted the role of bacteria in disease as a fact.

Germ theory proponents met with much resistance. Most doctors stayed loyal to medical teachings formulated in Antiquity, whilst sanitarians continued practices handed down through generations. Only by introducing new hygienic routines, such as washing hands and cleaning equipment, did the promoters of the new theory make inroads. Their standing rose in the communities they served by demonstrating lower mortality rates in sanatoria.

The Catholic and Protestant churches proved far more formidable obstacles. Through their religious teachings and prayers to ward off calamities and offer comfort whenever they befell a community, the clergy could maintain the original medical teachings Christianity had embraced since its emergence and spread amongst the population. In the early 1800s, European societies were generally structured according to the feudal system of three separate estates: the clergy, nobility and the commoners. Advancing the germ theory required physicians, individually and as a professional class, to gain prestige and social influence relative to the first estate. As noted, their social standing rose with dropping infection and mortality rates. The increasing social and economic transformations of the first industrial revolution and the redistribution of wealth in favour of entrepreneurial commoners diminished the church's hold on society. The development opened the path towards forward-looking laws, including in the public health domain. Like other emerging sectors seeking to influence decision-making by claiming specialist knowledge, medical doctors advanced their cause through professional associations.

Germ theory had to challenge 'spontaneous generation', a concept consistent with the Bible's divine creation narrative. To validate the new proposition, exponents had to apply greater rigour to their experimental research, including understanding and avoiding contamination in the laboratory. Ultimately, they had to demonstrate that an agent causes infectious disease, can be transmitted from one living organism to another, and that one agent is responsible for one disease only. These requirements eventually led to Robert Koch's postulates in 1890.[4] Researchers thus established causal relationships between multiple microbial agents and specific diseases, confirming the germ theory's foundation. From this point on, for biological warfare, there was an agent to manipulate.

Meanwhile, physicians and surgeons accepting the new explanation of disease causation also came to appreciate that certain bacteria released poisonous substances. The word 'toxin', coined in 1888, replaced earlier denominations such as septic or putrid poisons.[5] Together with evolution theory and new geological insights into the earth's history, the germ theory's success over creationist science helped to solidify social transformations in which S&T were to play central roles.

2.2.2 The Impact of Chemistry on the 'Poison' Concept

The rise of chemistry as a science in the late 1700s caused 'poison' to undergo a different semantic shift. Germ theory established a new understanding of contagion that allowed 'disease' to extract itself from under the 'poison' umbrella whilst the identification of bacterial toxins preserved an association between both. In contrast, advances in chemistry initiated semantic specialisation under the 'poison' header, the most consequential being the early conceptual differentiation between naturally occurring poisons (as captured in the *Encyclopaedia Britannica*) and human-made toxicants. However, the divergence proceeded gradually and not always in a clear direction.

The poisons from the mineral kingdom corresponded with inorganic chemicals. Orfila united the animal and vegetable poisons in the 'organic kingdom', yet organic chemistry still had to emerge as a subdiscipline.[6] Up to that point, 'vitalism' covered organic chemicals. The theory posited that anything alive or derived from it required a vital force.[7] In 1828, the German chemist Friedrich Wöhler synthesised urea from an inorganic compound, disproving vitalism. As a toxicologist, Orfila bypassed the contention by grouping poisonous substances according to their action on humans and animals. The six classes were: (i) corrosive or escharotic poisons, (ii) astringent poisons, (iii) acrid poisons, (iv) stupefying and narcotic poisons, (v) narcotico-acrid poisons and (vi) septic or putrefying poisons. The latter four classes comprised the so-called organic kingdom; the last captured the then-unknown bacterial toxins.

In another book published in 1826, Orfila explained the consequences and treatment of exposure to organic chemicals for a broader audience. He also addressed 'asphyxies', which he understood to be a suspension of the pulse and respiration, or 'suspended animation'. The types of asphyxiation ranged from toxic chemical exposure to strangulation, drowning and suffocation suffered by newborn children.[8] At the same time, both books by Orfila also reveal that industrial poisoning was not yet a source of societal concern. Apart from some references to chemical asphyxia from gas in coal mines and lime kilns, the sections on suffocation did not mention manufacturing processes.

Still, industrialising societies became exposed to new toxicants through factory emissions, waste and environmental degradation and their use in household products and pharmaceutical preparations. Increasingly, investigations began linking toxic chemicals to more ailments and premature deaths. Governments and politicians sought to draft and implement health regulations for workers and the public. However, the products, in increasingly common use, were economically and socially challenging to control. Chemist and druggist enterprises set up associations to protect their profession from 1813. Apothecaries in Britain united in the Pharmaceutical Society in 1841 to defend their economic interests. Both sectors sought and acquired higher political influence based on exclusive scientific knowledge. They also competed with each other as their interests did not always overlap, often resulting in muddled messages for politicians.

Halfway through the century, the context for considering poisonous substances was changing. Many industrialising countries recorded their growing role in murders, suicides and accidents. Forensic and medical jurisprudence required enhanced capabilities to distinguish between deliberate or accidental toxicant exposure. Moreover, the health conditions of labourers exposed to toxic chemicals also drew the attention of toxicologists. Poisonous gases thus became a separate concern, with medical scientists now differentiating between chemical asphyxiation and mechanical forms of injury caused by deleterious fumes.[9]

By the end of the nineteenth century, toxicologists had described industrial poisoning in great details. They identified multiple families of harmful manufactured compounds and outlined treatments. Industrial poisoning, however, was treated similarly to industrial accidents. Resistance to prevention remained strong until the final quarter of the century when the nascent socialist parties set about framing national and international regulations to protect workers. These regulations detailed the families of toxic compounds produced or consumed by the industry, the nature of manufacturing processes that released noxious fumes and other possible circumstances under which workers might get exposed to toxicants. Several West European countries foresaw special protection for children, young persons and (pregnant) women.[10,11]

2.3 Capturing Evolving Concepts of Disease and Toxicants in Restraining Warfare

When the world's foremost powers convened in The Hague for the first Peace Conference (18 May–29 July 1899), armament limitations were formally on the agenda. For many decades, diplomats and lawyers had expressed concern about the rising levels of armaments and associated demands on national budgets to equip expanding standing armies with the latest military equipment. Technology and industrial development pushed the armaments drive, which reached unprecedented levels in the century's final decade. Agrarian societies could not keep up with the pace set by industrial powers. Those considerations led Tsar Nicholas II to propose the international conference. The circular of December 1898 also included the codification of the customs of war as an agenda item.[12]

The Peace Conference failed in its main arms limitation goals. The 26 participating States reached no agreement on qualitative or quantitative limits on national armies or armaments but issued three declarations touching upon novel weapon developments and their use in combat: (IV, 1) Declaration relative to the prohibition of the discharge of projectiles from balloons; (IV, 2) Declaration relative to the prohibition of the employment of asphyxiating projectiles; and (IV, 3) Declaration relative to the prohibition of balls which expand in the human body. Under Declaration (IV, 2), the Contracting Powers agreed to abstain from using projectiles, the sole object of which is the diffusion of asphyxiating or deleterious gases. Even so, they codified the customs of war for the first time. Article 23 of the Convention (II) with respect to the Laws and Customs of War on Land, listed poison and poisoned weapons as one of the outlawed means of combat.

Delegates thus considered 'poison' as part of long-standing customs of war *and* as a novel type of weapon. More significantly, the same diplomats adopted the article banning poison use without any discussion, whereas they discussed the draft declaration extensively. The United States ultimately refused its signature, as did Great Britain because the document lacked consensus.

From a toxicological viewpoint, poisons and asphyxiating or deleterious gases both harm living organisms through their direct toxic action. However, the negotiators held different conceptions of poisons and asphyxiating or deleterious gases: the former were naturally occurring substances; the latter the product of scientific and technological advancement. The US naval delegate motivated his opposition on the grounds of the weapon's new technological qualities and the possibility that because of those qualities, it may prove decisive in a future war whilst not causing superfluous suffering. The German delegate also contemplated the potential contribution of toxic chemicals to the humanisation of warfare but did not block the Declaration.[13]

As described in the previous section, the semantic bifurcation of 'poison' originated in the evolvement of S&T and the rising hazards and regulation of exposure to industrial toxicants. But in the early nineteenth century, chemists were already unwittingly laying the foundations for modern chemical warfare. Three and a half decades after the isolation of chlorine in 1774, it was determined to be an element. The first synthesis of phosgene took place in 1812. The first preparation of mustard agent followed barely a decade later. None of the chemists involved looked at those chemicals as possible instruments of warfare.

In contrast, some individual military thinkers began contemplating toxic chemicals to overcome evolving defensive technologies. In particular, engineering advances in fortification design and construction significantly reduced the effectiveness of artillery in the first half of the nineteenth century. During the American Civil War, the North experimented with toxic clouds and munitions to dislodge confederate troops from trenches. Most ideas surfaced in Great Britain and the United States, but evidence of similar interest in other industrialising countries would not surprise.

Intriguingly, officials did not reject the proposals outright on the grounds of the customs of war. Instead, they submitted them to committees for consideration. It signified the scientification of policymaking. Moral or legal considerations entered the deliberations only when science or technology offered marginal benefits at best or longer-term advantages were in doubt. This trend was the clearest in Great Britain. Government bureaucracies forwarded successive submissions of ideas to evaluation committees. Ultimately the ideas failed to gain any traction because the science or technology was still immature or because of age-old bureaucratic prejudices. A case in point was the rejection in 1914 of a proposal to use noxious fumes in land warfare because the idea came from a naval officer.[13]

A final element contributing to the disagreements about Declaration (IV, 2) was the abandonment of the original disarmament objectives of the first Peace Conference. Rather than banning the weapon, the Declaration proscribed its use based on humanitarian considerations. The shift opened the door for the argument that nobody had yet used the type of weapon and might still prove decisive on battlefields. Therefore, it was impossible to judge whether it was less humane than other types of arms relative to military objectives.

2.4 Further Development of the Control of Toxic Weapons

Under customary law, the constraint on using toxic substances and, to a certain extent, diseases continued under a generic notion of 'poison'. Scientific and technological developments caused semantic shifts that extracted disease from the concept and introduced more specific terms, such as 'asphyxiating' and 'deleterious', based on how toxic substances harm organisms. The latter change led in the context of the customs of war to an association of 'poison' with primitiveness and 'asphyxiating and deleterious' with civilisational progress. The S&T products thus gradually drifted away from the established ban on poison use. However, placing them under a semantically updated prohibitory rule became controversial because of evolving technology-driven understandings of superfluous suffering relative to military objectives in war.

Declaration (IV, 2) thus emerged when fast-evolving relationships between military art, S&T and industrial capacity were likely to challenge the relevancy or interpretation of new rules on the battlefield, which is what happened 15 years later. World War I also exposed another flaw: the Declaration's scope was too narrow as it applied to projectiles with no other purpose than diffusing asphyxiating or deleterious gases rather than the toxic substances themselves. Poisonous clouds released from cylinders the document did not explicitly cover.

2.4.1 Confirming the Semantic Bifurcation

After the Armistice, international law banning poison weapons developed along the two tracks that emerged from the First Hague Peace Conference (Figure 2.1). At the Second Hague Peace Conference in 1907, the Regulations concerning the Laws and Customs of War on Land annexed to Convention (IV) respecting the Laws and Customs of War on Land repeated the ban on poison use. The International Military Tribunal of Nürnberg declared in 1946 that the 1907 Hague Convention, and therefore the prohibition on poison use, as universal, irrespective of whether a State was a party to them. The 1998 Rome Statute of the International Criminal Court confirmed a violation of the poison ban as a war crime in international and non-international armed conflicts.

The other track laid the foundations for today's disarmament treaties, the BTWC and the CWC. Declaration (IV, 2) did not survive World War I. Yet, the agreement had subtle influences during and after the armed conflict. Firstly, it consolidated the conceptual separation of asphyxiating and deleterious gases from poison. A mere two days after the first chlorine attack in April 1915, an Allied commission investigating war crimes already listed four German means of delivery: toxic fires, hand grenades, cylinders and shells. After the Armistice, the Commission on the Responsibility of the Authors of the War and on Enforcement of Penalties charged the Central Powers with 32 specific violations, including 'Use of deleterious and asphyxiating gases' and 'Poisoning of wells'. Second, the 1919 Versailles Treaty subjected Germany to coercive disarmament.[14] Article 171 captured the widening scope of a toxic weapon (emphasis added):

> The use of *asphyxiating, poisonous and other gases and all analogous liquids, materials or devices* being prohibited, their manufacture and importation are strictly forbidden in Germany.
>
> The same applies to *materials specially intended for the manufacture, storage and use* of the said products or devices.

New was the linkage of the warfare agents to delivery means and ancillary equipment or facilities *and* the industry and the arms trade. The prohibition went beyond the traditional restriction on battlefield use. It recognised that a gas weapon implied more than just the agent. It encompassed development and production capacities and the international chemical trade.

The 1925 Geneva Protocol drew on Declaration (IV, 2) and the Versailles Treaty. It banned 'the use in war of asphyxiating, poisonous or other gases, and of all analogous liquids, materials

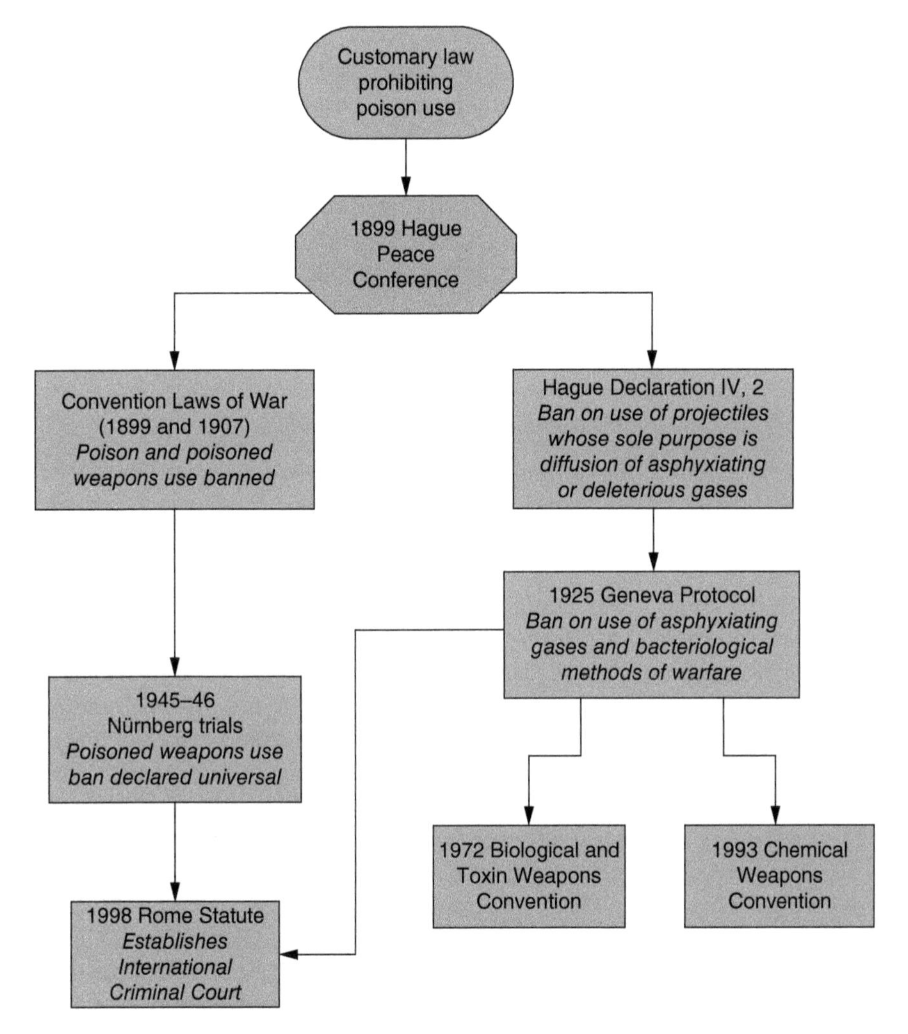

Figure 2.1 Development of international law banning poison weapons.

or devices'. From the former document, it took the prohibition as part of the laws of war. From the latter, it utilised the technical characterisation of the gas weapon to make the scope of application as comprehensive as possible. The mention of 'poisonous gases' links back to the 1899 and 1907 Conventions on the laws of war but without mentioning either explicitly. Another dimension of how the Geneva Protocol sought to capture scientific developments was the incorporation of 'bacteriological methods of warfare'. The phrase recognised the germ theory as a scientific fact and alluded to German wartime attempts at sabotage with infectious agents against animals. The 1918–1921 influenza pandemic may have also played a role.

Delegates called the agreement a 'protocol' because they viewed the prohibition on CW use as a necessary precondition for disarmament, i.e. the elimination of the weapon itself. They were already preparing for the disarmament conference under the auspices of the League of Nations, which was to take place in the 1930s. That conference eventually failed due to worsening geopolitical relations in Europe and Asia. Ultimately the Geneva Protocol fulfilled that role with the successful negotiation of the BTWC and the CWC many decades later. Both disarmament treaties define the weapon categories they govern for which, interestingly, they took up the three-part structure comprising the warfare agent, delivery means and ancillary equipment first seen in the Versailles Treaty. The BTWC reflected the discovery of other types of pathogens (including viruses) by adding 'biological' to the prohibition whilst keeping 'bacteriological' between parentheses as a link to the Geneva Protocol. Finally, to criminalise BW or CW use in armed conflict, the Rome Statute uses the phraseology of the Geneva Protocol rather than the wording in the BTWC and CWC (because some States feared that drawing on disarmament treaties might lead to the abolishment of nuclear weapons).

2.4.2 Of Humanitarian Foundations and a Dual-Use Quandary

Despite the apparent smooth evolution of the CBW prohibitory regime from the laws of war to disarmament, industrialisation and S&T posed unexpected challenges. The most important was the dual-use problem, an unrecognised impediment in the early 1920s. Diplomats confronted the issue for the first time during the Washington Naval Conference (12 November 1921–6 February 1922). The meeting's outcome was meagre because participants could not agree on curbing national armaments, and new geopolitical competition was already rearing its ugly head. It also had a forum on Rules for Control of New Agencies of Warfare comprising three subcommittees on CW, aircraft and the rules of international law. Just like The Hague Peace Conference almost a quarter of a century earlier, failing to make headway on limiting emerging military technologies, the conference shifted to humanitarian considerations.

Noting the increasing peacetime use of warfare gases, the subcommittee on chemical warfare noted (i) that research that may discover new warfare gases cannot be prohibited, restricted or supervised; (ii) the impossibility of restricting the manufacture of any particular gas and the impracticality of limiting the quantities of certain gases being manufactured; and (iii) that the kinds of gases and their effects on humans cannot be taken as a basis for limitation. It concluded that the only practical restriction is prohibiting use against cities or other large bodies of non-combatants like with high explosives. There could be no limitation on the use of toxic gases against enemy armed forces ashore or afloat. The subcommittee's recommendations got scuppered because, amongst other reasons, the five main participants – the US, Great Britain, France, Italy and Japan – were party to the Versailles Treaty, including its Article 171 (even if it did not strictly apply to them), or one of the peace treaties with Germany's allies. The Washington Naval Conference eventually adopted a text that became the direct precursor to the Geneva Protocol.[15,16]

The Washington Naval Conference was the first time that diplomats confronted the dual-use characteristics of a technology they wanted to control. They realised the quasi-impossibility of regulating research or production of toxic chemicals with industrial or commercial utility without a specific ban on their battlefield use. Their conclusion was to fall back on a fundamental principle of the laws of war, namely the protection of non-combatants.

Three years later, the League of Nations convened the Conference for the Supervision of the International Trade in Arms and Ammunition and in Implements of War in Geneva

(4 May–17 June 1925). At the outset, the US representative also proposed prohibiting the export of toxic chemical agents intended for use in warfare. Other delegates welcomed the suggestion but noted the dual-use nature of such compounds. They thought that the characteristics of the gases that could not be utilised in war and those with warlike and non-warlike purposes needed to be defined. Conventional weapons, in contrast, did not require a diplomatic assessment of the nature of the technologies. Despite work in committees and the involvement of chemical experts, participants could not resolve within the time frame at their disposal the core issues they had uncovered. They also realised that they could not regulate a specific aspect of chemical warfare preparations without an overarching prohibition on CW use. Hence, their formulation of the multilateral ban on chemical warfare.

The substance of the debates that led to adopting the Geneva Protocol differed entirely from those in The Hague or Washington. The US proposal to outlaw the trade in CW shifted the focus from humanitarian considerations to weapon technology and everything such technology implied from security, scientific, industrial and commercial perspectives. The consciousness of those various fields' interconnectedness not only gave diplomats and experts focus, namely the total elimination of the weapon category, but also helped them to frame the challenges with a precision that had eluded delegates in earlier deliberations.

From 1926, the preparatory commission for the disarmament conference addressed the previous year's technical questions and solved them by the start of the 1930s. Amongst other things, it defined the weapons based on their impact on living organisms rather than their chemical properties or degree of lethality. In this way, it avoided the problem of lists whose updating would have always posed technical and political challenges. To solve the dual-use matter, it proposed to base the prohibition on the goals to which the toxic chemicals may be applied rather than the compounds as such. This approach became known as the 'general purpose criterion' (GPC).[17] The definitions of a BW in the BTWC and CW in the CWC are founded on the solutions the experts came up with in the late 1920s.

2.5 Implications of Evolving Concepts and S&T Developments for Disarmament Law

The historical review of the emergence of the BTWC and CWC shows a clear relationship between developments in S&T and industry, on the one hand, and the design of legal structures and normative frameworks to counter security threats emanating from such developments, on the other hand. The link may appear straightforward, but it is entangled in multiple contexts through which participants and stakeholders interpret the challenges relative to their interests. It creates a dynamic environment that complicates the construction of new legal frameworks or the adaptation of existing ones.

Any international legal framework requires unanimity. Proponents will strive for consensus at the highest possible level. Other stakeholders may try to derail the process, aim for vague obligations or decide not to participate in the treaty. Such processes take place not just on the treaty level but also affect the consideration of every treaty component. Ultimately, a successful conclusion of negotiations reflects the broadest possible consensus amongst participants at a given time. Meanwhile, the world moves on.

2.5.1 Institutional Interests

A pristine moment for designing weapon control has never existed. The historical narrative indicates that scientific and technological progress preceded consideration and regulation of potentially new weapon threats. In itself, this is unsurprising. Formal arms control and disarmament are commonly perceived to trail scientific and technological innovation.

More important is how innovation affects society and social organisation and, consequently, how people frame and therefore label their understanding of the world. By the 1850s, the agents driving scientific and technological change organised themselves to effect the desired social changes. They sought to promote and protect their interests from undesired control and regulation. Those interests were much broader than the issues arising from a weapon control treaty but still influenced the approaches to designing, implementing and evaluating weapon control treaties.

Similar processes have continued until today. The decisions by chemical industry associations to support the CWC in the late 1980s contributed to the successful conclusion of the negotiations in 1992. It also allowed them to contribute to the design of the verification machinery to optimise the balance between compliance assurance and impact on economic processes. Today they advise and educate their members on treaty compliance and interact with the Organisation for the Prohibition of Chemical Weapons (OPCW) on the latest developments in products and processes and optimising verification. The biotechnology sector is far more decentralised than its chemical counterpart. It remained mostly absent during the negotiations on a legally binding protocol to supplement the BTWC between 1997 and 2001. In some instances, national industry associations strongly opposed any form of international verification, which then informed government positions. The negotiations failed to a certain extent due to the lack of stakeholder support.

Vested institutional interests play out on a different level too. International treaties or amendments to such treaties require parliamentary consent in many countries. The ratification process offers many opportunities to endorse or block treaty participation. For example, after World War I, the Chemical Warfare Service (CWS) mobilised the US chemical industry in support of its campaign for institutional survival. It lobbied hard for law enforcement agencies to adopt lachrymatory agents as riot control agents (RCAs). To this end, it sought and succeeded in blocking US ratification of the Geneva Protocol. Almost a century later, the impact of the CWS's actions continues today.[18] Many US officials and citizens still draw a sharp distinction between lethal and non-lethal CW, allowing them to argue that the United States' use of RCAs and herbicides in Vietnam during the 1960s and early 1970s was not chemical warfare. Industrial lobbying for so-called non-lethal or less-than-lethal weapon technologies and dispersal devices has significantly increased. It contributes to the difficulties in regulating the group of central nervous system-acting chemicals (the so-called incapacitating agents) under the CWC.[19] Assessment of the domestic public and stakeholder support levels influences a delegation's positions on negotiation items and the compromises it is willing to accept.

2.5.2 Semantic Shifts as Indicators of Scientific and Technological Advancements

Vocabulary tends to reflect the rapid changes in society. Multiple words get coined to delineate specific phenomena. To a large degree, they will be synonymous, yet they may also display significant semantic differentiation. Certain groups with a political interest in maintaining a sharp characterisation of their field of activities will eventually challenge particular emerging referent meanings. In contrast, some terms, whilst entirely synonymous, will claim an exclusive space because groups wish to isolate their ventures from broader debates to avoid political interference or, on the contrary, inject their specialist knowledge and expertise into political debates when seeking a more conducive environment for their activities.

The historical overview reveals that the semantic trajectories of pathogens and asphyxiating and deleterious gases relative to the general 'poison' concept in the early 1800s shaped the later framing of the CBW prohibitions. They reflected the new or more specialised understandings arising from the advances in medicine and chemistry, their increasing impact on society and the responses to societal challenges. From this perspective, Declaration (IV, 2) and subsequent international agreements derived from national and international regulations to protect workers and the public from regular or long-term exposure to industrial toxicants rather than the older restrictions on poison use in war.

Understanding of concepts evolves continuously, and new terminology reflects the changes and how societies absorb them. Industrialisation prompted national and international regulations in medicine and toxicology for public health and economic gains, thereby contributing to the early spread of new concepts and terms. Later, 'asphyxiating and deleterious gases' became 'chemical weapons' partly to include liquid and solid agents. 'Biological' superseded 'bacterial' as the pathogen spectrum widened. Both the Geneva Protocol and the BTWC reflect a pathogen-based understanding of infection. As science increasingly looks at disease as a function of biochemical processes inside a cell, novel modes of biological warfare become conceivable.

The five-yearly BTWC and CWC review conferences weigh the new understandings in S&T and conclude with agreements adjusting the scope of the prohibition. However, with the last two BTWC review conferences (2016 and 2022) ending without an evaluation of the functioning of the Convention, State Parties have not updated the common understandings since 2011. The 2018 and 2023 CWC review conferences also failed. Here too, the international community did not draw any political conclusions about the impact of S&T.

2.5.3 The Future Dimension of Disarmament

The BTWC and CWC are disarmament treaties. They focus on eliminating technologies – agents, delivery systems and specialised equipment associated with the preparations for and conduct of biological or chemical warfare. The CWC goes a few steps further by also demanding the destruction of infrastructure associated with CW development, production or stockpiling.

In parallel, both Conventions also aim to prevent the re-emergence of CBW. Contrary to weapon destruction, this objective has no finality. It springs from the stipulation that the Conventions shall be of unlimited duration (BTWC Art. XIII and CWC Art. XVI). In addition, both treaties express in their preamble the determination to completely exclude the possibility of CBW use for the sake of all humankind. However, neither document elaborates on this long-term ambition nor spells out whether State Parties can or should act on the provision.[20] This void poses a challenge. Like any other weapon control treaty, the BTWC and CWC reflect an understanding of the weapon-specific and general security challenges during their negotiation. They also testify to then-existing capabilities – political will, procedures and technologies – to detect, monitor and, if required, restore treaty compliance. Scientific and technological progress (e.g. novel design, synthesis and production processes in the biological and chemical fields and new types of agents), the emergence of new non-State security actors and systemic transformations in global geopolitics have pushed State Parties to reappraise the CBW threats and challenges to the treaty regimes.

Both accords allow amendments. However, the process is cumbersome, requiring negotiation of new language and national ratification. Moreover, a straightforward and necessary amendment proposal risks leading to additional requests for treaty changes, possibly affecting core building blocks that few State Parties would want to contemplate. Therefore, States have tended to adjust to new circumstances through the interpretation of treaty provisions at review conferences or (in the case of the CWC) consensus or majority decisions in the appropriate decision-making organs after substantive preparation by the OPCW Technical Secretariat, the Scientific Advisory Board (SAB) or temporary working groups.[21,22] Notwithstanding, preparatory work may yet be protracted.

The BTWC and CWC function through a division of labour between the international community and the individual State Parties. This partition follows from the obligation to transpose the treaty prohibitions, which govern inter-State behaviour, into domestic law so that they become applicable to natural and legal persons operating on a State Party's territory. At the time of writing this chapter, no State is known to maintain an offensive biological warfare programme. With the completion of the US CW destruction programme in July 2023, all declared CW stockpiles have been eliminated under the OPCW's supervision. Threats of terrorism, crime and assassinations with CBW have replaced major warfare scenarios contemplated during the Cold War, which means responsibilities for their prevention and mitigation now lie with multiple national and local agencies, institutions and individuals. Especially, since the 9/11 terrorist strikes against New York and Washington in 2001, internationally coordinated measures, such as UN Security Council resolution 1540 (2004), have emphasised national actions to prevent acts of terrorism with non-conventional weapons.

These developments have contributed to establishing international and bilateral assistance and cooperation programmes with significant funding from individual countries, associations of States such as the Global Partnership and regional organisations like the European Union. Many global and regional organisations also devote significant resources to counter the new threats.

Concerning the BTWC and the CWC, assistance programmes involve actively engaging international and national stakeholder communities. These include government agencies and parliaments, industry, research and academic institutions, civil defence and protection agencies, civil society, etc. The educational and training activities cover dual-use technologies and activities, domestic and international dual-use technology controls, awareness-raising about the international legal norms and risks associated with research, technology development and production in the biotechnological and chemical sectors, capacity-building for emergency response on the national and regional levels and implementation of biological and chemical safety and security measures, amongst other things.

Practice rather than design has replicated the division of labour between the international community and the individual State Parties concerning S&T. The BTWC and CWC have an evident interest in monitoring S&T advances and their commercial applications to understand how they may affect the core prohibitions and, in the case of the CWC, the efficacy of the verification machinery. But the legitimacy of both treaties also benefits from promoting cooperation and exchanges for peaceful purposes, which support economic, scientific and technological development and contribute to public health security. OPCW members receive and consider S&T assessment reports prepared by the SAB and decide on recommendations that affect the implementation of the CWC. The BTWC lacks such an institutional setup but has compensated this deficit in some measure by addressing S&T during the annual meetings of experts in between review conferences whose reports State Parties take up later in the year. Even if those report outcomes are not acted upon, their contents may influence assistance priorities funded by State Parties.

The assorted assistance programmes are implemented on the national or regional levels. Participants receive briefings and discuss the implications of S&T advances for their respective stakeholder communities. Through educational and training programmes, they learn or ameliorate mechanisms and procedures to mitigate threats and risks on the policy development and implementation levels or in their workplace. Even whilst the focus of the assistance programmes may be narrowly on preventing the re-emergence of CBW, the content links up with other security and safety issues and the objectives of other treaties and international organisations.

Through this division of labour, whilst treaties are slow to adjust to S&T developments, the communities with stakes in upholding the norm against CBW can remain up to date with the latest advances and agilely adjust their practices in line with State Party deliberations, even if these do not result in concrete decisions. The assistance programmes also foster national and regional networks of expertise. International organisations also increasingly cooperate and coordinate their assistance activities and programmes. The net result is the emergence of a general course of action to prevent the re-emergence of CBW embraced by stakeholders on the institutional, national and international levels across regions and on different continents.

2.6 Conclusions: Responding to S&T Developments

The relationships between disarmament and S&T are intricate, and their identification in today's armament and disarmament processes would require much effort. The history of modern CBW goes back almost a century and a half to when chemistry and biology arose as scientific disciplines, and S&T laid the foundations for medicine and industrialisation. The overview in this chapter allows the description of how certain currents affected societal attitudes and shaped certain conceptions about technologies and their potential roles in armed conflict. Particularly the efforts starting at the end of the nineteenth century to constrain the impact of new armaments on the battlefields took into consideration the evolving societal conceptions of technologies and the contributions of industrial manufacturing processes to those conceptions and national military capacities. As soon as the control of CBW got framed as a technical problem rather than as a humanitarian issue in warfare, diplomats and experts were able to phrase the types of questions that eventually led to the resolution of how to define a CW and resolve the dual-use problem posed by many commercial toxic chemicals. By formulating the GPC, they resolved a significant challenge caused by accelerating S&T developments: the definition of a CW would apply to all past, present and future toxicants. The GPC is a central building block of the BTWC and CWC. It contributes to their unlimited duration, leading to the need for monitoring and assessing advances in S&T and adjusting understandings of treaty provisions whenever necessary.

Many discussions assume a direct relationship between the CBW disarmament treaties and S&T, placing the BTWC and CWC at the centre. This is akin to arguing a geocentric model of the universe. Even with heliocentrism, the solar system is but one tiny part of the universe. The metaphor is appropriate to understand the complexities of disarmament today. With the end of the Cold War, the bipolar global system gave way to – depending on one's viewpoint – a unipolar or multipolar order. Evidence of a much more complex system has grown over the past 10–15 years. The polycentric governance model fits the rise of globalism with geographical decentralisation of research, business and industry activities in many aspects of international interactions and challenges earlier hegemonistic models. Polycentrism states that multiple and interconnected geographical centres of decision-making coexist and that decisions in one such centre do not necessarily benefit the interests of other centres.

The present author views polycentrism as bi-dimensional. Besides the geographical diffusion, decision centres can also be listed vertically from companies, institutions and associations over States to international organisations and multi- and transnational enterprises. This representation recognises the influence of new security actors and stakeholders, including non-State national or transnational players, on international processes. In a polycentric system, the role of States in shaping developments has declined. They may still regulate, but the pace of transnational events can easily overwhelm their capacity for agency. The BTWC and CWC, both governed by States, are perforce exposed to these trends. Still, they may help form quasi-global attitudes towards certain technologies, even if their influence is essentially limited to the dynamic interactions between S&T-related and security-related processes.

Despite the pace and scope of those changes that would have rendered any substantive formal modification to the BTWC and CWC obsolete before the finalisation of negotiations, State Parties have managed to set certain processes in motion whereby the focus of treaty reinforcement came to lie more with individual States. Through diverse assistance programmes, many stakeholders became involved in their implementation, thus leading to a multi-layered and multi-sectorial governance model well-adjusted to the polycentric environment. The following list presents a sample of actors working to prevent the re-emergence of CBW:

- *Weapon control*
  - Multilateral agreements (Geneva Protocol, BTWC, CWC)
  - Proliferation prevention arrangements (Australia Group, Proliferation Security Initiative, Global Partnership, etc.)
  - UN agencies: UN Security Council, UN Office for Disarmament Affairs, 1540 Committee, UN Environmental Programme, etc.
  - National laws and regulations (criminal, penal, trade, safety and security, etc.)
- *Disease prevention*
  - World Health Organisation, Food and Agricultural Organisation, World Organisation for Animal Health and their regional organisations and initiatives

- *Crime and terrorism*
  - UNSC Resolutions (UN resolution 1540, various terrorism resolutions, etc.)
  - Interpol, Europol, etc.
- *International transfers*
  - World Trade Organisation, World Customs Organisation, etc.
- *Economic development*
  - UN Development Programme
  - Organisation for Economic Cooperation and Development
  - European Union
- *Economic actors*
  - Companies (national, multinational and transnational)
  - Academic and research institutions
  - Individuals
- *Instruments of collective and individual governance*
  - Codes of conduct; Professional codes; Ethics
  - Awareness-raising and education
  - Whistle-blower protection schemes

The activities of these and other stakeholders create overlaying and partially overlapping layers of norm enforcement. At the heart of all these efforts lie the norm and prohibitions central to the BTWC and the CWC.

In shaping the respective S&T monitoring practices, the parties to the BTWC and CWC must remain aware that how they frame the issues will significantly impact their assessments and recommendations. At The Hague Peace Conference in 1899, dropping the original aim to limit armaments resulted in an agreement that proscribed using one type of ammunition rather than its contents, asphyxiating and other deleterious gases. In 1922, The Washington Naval Conference's consideration of CW control as a question of humanitarian principles almost led participants to legalise chemical warfare, except against non-combatants, when confronted with the dual-use characteristics of many modern toxicants. In contrast, the diplomats meeting in Geneva in 1925 considered banning the trade in CW, which raised many technical questions that eventually helped to legally define a CW and find a solution to the dual-use problem.

Similar risks still surface today. Many stakeholders try to insert a controversy in their field of work into BTWC or CWC deliberations. Through the inevitable securitisation of the topic, they may hope to engineer an outcome that serves their interests. The problem may be less acute for the CWC, given the roles of the Technical Secretariat, the SAB and the OPCW decision-making organs. The BTWC lacks a similar technical interface to prepare and accompany S&T decisions. Consequently, many issues like disease outbreaks, epidemics, novel research methodologies or techniques or particular experiments become the focus of S&T debates at meetings in Geneva without pinning down their pertinence for the Convention's disarmament objectives.

The organisation of assorted activities supporting the BTWC and CWC along the principle of the division of labour between the global community and the individual State Party has likely made both treaty regimes more resilient to the impact of S&T developments. Moreover, the involvement of multiple stakeholders means that many institutions actively reinforce the norm and prohibitions. Preventing the re-emergence of CW or BW no longer depends on a single international legal instrument. Perhaps the bigger challenge for State Parties will be to design a shared understanding of what 'preventing the re-emergence' means and then assess S&T depending on their future vision for the respective treaties.

Author Biography

Dr. Jean Pascal Zanders (Belgium) is an independent researcher/consultant on disarmament and security questions. He heads The Trench, a research initiative dedicated to the future of disarmament. He is also a Senior Research Associate at the Fondation pour la Recherche Stratégique (Paris). He holds Master's Degrees in Germanic Philology-Linguistics (1980) and Political Sciences (1992) and a PhD Degree in Political Sciences (1996) from the Free University of Brussels. He was Project Leader of the Chemical and Biological Warfare Project at the Stockholm International Peace Research Institute (1996–2003); Director of the Geneva-based BioWeapons Prevention Project (2003–08) and Senior Research Fellow at the European Union Institute for Security Studies (2008–13). He has participated as an expert to the Belgian and EU Delegations in the BTWC and CWC meetings since 2009. From January 2016 until December 2019, he chaired the Advisory Board on Education and Outreach (ABEO) of the Organisation for the Prohibition of Chemical Weapons (OPCW) and remained a member until December 2021.

References

1 Orfila, M. J. B. (1817) *A General System of Toxicology: or, a Treatise on Poisons Found in the Mineral, Vegetable, and Animal Kingdoms*, Carey & Son, Philadelphia, 1.
2 Bell, A. and MacFarquhar, C. (Eds.) *Encyclopædia Britannica; or, a Dictionary of Arts and Sciences (1711)*, Vol. III, Edinburgh, 152–153.
3 Fidler, D. P. (1999) *International Law and Infectious Diseases*, Oxford Monographs in International Law, Oxford, 26–28.
4 Gaynes, R. P. (2023) *Germ Theory: Medical Pioneers in Infectious Diseases*, Second ed., American Society for Microbiology Press and John Wiley & Sons, Washington, DC, 164.
5 Cavaillon, J.-M. (2022) From bacterial poisons to toxins: The early works of Pasteurians. *Toxins (Basel)*, 14 (11), 759.
6 Orfila, M. J. B. (1817) *General System of Toxicology*, M. Carey & Son, Philadelphia.
7 Wentrup, C. (2022) Origins of organic chemistry and organic synthesis. *European Journal of Organic Chemistry*, 1–12.
8 Orfila, M. J. B. (1826) *Treatise on Poisons and Asphyxies*. Translated from French by Stevenson, *J. G.* Hilliard, Gray, Little, and Wilkins, Boston, pp. 176–205.
9 Christison, R. A. (1845) *Treatise on Poisons in Relation to Medical Jurisprudence, Physiology, and the Practice of Physic*, Barrington & Geo. D. Haswell, Philadelphia, 611–637.
10 Brooke, E. A. (1898) *Tabulation of the Factory Laws of European Countries in So Far as They Relate to the Hours of Labour, And to Special Legislation for Women, Young Persons, and Children, Grant Richards*, London.

11 Rambousek, J. (1913) *Industrial Poisoning: From Fumes, Gases and Poisons of Manufacturing Processes, Edward Arnold*, London, 219–221.

12 Carnegie Endowment for International Peace. (1921) *Documents Relating to the Program of the First Hague Peace Conference. Pamphlet No. 36*, Clarendon Press, Oxford, 1–3.

13 Zanders, J. P. (2015) The road to The Hague. in Zanders, J. P. (Ed.) *Innocence Slaughtered*, Uniform Press, London, 22–45.

14 Zanders, J. P. (2015) The road to Geneva. in Zanders, J. P. (Ed.) *Innocence Slaughtered*, Uniform Press, London, 245–251.

15 (1922) *Text of the Conference Discussions, New York Times*, 7 January, 3.

16 Schindler, D. and Toman, J. (1988) *The Laws of Armed Conflicts, Martinus Nijhoff Publishers*, Dordrecht, 877–879.

17 Zanders, J. P. (2015) The road to Geneva. in Zanders, J. P. (Ed.) *Innocence Slaughtered*, Uniform Press, London, 265–267.

18 Feigenbaum, A. (2017) *Tear Gas*, Verso, London, 25–45.

19 Crowley, M. (2016) *Chemical Control*, Palgrave Macmillan, Basingstoke; OPCW. (1 December 2021) Decision Understanding Regarding the Aerosolised Use of Central Nervous System-acting Chemicals for Law Enforcement Purposes. Conference of the States Parties, Document C-26/DEC.10.

20 Sims, N. A. (1988) *The Diplomacy of Biological Disarmament*, Macmillan, Basingstoke, 27–28; Krutzsch, W. and Trapp, R. (1994) *A Commentary on the Chemical Weapons Convention*, Martinus Nijhoff Publishers, Dordrecht, 248–249.

***21** OPCW. (2011) *Report of the Advisory Panel on Future Priorities of the Organisation for the Prohibition of Chemical Weapons*. Note by the Director General, Document S/951/2011. OPCW, The Hague. 25 July.

22 OPCW. (2015) *The OPCW in 2025: Ensuring A World Free of Chemical Weapons*. Note by the Technical Secretariat, Document S/1252/2015, OPCW, The Hague. 6 March.

3

A Multifaceted Threat

Gemma Bowsher

School of Security Studies, King's College London, London, UK

Key Points

1) Biological weapons are not solely weapons of mass destruction – they may be used by different actors for diverse objectives and on differing scales ranging from political assassination, crime, influence operations and agroterror.
2) Examples of biological weapons use illuminate the evolving strategies of various State, non-State and individual actors.
3) Scientific advances such as those in synthetic biology may facilitate the development of new agents, targets, means of production and dispersal and delivery systems of biological agents.
4) Evolving risks such as cyber, artificial intelligence (AI) and disinformation are altering the risk landscapes in which biological weapons reside.

Summary

This chapter examines the multifaceted nature of biological weapons through their use by a range of State, non-State and individual actors to achieve objectives other than mass destruction. It considers both the use of biological agents in historical timeframes, and it explores the evolving strategic and operational utility of their deployment. Using case studies, the role of biological weapons in the achievement of alternative objectives beyond mass destruction will be outlined to highlight a range of themes such as political influence, insider threat, hybrid warfare and assassination. The chapter addresses the role of novel biotechnologies in the evolution of this threat paradigm, in particular the role of gene editing technologies and gain of function research. Finally, it touches on additional amplifiers of this particular form of biological risk – particularly, hostile disinformation to highlight the evolving, multifaceted nature of this threat domain.

Essentials of Biological Security: A Global Perspective, First Edition. Edited by Lijun Shang, Weiwen Zhang, and Malcolm Dando.

3.1 Introduction

The concept of weapons of mass destruction (WMD) encapsulates a spectrum of threats responsible for mobilising many of the profound geopolitical shifts of the late twentieth and early twenty-first centuries. From the first Gulf War in the 1990s to the US-led invasion of Iraq in 2003 and the Syrian Government's repeated use of chemical agents in the conduct of its protracted civil war, the alleged and confirmed use of WMD by 'rogue' actors has drawn together international security communities around a common threat concept.

WMD is defined by the United Nations as:

> ... atomic explosive weapons, radioactive material weapons, lethal chemical and biological weapons, and any weapons developed in the future which might have characteristics comparable in destructive effect to those of the atomic bomb or other weapons mentioned above.

This definition draws a direct link between the destructive impacts of the atom bomb and the scope and scale of biological agent attacks, likely familiar to public health and biological research professionals, as well as arms control actors. Biological weapons are somewhat unique within the WMD threat landscape since the problem of dual-use rises most prominently to the foreground in this domain. The challenge of differentiating legitimate biological research activities from weapons development programmes is well documented given the ubiquity of naturally occurring pathogens with the potential to cause mass destruction and the need to carry out research at the leading edge of biological innovation. Potential biological threats are highly diverse, encompassing natural and engineered pathogens, genome-editing tools, synthetic biology and meta-genomic technologies. Both contemporary and historical evidence clearly demonstrate that as technological innovations in the biosciences proliferate, so too does the risk of manipulation for harmful objectives by hostile actors.

Gregory Koblentz, the scholar of biological arms control, has reflected on the challenge of the WMD framing when it comes to bioweapons arguing that:[1]

> ... the widespread use of these labels has hindered our understanding of the international security dimensions of biological weapons ... obscure[ing] important differences between these different weapons and the strategic consequences of their proliferation.

This position reflects the diversity of biological agents and their often indistinct status at the boundary of public threat, strategic advantage and scientific progress. The porous nature of these boundaries renders biological weapons an appealing product for malign actors since such agents amplify the asymmetric advantage of the 'first-belligerent', and the challenge of verification can afford a degree of impunity. This perspective also opens up the understanding that biological weapons are multifaceted and may be used by different actors for diverse objectives and on differing scales – not solely for *mass destruction*. The achievement of alternative targeted objectives such as political assassination, crime, influence operations and agricultural damage demonstrates alternative strategic imperatives beyond the confines of mass destruction alone. In addition, the role of other contemporary dual-use technologies such as cyber, artificial intelligence (AI) and disinformation are shifting the wider emerging threat ecosystem, which in turn, magnifies the scope and scale of biological weapons' multivalent threat surface.

This chapter explores the multifaceted nature of biological weapons, considering both their increasing diversity and versatility as a threat paradigm, and their evolving scope for aiding the achievement of a range of strategic objectives beyond mass destruction alone. Drawing out this complexity aids the appreciation of the potential for novel innovations in the biosciences to be misappropriated. Rather than acting as a handbrake on cutting-edge research, these insights highlight the perennial need to balance scientific advances with pragmatic policy-making in order to protect a range of communities and sectors from the threat of malign actors.

3.2 Assessing the Utility and Scope of Biological Weapons at Various Scales

The term 'weapon' describes an enormous array of offensive and defensive tools used during conflict and hostility. Determining what constitutes a weapon is generally relatively straightforward; a bullet plainly only serves the purpose of a weapon, and so too does a chemical agent such as the nerve agent Sarin, also designed for offensive purposes and only producing harmful indiscriminate health effects on target groups. Biological entities, however, are far more challenging to define in these terms. For example, the bacteria *Bacillus anthracis*, commonly known as anthrax, is a bogeyman of the biological weapons field given its recent use in the 2001 *Amerithrax* events and its high associated lethality. Nevertheless, anthrax, like so many other weaponisable pathogens, is a naturally occurring bacterium found in soil and livestock around the world and implicated in naturally occurring disease outbreaks in humans and animals. Stocks of anthrax in medical laboratories are legitimate objects of research and vaccine development and do not in themselves provide a 'smoking gun' to suggest a bioweapons development programme. In this circumstance, the notion that the use of such agents to cause harm is *deliberate* is an important dimension of the determination that a naturally occurring pathogen constitutes a weapon.

A key question consequently arises regarding not just the assessment of whether an agent itself is indeed a weapon, but what the scope and impact of such entities are when used in this manner. Traditionally, defence organisations have assessed these impacts along the axes of *lethality* and *destruction*. These categories deliver important insights in the assessment of the scale of effect of a singular weapons deployment and reflect the need to understand the frequently multi-order effects of a particular technology. Assessments of destructive effects of biological weapons do signal devastating consequences for targeted populations. A 1969 report by the United Nations assessed that a 10-t biological bomb released in the appropriate weather conditions could reach a population spread across 100,000 km^2, an area far greater than a one-megaton nuclear payload.[2] There is of course more to the deployment of weapons than simply the desire to effect maximal destruction in a singular event (although some systems are designed for this objective). Other lenses through which to assess the scope of weapons systems, including biological weapons, are the *tactical, operational* and *strategic* factors driving their use, which help illustrate the specific motivations and differential benefits of particular weapons use at varying scales of effect.

In this context, biological weapons can be seen as an immensely heterogeneous collective with scope for producing a variety of harms – in terms of lethality and destruction, as well as in terms of the specific objectives of their deployment. It has generally been assessed that obstacles to the use of biological weapons have limited the more widespread deployment of these entities in favour of other categories such as chemical weapons. Regularly, cited challenges centre on the difficulties inherent in culturing sufficient quantities of pathogen to deploy at scale, as well as the difficulties in producing consistent effects either through engineering its intrinsic biological properties or ensuring consistent transmission effects in specific environments. A further challenge is the need to protect the operator from the harmful effects of the weapon deployed. These obstacles are a challenge for even relatively well-resourced actors; for example, the Syrian Government led by Bashar Al-Assad is known to have developed an illicit chemical and biological weapons programme, producing industrial quantities of agents such as Sarin and Chlorine Gas, which is alleged to have been used in the conduct of hostilities towards opposition groups. However, its endeavour to produce ricin, a biological agent derived from the seed of the castor plant, was less successful due to the biochemical challenges of stabilising it during synthesis. Nevertheless, although these challenges may constrain the wider adoption of such weapons as an agent of mass destruction at the scale of an atomic bomb, they have shaped the modes of use over time and will undoubtedly do so for the biological risk landscapes of the future.

3.3 Diverse Objectives of Bioweapon Use: Past and Present

The use of biological weapons has an ancient history. One of the earliest documented instances in war is from the sixth century BCE, when the Assyrian army contaminated the wells of their enemy, the Israelites, with the fungal disease rye ergot. This psychoactive fungus induces convulsions, muscle spasms, vomiting and hallucinations and can ultimately lead to gangrene and death in its victims. Further examples include the army of the Tartars throwing plague-infected corpses over the city walls of Kaffa, a Crimean Seaport during its bombardment in 1346 to spread the disease within the besieged city. Famously, British and American settlers and pioneers provided smallpox-infected blankets to Native American communities, leading to large disease outbreaks and the extinction of several native groups. More recently, the Imperial Japanese Army ran a biological weapons campaign throughout the Sino–Japanese (1937–1945) and the Second World War, which involved the aerial release of plague-infected fleas over selected Chinese cities. The development of porcelain bomb casings to hold infected fleas allowed the Japanese to target critical infrastructure such as dams and reservoirs with water-borne infections such as cholera. During the Korean War between 1950 and 1953, the governments of the Soviet Union, North Korea and China made contested allegations that the US government used offensive germ warfare, alleging that it had intentionally infected Korean troops with smallpox and plague by dropping infected insects from aircraft in collaboration with the Japanese.

These examples highlight the long history of biological weapons implicated in military campaigns to leverage 'traditional' kinetic combat action with the enfeebling effects of widespread disease. Co-opting the natural 'weapons' provided by nature – bacteria, viruses and fungi – and redirecting them towards opposition factions is nevertheless a high-stakes game with only a limited amount of control afforded to the aggressor over the onwards transmission pathways of the pathogen. Indeed, infection of aggressor forces themselves has always been a prominent risk of such campaigns. Military uses of biological weapons, therefore, form only one dimension of the wider scope of application. The European External Action Service (EEAS), the Foreign Policy arm of the European Union (EU), has identified this widened remit saying:

> ... in addition to strategic or tactical military applications, biological weapons can be used for political assassinations, the infection of livestock or agricultural produce and cause food shortages and economic loss, the creation of environmental catastrophes, and the introduction of widespread illness, fear and mistrust among the public.

These examples reflect the historic and evolving potential for such weapons to deliver focused objectives imperilling individuals, communities and societies.

Political assassinations likely come second to military use in the hierarchy of biological weapons deployment. Again, ancient examples show the use of biological poisons such as belladonna and hemlock by individuals such as Agrippina and Cleopatra, against political rivals. In more recent history, the Soviet Union directed its biological weapons programme towards assassinations of prominent dissidents. One such incident involved the Bulgarian journalist and writer Georgi Markhov, who worked for the BBC World Service and Radio Free Europe, publishing materials critical of Soviet policy and Bulgarian political activity. On 7 September 1978, whilst crossing Waterloo Bridge on his way to the BBC, Markov's leg was pierced by an umbrella carried by an individual connected to the Bulgarian secret police and KGB. Markov died four days later, and afterwards a tissue segment was sent to the UK's Defence Laboratory Porton Down, where it was established that he had been poisoned by a pellet of ricin introduced by the umbrella. The Central Intelligence Agency (CIA), too, has been implicated in historic attempts to target political opponents with biological agents. Prior to the advent of the Biological Weapons Convention (BWC) in 1972, the United States, in the same way as several other Western nations, possessed a biological weapons programme. In the 1960s, many African nations were moving to independence from colonial states and countries such as Britain and France, and the new superpower of the United States was jostling for influence and control. The left-wing leader of the newly independent Congo, Patrice Lumumba, was a controversial figure in the eyes of communist-fearing United States. In September 1960, the CIA, through the efforts of its poison expert, Sidney Gottlieb, prepared an assassination bundle containing pathogens, including tularaemia, brucellosis, anthrax, smallpox, tuberculosis and Venezuelan equine encephalitis. These were designed to be mixed with Lumumba's toothpaste so that he would infect himself whilst brushing his teeth. Before this package could be used, Lumumba was assassinated by Belgian-backed Congolese rivals in January 1961, allowing several interested States, such as the United States to sideline their existing assassination plans.

The nature of assassination does favour the use of certain types of biological agents, due to the ambiguity of the natural versus deliberate spectrum of pathogens, which can help aggressors evade attribution. There were some concerns during the COVID-19 pandemic that individuals would intentionally spread infection to others with the goal of murder. Cases where human immunodeficiency virus (HIV) has been intentionally spread highlight the potential for a naturally occurring pathogens to be exploited *in* and *ex vivo* to co-opt existing transmission networks for malign ends.

The risk of agroterrorism is a further concern when it comes to the use of biological weapons. This particular form of terrorism involves the deliberate introduction of a disease agent into livestock or food to cause harm to the public, societal upheaval and fear. Agroterrorism is an oft-cited but relatively poorly addressed field by law enforcement and counter-terror actors. In terms of lethality, an attack targeting the food chain has the potential to cause significant disruption to affected communities. The natural disease outbreak of bovine spongiform encephalitis (BSE) in cattle decimated the British beef industry in the 1980s and 90s and demonstrated the major economic and social consequences of contaminated food supplies and the long-term challenge of restoring public confidence in the integrity of the agricultural sector. Examples of intentional food supply contamination do exist on smaller scales[3] – in 1952 a group of Mau Mau rebel fighters in Kenya were found to have poisoned a herd of cattle at the British Mission, killing 33 cows. The relative ease with which this was achieved, by making small incisions in the legs of the cows and introducing the sap of the African milk bush, shows the relative fragility of food chains, which remains the case to this day. Even the threat of agroterror is a serious problem for law enforcement. In 2011, a South African man was arrested for threatening to infect British and American livestock with foot and mouth disease unless he was paid a multi-million-dollar ransom. Agroterrorism operates along 3 major trajectories: (1) as in the Mau Mau example, by eliminating a source of food for a specific community; (2) contaminating a food source such that human consumers experience deleterious and potentially lethal side effects; and (3) introducing fear and mistrust in the food system – with the potential for wider political disruption and public disorder. Agriculture is a poorly protected critical infrastructure, with several vulnerabilities such as the intensity of contemporary farming practices, the increased susceptibility of farmed animals and plants to disease and poorly connected disease surveillance networks. In addition, the nature of agriculture is diffuse, and potential aggressors span well-resourced nation-states, lone-wolf actors and non-State groups.

3.4 Evolving Biotechnologies

Developments in the biosciences have led to the twenty-first century being characterised as the 'Age of Biology'. Progress in our understanding of genetics and molecular techniques has fundamentally altered the potential for research and biotechnology to deliver breakthroughs in fields such as medical research, agriculture and climate change mitigation. Of course, with these developments appear accompanying risks, as predicted by the US National Intelligence Council in its 2004 horizon scanning project *Mapping the Global Future*, which stated that in 2020:

> ... major advances in the biological sciences ... probably will accelerate the pace of BW agent development, increasing the potential for agents that are more difficult to detect or defend against.[4]

Evolutions in the field of synthetic biology have also democratised bioscience work, with simplified genome-editing systems becoming more accessible to non-research actors. There is growing concern that these converging trends may permit hostile actors to weaponise biological entities through their use of novel and accessible biotechnologies, producing a new landscape of biological risk for authorities to grapple with.

The US Department of Defense (DOD) outlined its major bioweapons concerns with regard to synthetic biology, including the recreation of a human pathogenic virus, for example, the extinct smallpox virus. In 2018, an American biotechnology organisation requested a German firm to resurrect the extinct, related horsepox virus from Deoxyribonucleic acid (DNA) sequence information. Concerns mount that similar techniques could be used to revive viruses such as the eradicated smallpox virus, and this could be released intentionally by hostile actors into populations with no existing immunity in order to generate a lethal outbreak. Significant concerns centre on the risk of actors using gene synthesis techniques to produce entirely new biochemical products, which can be used as weapons, and which may evade current biothreat detection capabilities. These kinds of technologies might involve exploiting current gene therapy techniques to engineer genetic elements for introduction within viral vectors into victims' bodies. These novel elements, rather than being therapeutic, could induce specific changes within the host germline such as the emergence of cancer. There is also a risk that a so-called 'stealth virus' could lay dormant in host populations until a predetermined external signal activates it in a particular individual or population. Lastly, there is mounting anxiety that existing bacteria could be manipulated to make them more dangerous – for example, the current explosion in drug-resistant bacteria is rendering several existing medical countermeasures useless. Targeted engineering of this phenomenon, either to induce drug resistance or to enhance transmission attributes, for example, could spread widely with limited countermeasures available to authorities. The Soviet Union's offensive biological weapons programme is known to have worked on these approaches, by engineering a 'new and improved super-plague', by altering existing plague bacteria to be resistant to known antibiotics.

The diffusion of biotechnology has enormous promise to deliver important benefits for humanity – however, the scope for misappropriation grows significantly with the expansion of technologies. As costs decline, and emerging bio-economies scatter expertise more widely, the need for targeted biosecurity strategies in the field also grows. Both the United Kingdom's 2023 Biosecurity Strategy and the United States' 2022 National Biodefense Strategy express this imperative clearly and together signal the commitment to connecting researchers, policymaker and defence actors in this important and emerging field.

3.5 Changing Biothreat Landscapes

Biological research is not the only source of emerging technologies implicated in shifting bioweapon risk landscapes. Rapid developments in cyber and AI are coupling with biological risk to produce novel forms of biological threats, meriting new kinds of expertise.

The nascent field of cyber–biosecurity faces these converging threat dimensions in its assessment of how material at the interface of life science and digital ecosystems can be misused, exploited or weaponised for malicious purposes. The increasing digital dependence of critical sectors in the health and biosciences is one such risk, opening the door to foreign and hostile actor interference in the manipulation of digital biomaterials. The phenomenon of biological disinformation has been characterised as the future of biowarfare due to its ability to perpetuate and propagate pathogen transmission through secondary effects, as well as engendering hostility and violence towards healthcare and other governmental and private actors.

3.5.1 Cyber-Dependency

The cyber-dependence of health and biological sectors is increasingly appealing as a target to malign actors who are able to use this novel threat surface to leverage geopolitical influence, extort large sums of money or simply cause chaos and disruption.[5] The COVID-19 pandemic has revealed the central role of health and biological systems within critical national infrastructure and revealed a host of vulnerabilities, never previously stress-tested at the scope and scale of the pandemic. Natural events such as pandemics strain the integrity of the cyber-bionexus by exposing the vulnerabilities of systems already under stress and by providing opportunities for their exploitation. So too do increasing numbers of hostile cyberattacks such as the 2017 WannaCry ransomware attacks that crippled the United Kingdom's national health service, caused failures of critical clinical services, and cost the UK Government approximately £92 million. The Waikato District Health Board (DHB) cyberattacks on New Zealand's oncology and COVID-19 vaccination services necessitated the transfer of cancer patients out-of-country to receive urgent cancer therapy. A death attributed to a hospital cyberattack in Germany signals the scope for these sorts of disruptions to cause patient harm and death, and cyberattacks on COVID-19 vaccine trials reinforced the susceptibility of the disease control apparatus to hybrid threat actors.

Bioscience data is also vulnerable to exfiltration and weaponisation. The US National Medical Intelligence Center has warned that Chinese firms are accessing global health data by establishing laboratories to support COVID-19 testing. Population-wide genetic data sourced through these initiatives or stolen via State-sponsored hacking campaigns poses a security risk to nations and communities from whom the data is sourced. AI algorithms are permitting the processing of such data at enormous scale. Biosector reliance on legacy digital systems and network fragmentation has resulted in vulnerable cyber-terrains, to which State and non-State actors are increasingly turning their attentions with the threat of and actual generation of harmful biological consequences. As biological research increasingly allies with innovations in cyber, it is essential that leaders invest in strengthening existing resources and take seriously the need to horizon scan for future cross-cutting cyber-biological risks.

3.5.2 Disinformation

Disinformation is an updated version of the age-old phenomenon of the calculated spread of falsehoods to amplify public fear and alter public behaviours during crises. The COVID-19 pandemic is just the latest episode in the developing story of false narratives and factual distortion in critical health events. From Ebola to Measles the advent of 'fake news' has implicated hostile actors from the Kremlin to North Korea in the weaponisation of health and biological information. Disinformation is a 'deliberate falsehood', shared in full knowledge of its deception and with malicious intent.[7] The ease of information spread through the internet has profoundly altered the accessibility of publics for actors interested in strategic messaging. These campaigns have been so significant during the course of the COVID-19 pandemic that the World Health Organisation (WHO) took the novel step of declaring an 'infodemic' – or a crisis of information detracting from disease control efforts.[6] It has been suggested that disinformation is an evolving battleground for the propagation of the malign interests of criminals, terror groups and hostile state actors. This phenomenon is a consequence and driver of the changing character of hostilities, with regions such as Eastern Europe and the Indo-Pacific ascending in importance in the global sphere of disinformation operations. Beyond health and critical infrastructure, disinformation operates to amplify the strategic influence of hostile actors through 'soft power' effects. Examples range from campaigns relating vaccine diplomacy, COVID-19 origin narratives and false allegations asserting the existence of bioweapons laboratories.

Biological disinformation, particularly the campaigns waged in recent years by the Kremlin, has already compromised essential health and scientific programmes internationally.[7] At the onset of the Russian invasion in February 2022, Ukraine was the least vaccinated country in Europe against COVID-19 after years of targeted Russian information operations levelled against immunisation programmes for measles and other vaccine-preventable diseases. A second epidemic of Polio amongst Ukraine's children before the war was described as a 'biological emergency on a regional scale' by Ukraine's Ministry of Health – noting that the country experienced uniquely poor immunisation rates across the range of preventable childhood diseases. The harmful effects of disinformation are only recently being assessed by academics and policymakers; however, it is becoming clear that these campaigns do more than simply sow distrust, they have consequential effects on actual health and biological outcomes. A clear focus on disinformation and its destructive consequences is a matter of urgency for scientists, public health professionals and decision-makers alike.

This new cyber-biological frontier is in some ways more threatening than an intentionally engineered organism. Without the technical constraints that generally impede the dissemination of engineered pathogens, as well as the absence of international sanctions related to contraventions of the BWC, its temptations to those interested in causing harm are clear and escalating. Unlike conventional biological weapons, none of the usual constraints apply, and the scope for population coverage is significantly greater due to the ease of digital data transmission. Regulatory processes remain weak, and self-governance efforts by technology companies are only recently becoming more visible. Global campaigns to expand the bioeconomy in critical regions place development objectives in tension with the need to protect populations from the real-world biological threats associated with increased digital connectivity.

3.6 Conclusion

Biological weapons continue to diversify in range and scale, constituting a multifaceted threat of enormous consequence for the global publics, security actors and biological researchers of the twenty-first century. Historic examples demonstrate the ingenuity of aggressors using natural outbreaks to achieve their own strategic ends in military campaigns, assassinations, poisonings and food supply disruption. The incredible progress being made in the biosciences heralds promise for tackling many obstacles to human health and livelihoods, from genetic disease, cancer, environmental degradation and food insecurity. Nevertheless, these evolutions have always presented the problem of 'dual use', the ability for new innovations to be reformulated into a weapon.

The growing entanglement of digital and biological technologies catalyses new opportunities for biological warfare carried out via cyber networks. The hacking of an immunisation centre can cripple disease prevention for specific diseases to produce the same effects as those ancient armies throwing disease-ridden corpses over city walls. Hostile biological disinformation operations operate through the age-old human impulses of fear and mistrust – yet the vectors used to propagate these today, such as AI and cyber, are at the forefront of innovation.

These collective facets of contemporary biological weaponry will continue to be important concepts for professionals working with biological entities in any capacity, whether it be research, industry, agriculture or medicine. As calls grow to expand the bioeconomy, economic thinkers and policymakers must also be driven to engage with the evidence emerging at the forefront of bio-innovation. The risks of 'business as usual' research only escalate as innovation in the biological sciences is harnessed by adversaries with no regard for the norms and safeguards enshrined in the 1972 BWC.

Author Biography

Dr. Gemma Bowsher is the lead for Global Health Security at the Centre for Conflict and Health King's College London, where she is also appointed at the Centre for Science and Security Studies at the Department of War Studies. Her research interests lie at the intersections of global health and biosecurity. Dr. Bowsher engages with governments, security and health sector organisations to develop approaches to anticipate and prepare for emerging threats across the CBRN, health and cyber domains. Dr. Bowsher is a social scientist by background and holds medical and postgraduate degrees from King's College London and Harvard University. Dr. Bowsher has been awarded a range of fellowships and external appointments, including Fellowship of the Royal Society of Public Health, the Johns Hopkins Emerging Leaders in Biosecurity Fellowship and an appointment as a senior research associate at the Intellectual Forum of Jesus College, Cambridge University. Dr. Bowsher is also a practising doctor in the United Kingdom's National Health Service.

References

*1 Koblentz, G. D. (2010) *Living Weapons: Biological Warfare and International Security*, Cornell University Press, Ithaca.

2 Meselson, M. S. (1970) Chemical and biological weapons. *Scientific American*, 222 (5), 15–25.

*3 Keremidis, H. *et al* (2013) Historical perspective on agroterrorism: lessons learned from 1945 to 2012. *Biosecurity and Bioterrorism: Biodefense Strategy, Practice, and Science*, 11 (S1), S17–S24.

4 Rose, G. (2005) Mapping the global future: report of the National Intelligence Council's 2020 project. *Foreign Affairs*, 84 (3), 133.

***5** Bernard, R. *et al* (2020) Cyber security and the unexplored threat to global health: a call for global norms. *Global Security: Health, Science and Policy*, 5 (1), 134–141.

***6** Zarocostas, J. (2020) How to fight an infodemic. *The Lancet*, 395 (10225), 676.

***7** Broniatowski, D. A. *et al* (2018) Weaponized health communication: twitter bots and Russian trolls amplify the vaccine debate. *American Journal of Public Health*, 108 (10), 1378–1384.

4

Biological Weapons from the Ancient World to 1945

Brett Edwards

Department of Politics, Languages and International Studies, University of Bath, Bath, UK

Key Points

1) Long before scientific understandings of poisons and pathogens humans have employed them as part of warfare.
2) In every era biological weapons (BWs) have occupied a small niche.
3) The BWs niche is shaped in each era by a number of historical, cultural, political and technological factors.
4) Humans have experimented with and employed a wide range of pathogens and poisons as part of warfare since prehistoric times.

Summary

In this chapter, a review of the history of biological warfare (including toxins) is presented. This involves a review of the open literature on the topic and a presentation of the chronology of biological warfare, distinguished into key eras. It is argued that in each era the development and use of biological weapons (BWs) have been determined by a range of historical, cultural, political and technological factors. Together, these factors have shaped the specific niches which BWs have come to occupy in warfare. The paper demonstrates a general trend towards the possibility of increasingly versatile, sophisticated and effective weapon systems in the era studied, which were increasingly developed through research based on scientific principles. It also demonstrates the extent to which developments in biological warfare accelerated rapidly in the twentieth century. However, it also highlights that the development and use of BWs occupied only a very narrow niche in warfare during the period studied.

Essentials of Biological Security: A Global Perspective, First Edition. Edited by Lijun Shang, Weiwen Zhang, and Malcolm Dando.

4.1 Introduction

The aim of this chapter is to provide a survey of the history of biological warfare before 1945. The scope of this chapter then is expansive in two key senses. Firstly, it covers a vast period of human history – with the earliest time periods marked out in millennia and centuries. Secondly, while more recent history is organised into narrower periods, it covers the growth of numerous modern programmes during the twentieth century. The question a reader should ask then, is why read a chapter covering such a broad sweep of history – especially when there are certainly more narrowly focused and detailed histories covering the early historical eras as well as specialist studies covering twentieth-century programmes. The first answer, of course, is that it allows for a quick skim and survey of the scope of this history writ large. The second, I believe, is that in attempting to present such a broad history, we are forced to appreciate a few useful things about contemporary preoccupations and concerns about biological warfare and biological terrorism. Not least, the niches biological weapons (BWs) have come to occupy in the past, and why, and consider the possible implications of this for the future. The term niche, in the most basic sense, refers to the idea of something coming to occupy a space in a broader context. In more technical senses, the term has also found its place in the discussion of ecological systems and technological change. And, it is a term which is usually related to the broader idea of evolutionary processes. In the history presented here, we are interested in the roles in which BWs have been considered and employed – and the broader context in which these sat. This extends to the character of warfare, the broader technological context and political and military doctrinal factors that shaped the development and use of these weapons.

In this chapter then, the aim is to point to how both the broader historical context, processes of development and production and understood applications and potentials of BWs have changed through history.

The use of poisons, toxins and infectious diseases in warfare, certainly predates our modern scientific understandings of the chemical and biological mechanisms through which these agents exhibit effects. This is not surprising. Humans used yeast in brewing, for many millennia before they could watch *Saccharomyces cerevisiae* (ale yeast) at work under a microscope. But, this did not stop them from discovering and then retaining the ability to harness such abilities. It is apparent then, that biological warfare may have ancient roots, our contemporary understandings of BWs are decidedly modern in several important senses, and this raises some interesting questions about the relationship between modern and historic practices, as well as the scope and preoccupations of contemporary framings.

Pathogens, toxins and poisons are today understood with reference to scientifically derived typologies. They are also either grouped, or distinguished, dependent on the context humans seek to study, exploit or protect people or agriculture. Historically, however, such distinctions were often entirely lacking, ad hoc or made in ways which seem counter-intuitive today. For example, before the establishment of germ theory in the nineteenth century, a host of explanations for pathogenic disease existed in folk law and medical practice (see Chapter 2). And while sometimes practitioners struck upon cures which worked, the explanations provided appear naive and misguided to us. Miasmic theories of disease, which were pre-modern explanations for the cause of disease centring on invisible emanations which could travel through the air make for an excellent case in point. They were technically incorrect, and at times served to hinder the emergence of germ theory in later centuries – but may also have at times led to improvements in sanitary conditions which would have conceivably helped reduce the prevalence of some diseases.

The military programmes which emerged in the first half of the twentieth century, and since, also continue to inform our contemporary understanding of biological and chemical warfare. These extended to the development of *bacteriological* (often used historically as an all-encompassing term to include the effects of pathogenic organisms, including bacteria, viruses and fungi) as well as toxins derived from microbes, plants and animals, which have usually found themselves bundled into one of the above two categories since this time. As we will see, the State programmes which emerged in the twentieth century were diverse and tested and developed a wide range of agents for a wide range of uses – and also spurred the development of a range of capabilities and strategies to defend both militaries and broader society against them. A common refrain among bioweapon experts today is that too much recent discussion of BWs has focussed on high-lethality mass-casualty weapons targeted against humans. And indeed, BWs were thought to potentially occupy a much broader range of niches during the later periods studied in this chapter. For this latter reason alone, the pre-1945 era certainly merits revisiting if we are to appreciate and anticipate contemporary threats.

4.2 Map of the Literature

In putting together this study, I have been reliant on two key types of work. The first has been general surveys of the history of biological warfare – which extend back and beyond the First World War. There are only a handful of substantive surveys conducted and available in the open literature which I am aware of, which have relied on an explicit set of criteria for inclusion and assessment of claims of development and use.

The first is an edited volume produced by Geissler and Moon which covers the history of biological warfare up until 1945.[1] Published some 20 years ago, this collection continues to be a touchstone in the field. The chapter on biological warfare before 1914, produced by Mark Wheelis[2] reflects a seminal review of pre-twentieth century history and remains a useful resource for those compiling lists of early allegations of use, as well as those critically assessing such claims. Likewise, there are also chapters dealing with biological warfare programmes between 1914 and 1945. Although some of this work has been superseded by more recent historical research and reviews, this latter part of the collection remains an excellent sketch of the biological warfare landscape in this era. Two general histories produced by Seth Carus have also been central to compiling this review.[3,4] Both works involve a critical appraisal of existing literature on the topic, as well as fresh historical insights, and have been particularly valuable in this study, as they reflect the most substantial survey of broader literature relevant to pre-nineteenth century biological warfare. In addition to these surveys, there are also numerous additional peer-reviewed surveys of the history of biological warfare which should be used with caution. This is because they occasionally reproduce earlier claims about the use (particularly in the pre-twentieth century) which have long since been debunked or heavily caveated by relevant experts. Such reviews, while often useful and thought-provoking have been written for a diverse range of purposes, and with differing levels of historical rigour. They also often lack a clear criterion for inclusion and critical appraisal of early allegations of use – concerning things such as technical feasibility and the evidence base which supports such claims.

In addition to these general surveys, there are also several other key studies which are of value to general reviews of biological warfare programmes in the first half of the twentieth century. This includes two further studies by Seth Carus. The first covers bioterrorism and other crimes involving pathogens since 1900,[5] and a more recent survey deals with biological warfare programmes since 1915.[6] In the remainder of the chapter, I present a brief characterisation of biological warfare in key eras of human history.

4.3 Historical Review

The review below is certainly not comprehensive and must be selective in terms of which threads of the history of biological warfare before 1945 are drawn out. As a rule, I have attempted to point to some key dimensions to help paint a general and introductory picture of each era, based on the current state of the literature on the topic. This includes both the technical character of such weapon systems, as well as their broader (and usually marginal) role in warfare. In terms of periodisation, a few methodological issues are worth bearing in mind.

Firstly, while periodisation is an essential means to construct wide-ranging historical reviews, such distinctions are in many ways arbitrary. For example, there are more continuations than distinctions between the pre-modern eras – not only when we consider the use of such weapons by larger civilisations but also when we consider the continual use of poison arrows and other rudimentary weapons by hunter-gather communities throughout this period. This review could have been arranged by other and more systematic general schema drawn from the literature on global history – centred on scientific and technological developments, warfare, economics or other cultural developments. However, I have tended to follow the periodisation adopted in previous studies which do still provide the best vehicle for an introductory account of the rich and varied history of this issue area. This approach also reflects certain inherent biases within the existent literature – particularly in the pre-nineteenth century world – in as much, as it is Western European civilisation centric in terms of both content and chronology. With such caveats in mind, a survey of this history is now provided.

4.3.1 Pre-history (72,000–500 BCE)

The early history of the use of either infectious disease or natural toxins as a precursor to modern biological and chemical warfare has been subject to only limited academic study. This is perhaps for good reason. Firstly, there was only a very rudimentary understanding of toxins and an even more limited appreciation of the mechanism of disease spread among early hunter-gatherer societies. Any would-be historian also faces a scarcity of evidence, in terms of identifying instances of the use of such agents or the existence of broader practices which could be understood to constitute crude forms of biological warfare. Instead, we tend to be reliant on inference from later practices. It might be assumed for example, that certain early communities familiar with the effects of poisons and toxins in the natural world – especially those employed as part of medicine or hunting, may have put such knowledge to use in conflict both within and outside their immediate community. For example, the earliest physical example, we have of a preserved poison arrow comes from somewhere around 2200 BC. In this case, from an arrow found on the East Bank of the River Nile in Egypt – near an ancient graveyard. Experiments in the 1970s appeared to demonstrate that the arrow was covered with a toxic substance.[7] However, poison arrow use stretched back even further. For example, in one recent study, which relied on examining the size and shape of around 400 arrowheads, a specialist determined that poisoned bone-tipped arrows may have been used in southern Africa for up to 72,000 years.[8] There remains disagreement, however, on the extent to which such weapons were used as part of warfare in the prehistoric era. What is clear, is that the use of poison and toxin weapons in warfare may have occurred ad hoc innumerable times in the prehistoric world, and at various times become part of primitive warfare in the evolving patchwork of loosely connected communities wherever humans lived.

4.3.2 Ancient History (500 BCE–1000 AD)

The ancient world reflected a period in which larger sedentary agricultural societies emerged. This resulted in a fundamental transformation in the relationship between these societies and the natural world which also led to changes in warfare. Ancient civilisations were familiar with a wide range of poisons and toxins. And in many civilisations, the threat of murder via poisoning appears to have been a significant concern among senior political figures. This included a range of materials which were available locally. This extended to mineral poisons, such as arsenical compounds, as well as toxins produced by plants – such as belladonna, henbane, hellebore and hemlock, as well as animal-derived toxins and venoms.[9] There were then, certainly motivations for elites to seek and support the development of defences against such weapons – which extended to the emergence of a range of practitioners ostensibly specialising in protective measures and cures.[10]

It is also clear that both the Romans and Greeks fostered mythological knowledge of poison arrows.[11] This perhaps reflected the memorialisation of practices which had all but died out within those societies, as well as interaction with local nomadic peoples in a European context. It seems likely that poison arrow use was already well in decline at the start of this era, but that these weapons were not given up by surviving nomadic and hunter-gatherer communities in any sort of uniform way. The last hold-outs in Europe may, for example, have been the Slavic Poles, as late as the fourteenth century.[12] The picture is more complicated in Asia, however, where it is reasonable to believe that both Indigenous communities, as well as the larger dynasties of that region, continued to employ arrow poison (such as aconite) in warfare until as late as the eighteenth century.[13] If this is the case, it is apparent that a range of cultural and environmental factors could have led to this divergence with Europeans. It is tempting, for example, to consider if the prevalence of certain large predators, which were often effectively hunted with poison, might be one such factor. Either way, poison arrows appear to have been a marginal concern, even a curiosity, within sources from the era. This is not surprising, not least because conventional armour would have also served as the best defence against such weapons. These two factors then seem to explain why there appears to be little (if any) discussion of the need to develop specialist defences against such weapons in this era within the Western military literature of the major civilisations at the time.

Water warfare also appears to have been an aspect of conflict, in the advent of larger settlements, which were increasingly dependent on human built systems to provide water to inhabitants, irrigation and to a lesser extent sanitation. There are numerous claims of the intentional poisoning of water supplies, often with toxic plants – and even pathogens in this era. All, however, suffer from a lack of evidence, and all but the crudest allegations seem technically unfeasible. However, it does seem likely that corrupting water supplies with human or animal carcasses did occur – particularly in arid areas, where communities were dependent on perhaps a single or handful of wells.[14] A question which remains then, is the extent to which it is possible to discern if causing disease, rather than dehydration, was ever a primary of secondary intention in this era.

This era is also important in attempts to trace the history of the taboo against the use of poisoning in warfare. However, the evidence base remains sparse – and of course, such norms also varied between communities and eras and would also have been subject to processes of contestation at the time. Be that as it may, there has long been a tendency, which is still largely unsubstantiated, to frame claims about the immorality of poison warfare in this era as precursors to the prohibition norms which emerged in much later centuries.

4.3.3 Medieval and Early Modern (1000–1750AD)

The scope of possible use of poisons and pathogens in the Medieval and Early Modern eras appears largely a natural extension of the types of use and allegations of use in earlier eras.

In terms of poisons, a comparable list of poisons known since antiquity was still in use. However, certain mineral poisons and toxins, appear to have been more widely available due to their applications in both medicines and in pest control. It is in this era, for example, that a range of arsenic-containing substances were widely in use in Western Europe by the sixteenth century. And indeed, advances in the refinement of mineral poisons for these purposes also contributed to the effectiveness and ease of use of these poisons for assassination and murder in this era. Indeed, concerns about assassination, and quests for antidotes even contributed to the emergence of human experimentation in regions such as Italy.[15]

In terms of the use of biological agent uses as part of siege warfare, there are also numerous allegations in this era. The majority of these have centred on claims that biological materials, such as cadavers and dead animals, were catapulted into besieged cities – with at least the secondary aim of causing disease outbreaks. However, the evidence of such claims is sparse, and the motivations for such actions, if they did take place, are hard to discern.[16] It is clear, that the purposeful spread of disease was not beyond the scope of imagination within cultures globally at this time – but the means were lacking due to an absence of understanding of how to manipulate infectious agents and understanding of the role of vectors such as flees. This then goes to show why there are allegations of purposeful disease spread as well as some indications of stigma against the purposeful spread of disease, as well as poisons – which were not distinguished. For example, from this era, and indeed, early eras there is occasion reference to stigma against the poisoning of water supplies, in both peace treaties and other works on the norms of warfare. However, while disquiet around the use of poison and disease as an aspect of warfare has long antecedents, it is only in retrospect that this can be understood to form part of a broader trend towards a general prohibition in warfare.[17]

4.3.4 Late Modern (1750–1915 AD)

The late modern era witnessed significant developments relevant to the potential and understanding of biological and toxin weapons. These advancements included progress in the isolation, characterisation and production of pathogens, which played a crucial role in the development of vaccines against diseases like smallpox, rabies, cholera and plague. Alongside a general trend towards large-scale industrial warfare, this era also saw a patchwork of colonial and asymmetrical conflicts occurring at various scales. Notably, anti-establishment movements increasingly employed terror strategies. Added to this, the prospect of biological warfare began to capture the imaginations of artists and science fiction writers.[18]

There are numerous claims of attempts to spread of agents such as yellow fever and smallpox throughout the eighteenth and nineteenth centuries. These can be understood in the context of a growing understanding of infection mechanisms through ingestion and physical contact. However, significant ambiguity remains regarding the occurrence and effectiveness of many alleged attempts.[19]

The use of poison arrows by native peoples also persisted during this era. The period also witnessed an increasing appreciation of Indigenous knowledge and utilisation of plant and animal-based toxins, including alkaloid poisons like curare from South and Central America. Several Indigenous communities employed these weapons in warfare. Reports on various European expeditions frequently mention poison arrow incidents, such as an account of Indigenous islanders in the South Atlantic archipelago using poison arrows that resulted in the death of a ship's master. The report noted that this was the third Fijian vessel master to be killed in such a manner in recent weeks.

The rapid advances in microbiology during this era captivated the public's imagination, and many of the developments would later prove fundamental to offensive and defensive programmes. However, instances of BW use, if they did occur, closely resembled previous patterns, and tended to be locally improvised on a small scale.

4.3.5 World War I (1914–1918 AD)

The first dedicated biological warfare programmes emerge during the First World War. The emergence of these programmes reflected the convergence of a wide range of factors. This included State investment in fields such as microbiology, epidemiology, vaccination and prophylaxis. These were important in as much as they would come to provide the technical principles and institutional capacities to consider and investigate the potential of BWs. The character of the First World War, and the broader historical moment it occurred, would drive battlefield innovation across air, sea and land and opened several potential niches for biological warfare.

A range of potential applications were considered and mooted in several States, and at various levels of the political and military leadership in this era, including aerially delivered agents. The character of which, at least in part, reflected the entrance of microbiology into the broader public imagination in the late nineteenth century. As well as the emergence of novel battlefield chemical weapons such as phosgene and sulphur mustard. However, it would be the clandestine use of pathogens against pack animals which would become the central focus.

The most significant programme emerged in Germany. The roots of this programme can be traced to a pre-war global intelligence network within the German Navy. Part of the function of this global network was the disruption of enemy shipping. At some point in 1915, a decision was made within the German command to launch clandestine biological attacks, utilising pathogens such as the causative agents of Anthrax and Glanders against livestock. Although much of the evidence for these programmes was destroyed at the end of the war, it is believed that shipping was targeted in several states, including Argentina, Finland (Russia), France and the United States.[20] The route of delivery for much of this work was animal ingestion – with the cultured pathogen being mixed with food and often glass – as to increase the likelihood of infection. The programme appears to have had little impact in terms of causing disease. However, the discovery of such plots, as well as later revelations about the existence of the programme, contributed to increased attention to the need to protect against such sabotage, as well as the broader spectre and potentials of biological warfare.

It also seems likely, that there were comparable if tentative explorations of the broader military potentials of toxins and pathogens in other States' programmes in this era. Exploration of toxin agents as means of warfare in other States. In the United States for example, work was conducted on the feasibility of shrapnel or cloud-delivered Ricin as a battlefield weapon.

In this era then, while a range of potential weapon systems, were being considered, and even tentatively explored, biological warfare would occupy only a modest and circumspect niche. It continues, however, to serve as prologue of the developments as part of more substantive research and development programmes in the decades which would follow. It is also worth remembering, however, that developments in this same era – including advances in chemical warfare defence, battlefield medicine, veterinary science and public health which would also later prove fundamental to States' assessment of and responses to the threat posed by BWs to soldiers and broader society.

4.3.6 Inter-War Years (1918–1939 AD)

State interest in the prospect of biological warfare grew during the inter-war years. This led to the establishment of several new State programmes, as well as attempts to forestall use by agreeing a new prohibition on use.

Seth Carus identifies five States in which dedicated programmes were first established during the period between 1918 and 1939. This includes France (1922), Hungary (1936), Italy (1934), Japan (1934) and the Soviet Union (1928).[21] In addition, Poland also established the basis of a biological warfare programme in this era.[22]

These programmes were driven by a range of factors. This included suspicion that other States were developing such weapons, with the line between defensive and offensive research usually being blurred from the outset. These programmes tended to start quite modestly, often centring on the research of a single team – but would come to vary in terms of both scale and focus. The Italian programme was perhaps the smallest – conducting only the most rudimentary research. On the other hand, the Russian and Japanese programmes had already grown substantially by the 1930s. In the case of Japan, this extended to a second larger site in Manchuria in which experiments were carried out on Humans. And in the case of Russia, by the outbreak of World War Two – its offensive programme boasted multiple dedicated research facilities, as well as three open-air test sites.[23] These programmes also varied in terms of the agents and delivery systems they considered and developed. In the case of Poland, the emphasis was on espionage by resistance forces. In France, there was exploration of various munition types, accompanied by field tests and associated meteorological work and vulnerability assessments. This included work with a range of agents, notably botulinum toxin as well as the field testing of aircraft delivered bombs, hand grenades and artillery shells.[24]

The Hungarian programme, which was smaller than the French, also explored the potential of a range of agents including the causative agents of anthrax, salmonella and dysentery – to be deployed on the battlefield, for acts of espionage against occupiers, and to allow retreating forces to contaminate territory. This extended to research and field tests with bombs and artillery shells, as well as infected projectiles. It also included attempts to increase the virulence of at least one pathogen.[25] In the Russian case, early work also focused on a range of agents and also extended to both laboratory as well as open-air field trials. And as with the Hungarian case, attempted to produce modified strains.[26]

These programmes emerged in the context of a drive to ban chemical weapons following the horrors of chemical warfare in WW1, and the long-term effects of those weapons on the soldiers that returned home. The issue of chemical warfare also became linked to the emerging societal cognisance of the threat posed by biological warfare to civilian populations. This contributed to the negotiation of a ban on use of both chemical and BWs. (1925 Geneva Protocol for the Prohibition of the Use in War of Asphyxiating, Poisonous or Other Gases, and of Bacteriological Methods of Warfare) This general prohibition was caveated by State reservations, such as the right to respond in kind to States that violated the treaty and non-signatories. The result then was essentially a 'no first use' agreement between signatories.

This was an era then, in which there was growing State interest in assessing and developing the potentials of BWs – with several states having demonstrated the technical feasibility of a range of weapon systems. However, by 1939, states did not possess useable capabilities much beyond what had been available during the First World War. It was also an era in which concrete attempts to deter use – through moral argument as well as the threat of like-for-like and chemical retaliation would emerge.

4.3.7 World War II

By 1939, there were five programmes. In three of these programmes there appears to have been little progress beyond what was achieved during the interwar years. The French and Italian programmes had ceased by the end of 1940 and the small Hungarian programme also appears to have continued up until it was destroyed in 1944. Although in all three cases, there is a scarcity of records in the public domain.[27]

In terms of programmes which continued to operate to any significant degree, by far the largest was the Japanese programme. By 1939, the programme involved a research institution in Tokyo, which focused primarily on defensive work, as well as a complex of research institutions and prison facilities in occupied Manchuria. At these latter sites, human experimentation was carried out on a massive scale as part of attempts to develop an offensive biological warfare capability, and the production of a range of pathogens had commenced. The 'fielding' of these agents had also begun, primarily against Soviet forces – through the contamination of a river, as well as the use of shells. In the following years, a barbaric programme of human experimentation continued – in which as many as 10,000 were killed. The programme also became the base for a regional terror campaign, directed at the Chinese. This involved the contamination of water-supplies and food supplies with the causative agents of diseases such as anthrax and dysentery, the arial delivery of plague carrying flees, as well as the purposeful infection of prisoners with typhoid which were then released in order to cause outbreaks.[28] Another significant wartime programme, which had already been established during the interwar years was that of the Soviet Union. During the Second World War this programme continued to progress in a number of areas, despite losing many of its best scientists in the Stalinist purges of this era. However, the extent this had resulted to systems which were available for use by the Soviets beyond sabotage operations remains unclear. Indeed, there are a number of uncorroborated accusations of this type of use against German troops.[29]

The Allies also cooperated in the establishment of three further programmes during the conflict: United Kingdom (1940), Canada (1940) and United States (1941). The British offensive programme emerged out of a smaller defensive programme, which was established, at least in part, due to concerns about the possible existence of an offensive German programme. The tri-State programme explored a range of weapons systems and agents. This included the development and stockpiling of 5,000,000 cattle-cakes laced with the causative agent of anthrax by the British, which were developed for use against German cattle. Work also included the adaption of chemical cluster munitions to deliver the same agent as well other shell-developed systems for a range of pathogens – although these were never produced at scale and would likely have been dependent on US production capabilities if they had entered service during the conflict. Many of these weapons, were tested in field-trials by the Canadian programme. The US programme would be the largest partner in this collaboration. In terms of research and development the United States explored a wide range of agents and delivery systems – producing a significant research base for later attempts to overcome the challenge of pathogen aerosolization. It also developed plans to become the main producer for weapon agents for its own arsenal and allies. Although, by the wars end, it had yet to begin building a usable stockpile, and agent production had not come online.[30] In addition, the Polish resistance movement employed both chemical and BWs as part of a sabotage campaign against the Third Reich.[31]

4.4 Conclusions

The niches occupied by BWs in warfare before 1945 were delimited by a range of factors. In the pre-modern era, the biggest obstacle to the effective use of pathogen weapons, even the most rudimentary ones, was lack of scientific understanding – particularly in relation to routes of transmission. The use of plant and animal toxins (primarily, if not exclusively) by nomadic and hunter-gather peoples in poison arrows was similarly bounded. They could only be produced at small scale, required local specialist knowledge and had limited effectiveness against increasingly well defended, organised and armed civilisations. From the nineteenth century onwards, increased understanding and practices of vaccination and the microbial causes of disease did lead to potentially effective attempts to spread contagious diseases by fomite transmission. However, in practice, it is difficult to discern the effectiveness of such actions, as they tended to be employed in contexts in which vulnerable populations were already suffering epidemics. In the early twentieth century, some of the basic conceptual obstacles began to erode. This made the production and dissemination of several pathogens, primarily through ingestion routes increasingly viable – and so it allowed pathogens to emerge as a potential form of sabotage. By the interwar period, increasing attention was being given to the delivery of pathogens and toxins as aerosols – through munitions and spray devices. However, by World War Two, even in the context of the emergence of dedicated programmes, the technical challenges of delivery had still to be overcome. It is notable, for example, that Japan, the largest and most advanced programme during that conflict, was still hampered in this regard. With regard to the prohibition on use in this era, it is clear that the Geneva Protocol was certainly a factor in State decision-making. However, it did not significantly constrain offensive development. And indeed, perversely, sometimes appears to have provided justification for the development of an offensive capability. What seems to have been more important was the threat of like-for-like or chemical retaliation. In Manchuria, the Japanese faced no such threat. However, in the European theatre, the threat of such retaliation, coupled with the very modest capabilities of Western States meant that the use of such weapons by States was both impractical and undesirable – beyond more modest deniable sabotage campaigns. This, of course, proved less of a deterrent to resistance groups in this era, but these groups had much more limited means available to them. This then kept the niche for BW small and, in many respects, comparable to the era that had preceded it.

Author Biography

Dr. Brett Edwards is Senior Lecturer in Security and Public Policy at the University of Bath. His research focuses on historical and contemporary dimensions of chemical and biological weapons and warfare – including disarmament and non-proliferation. He hosts a podcast series on this topic called 'Poisons and Pestilence'.

References

1 Geissler, E. and van Courtland Moon, J. E. (1999) *Biological and Toxin Weapons: Research, Development and Use from the Middle Ages to 1945*, Oxford University Press, Oxford.

2 Wheelis, M. (1999) Biological warfare before 1914. in Geissler, E. and Moon, J. E. V. C. (Eds.) *Biological and Toxin Weapons: Research, Development and Use from the Middle Ages to 1945*, Oxford University Press, Oxford.

3 Carus, W. S. (2017) *A Short History of Biological Warfare: From from Pre-History to the 21st Century*, Government Printing Office, Washington.
4 Carus, W. S. (2015) The history of biological weapons use: what we know and what we don't. *Health Security*, 13 (4), 219–255.
5 Carus, W. S. (2002) *Bioterrorism and Biocrimes: The Illicit Use of Biological Agents since 1900*, Fredonia Books, Amsterdam.
6 Carus, W. S. (2017) A century of biological-weapons programs (1915–2015): reviewing the evidence. *The Nonproliferation Review*, 24 (1–2), 129–153.
7 Borgia, V. (2019) Chapter 1 - The Prehistory of Poison Arrows. in Wexler, P. (Ed.) *Toxicology in Antiquity*, 2nd ed., Academic Press, Cambridge, Massachusetts.
8 Lombard, M. (2020) The tip cross-sectional areas of poisoned bone arrowheads from southern Africa. *Journal of Archaeological Science: Reports*, 33 (1 October 2020). 102477.
9 Wexler, P. (2014) *History of Toxicology and Environmental Health: Toxicology in Antiquity Volume I*, Elsevier Science, Amsterdam.
10 Wexler, P. (2014) *History of Toxicology and Environmental Health: Toxicology in Antiquity Volume I*, Elsevier Science, Amsterdam. Chaps. 4, 5.
11 Carus, W. S. (2015) The history of biological weapons use: what we know and what we don't. *Health Security*, 13 (4), 219–255. 225.
12 Carus, W. S. (2015) The history of biological weapons use: what we know and what we Don't'. *Health Security*, 13 (4), 219–255. 229.
13 Ibidem.
14 Del Giacco, L. J. *et al* (2017) Water as a weapon in ancient times: considerations of technical and ethical aspects. *Water Supply*, 17 (5), 1490–1498.
15 Rankin, A. (2021) *The Poison Trials: Wonder Drugs, Experiment, and the Battle for Authority in Renaissance Science*, University of Chicago Press, Chicago.
16 Zanders, J. P. (2022) 'Catapulting Cadavers: A Medieval Practice of Biological Warfare? Working Paper', *The Trench, Historical Note Series* 2 Retrieved from https://www.the-trench.org/catapulting-cadavers.
17 Price, R. M. (2018) *The Chemical Weapons Taboo*, Cornell University Press, Ithaca.
18 Carus, W. S. (2020) Perspectives on 'bioterrorism' in the nineteenth century: the philosophy of mass destruction, fake news, and other fictions. *The Nonproliferation Review*, 27 (4–6), 267–275.
19 Carus, W. S. (2015) The history of biological weapons use: what we know and what we don't. *Health Security*, 13 (4), 231–232.
20 Carus, W. S. (2017) *A Short History of Biological Warfare: From from Pre-History to the 21st Century*, Government Printing Office, Washington, 12–13.
21 Carus, W. S. (2017) A century of biological-weapons programs (1915–2015): reviewing the evidence. *The Nonproliferation Review*, 24, 129–153.
22 Carus, W. S. (2017) *A Short History of Biological Warfare: From Pre-History to the 21st Century*, Government Printing Office, United States. 12: 22.
23 Leitenberg, M. and Zilinskas, R. A. (2012) *The Soviet Biological Weapons Program: A History*, Harvard University Press, Cambridge, 22.
24 Rózsa, L. and Nixdorff, K. (2006) Biological Weapons in Non-Soviet Warsaw Pact Countries. in Wheelis, M. (Ed.) *Lajos Rózsa, and Malcolm DandoDeadly Cultures: The History of Biological Weapons since 1945*, Harvard University Press, Cambridge, 158.

25 Rózsa, L. and Nixdorff, K. (2006) Biological Weapons in Non-Soviet Warsaw Pact Countries. in Wheelis, M. (Ed.) *Lajos Rózsa, and Malcolm Dando Deadly Cultures: The History of Biological Weapons since 1945*, Harvard University Press, Cambridge, 158.

26 Leitenberg, M. and Zilinskas, R. A. (2012) *The Soviet Biological Weapons Program: A History*, Harvard University Press, Cambridge, 21.

27 Carus, W. S. (2017) *A Short History of Biological Warfare: From Pre-History to the 21st Century*, Government Printing Office, United States, 12–27.

28 Carus, W. S. (2017) A century of biological-weapons programs (1915–2015): reviewing the evidence. *The Nonproliferation Review*, 24 (1–2), 16–18.

29 Carus, W. S. (2017) A century of biological-weapons programs (1915–2015): reviewing the evidence. *The Nonproliferation Review*, 24 (1–2), 21–22.

30 Carus, W. S. (2017) A century of biological-weapons programs (1915–2015): reviewing the evidence. *The Nonproliferation Review*, 24 (1–2), 24–25.

31 Petersen, R. (2020) The covert battlefield: doctor Witaszek, the WKZO, and the polish use of biological and chemical warfare against the third Reich. *The Nonproliferation Review*, 27, 4.

5

Biological Weapons from 1946 to 2000

Brian Balmer

Department of Science and Technology Studies, University College London, London, UK

Key Points
1) Interest in biological weapons between 1946 and 2000 was relatively widespread but use was much rarer. 2) Not all programmes focused solely on battlefield, anti-personnel weapons. Historically, there were different concepts about what constitutes a biological weapon. 3) The development of biological weapons is not inevitable; history shows that we should avoid technological determinism.

Summary

Rather than presenting a country-by-country summary of State biological warfare (BW) programmes in the late twentieth century, this chapter instead explores critical themes from the history of BW. The chapter begins with a brief overview of the main State-sponsored BW programmes in this period. It then examines key topics: offensive aspects of programmes; human exposure and experimentation; non-State actors; and consideration of the main drivers and inhibitors of state programmes. The conclusion will use the concept of technological determinism to discuss how the history of BW should make us wary of seeing the development of biological weapons as unavoidable.

5.1 Introduction

The Second World War provided impetus to several State-sponsored biological warfare (BW) research programmes, several of which continued into the Cold War. This chapter provides an overview of BW activity in this period after 1945 until the end of the twentieth century. Rather than presenting a detailed country-by-country summary of State BW programmes in the late twentieth century, where there are already several easily accessible

Essentials of Biological Security: A Global Perspective, First Edition. Edited by Lijun Shang, Weiwen Zhang, and Malcolm Dando.

academic sources that readers can consult, the chapter will instead explore key topics from this period in the history of BW. Following a brief overview of the major State BW programmes, the chapter turns to examine the offensive aspects of several programmes. It then discusses human experimentation and exposure to experimental agents. State involvement was not the only BW activity and so the chapter will briefly consider the role of non-State actors and BW. The final topic in the chapter focuses on the main drivers and inhibitors of State BW programmes.

Three main points arise from this overview chapter. The first is that interest in biological weapons between 1946 and 2000 was relatively widespread but use was much rarer. Secondly, not all programmes focused solely on battlefield, anti-personnel weapons. Historically, there were different concepts about what constitutes a biological weapon. Finally, this chapter argues against technological determinism, which is the view that technology builds ineluctably on prior technology, with no role for social or political intervention that would otherwise shape and direct that technology along different pathways. Indeed, this overview indicates that development of biological weapons was by no means inevitable.

5.2 Overview of State BW Programmes

In his thoroughly researched overview of twentieth-century BW, Carus points to the difficulty of even determining what constitutes a BW programme.[1] He indicates that many attempts to create typologies of past programmes focus on organized, large-scale, science-intensive undertakings, at the expense of smaller, cruder or less co-ordinated activities. Carus then reviewed BW allegations, made between 1915 and 2015, about 42 states and categorised them as having: known programmes, probable programmes, possible programmes and none. For the purposes of this introductory chapter, we will focus on the known offensive programmes, which were: Canada (1940–1958); Egypt (1960s); France (various periods between 1916 and 1967); Germany (1915–1918); Iraq (various periods between 1974 and 1991); Israel (1948-?); Italy (1934–1940/43?); Japan (1934–1945); Poland (no dates); Rhodesia (1977); South Africa (1981–1993); Syria (1970s-?); United Kingdom (1940–1957); United States (1941–1971); and Union of Soviet Socialist Republics (USSR) (1928–1991). Carus also notes that China, Hungary, Iran and North Korea fall into the 'probable' category, and Russia after 1992 (i.e. not USSR) into the 'possible' category.

State-sponsored BW programmes shared the characteristic of secrecy. Depending on the country involved, this need for secrecy sprang from an awareness of widespread moral censure, international legal prohibition, and concerns about national security. Such factors ensured that unusual measures were taken to keep BW activities, together with their associated military strategy and government policy, hidden. In several programmes, research was hidden in plain sight, established as, or embedded in, ostensibly civilian research facilities (for example, Biopreparat in the USSR; in South Africa, the Roodeplaat Research Laboratories and Delta G Scientific). In the USSR, entire cities housing BW research facilities, such as the city of Stepnogorsk, were closed off. An important corollary of this secrecy is that histories of biological weapons are inevitably filled with absences and uncertainties, even for countries where relatively rich archival sources are available.

The Soviet BW project was by far the largest, most secretive and longest-lived research programme in the twentieth century. It has been estimated that during the Cold War, there were some 60,000 personnel, spread across 40–50 institutions, working on aspects of BW.[2] Leading Western experts on the history of the Soviet programme divide it into two major phases.[3] The first phase grew from the Soviet chemical weapons programme, lasted from around 1928–1971 and adopted an offensive policy in 1928. Work in this period concentrated on classical microorganisms but was inhibited by State-sanctioned Lysenkoism, a view which denied Mendelian genetics and Darwinian evolution, and which persisted until the mid-1960s. The next phase commenced at the behest of the Central Committee of the Communist Party and the USSR Council of Ministers. Consequently, the Soviet programme began an enormous expansion shortly before the USSR signed the 1972 Biological Weapons Convention (BWC). The BW programme eventually spread across numerous sites, undertaking research and development, agent production, testing and weapons design and production. During this second phase, scientists drew intellectually and practically on the burgeoning field of recombinant deoxyribonucleic acid (DNA) technology, or more colloquially genetic engineering. Leitenberg et al. pointed out that the policy guiding this research and future use of biological weapons remains shrouded in secrecy, although they infer that weapons were most likely being envisaged for longer-term strategic, rather than tactical, battlefield use.[4] In the years before the Soviet Union collapsed, and into the early days of a new Russia, several Western countries placed pressure first on President Gorbachev and later Yeltsin to close their BW programme and to offer greater transparency about their BW activities. Yeltsin, in 1992, publicly revealed that there had been a programme and issued a decree to terminate it. That said, a recent historical overview of the programme indicates several points of doubt, such as continued secrecy around some organisations formerly involved in the programme, that call the decree into question.[5]

The United States, United Kingdom and Canadian BW research efforts post-WWII were intertwined through a formal Tripartite collaboration between their programmes. This arrangement started during WWII and involved exchanges of information, regular meetings and visits between scientific research sites. In terms of scale and scope, the US programme was by far the senior partner in the alliance. Around 3400 personnel worked in the US programme during the early Cold War, compared with around 225 in the United Kindom.[6] Guided initially by a policy of 'no first use', which in 1956 shifted to a policy of use as decided by the President, the US offensive programme ran until 1969.[7] At this point, President Nixon publicly and unilaterally abandoned the US offensive programme. Throughout this period, BW research funding was often outcompeted for priority by alternatives such as chemical and nuclear weapons, with military interest only being roused by the Korean War and later in the search for alternatives to full-scale nuclear war.

The UK programme commenced in 1940, shortly after the fall of France to the Nazis, although its senior civil service architect, Maurice Hankey, had advocated for a project for some time.[8] After the end of WWII, the United Kingdom government invested considerable resources into its chemical and biological weapons research programmes. In accordance with UK policy, the BW research programme aimed to create a retaliatory capacity in anti-personnel weapons and, for a short period of time following the war, research was accorded the highest priority, equal to atomic weapons research. This status gradually eroded as the Cold War advanced. Economic constraints, the acquisition of an independent nuclear deterrent and the decision to join the race for thermonuclear weapons, all impacted detrimentally on the priority that had been accorded to the UK programme. Although there is no firm date in the archives of a decision to change policy, by the mid-1950s most, if not all, official documents were referring to the UK programme as defensive. Turning to the third Tripartite member, Canada did not pursue offensive anti-personnel weapons during this period but did make its basic research and large area outdoor test sites available to the offensively oriented UK and USA programmes' staff.[9] Canada did, however, pursue anti-livestock weapons, with its research in this field located at Grosse Isle on the St Lawrence River, Quebec.

A French BW programme ran between 1921 and 1940. When the programme recommenced in 1947, it commanded a substantial and well-resourced research and development effort until 1956.[10] As with the United Kingdom, the French military leaders at this time initially regarded biological weapons as holding equivalent potential to nuclear weapons. After 1956, the programme was overshadowed by nuclear weapons and policy, leading to a steady decline in priority and budget. Lepick notes that in 1967 the term 'aggression', used to denote offensive work, ceased to be used in the minutes of the main co-ordinating group for the programme, the Sous Groupe de Travail et d'Etudes Biologiques (SGTEB).[11] A few additional half-hearted attempts were made to revive interest in BW research, but although not formally decided on, 1969 marked the effective end of the offensive programme; French accession to the BWC in 1972 formalised that end.

Compared to chemical and nuclear weapons, Iraq's biological weapons programme was 'the poor cousin of Iraq's Weapons of Mass Destruction (WMD) programmes. It was the last to start, the least well-funded, the least well-staffed, the worst managed and the most covert' (p.114).[12] The programme ran in two phases. Apparently in response to the threat from Israel and Iran, a small research effort was launched but terminated in 1979 without producing significant results.[13] The programme restarted in earnest in 1983, in the midst of the Iran–Iraq war and ran until 1991, when Iraq surrendered at the end of the Gulf War and the UN Security Council made destruction of its WMD capability a condition of surrender.

Code named, Project Coast, the South African BW programme ran between 1981 and 1993. Unlike many of its larger State counterparts, research in the BW programme focused primarily on toxins and poisons, with little distinction made between chemical and biological agents. The main aim of the BW work was to produce agents for use in small-scale assassination and counter-insurgency situations as a support for the apartheid regime.[14] Work was also carried out on creating a fertility controlling vaccine, most likely intended for covert use to control the birth rate in the country's black population. South Africa also had a separate chemical weapons programme, which aimed to produce riot-control agents, protective equipment and detectors. Also with the goal of counter-insurgency, the Rhodesian programme was a combined chemical and biological effort.[15] It involved a small research team, with most effort concentrated on chemical weapons. There is evidence that the BW work involved culturing anthrax and cholera, attempts to disseminate agents in a liquid slurry, and some work on creating cigarettes poisoned with anthrax bacilli.

Carus' survey of BW programmes, and in a later historical overview, also mentions Israel, Egypt and Syria in his 'known' category.[16] Israel initiated a research programme in 1948 and shortly afterwards attempted to use biological agents against British, Egyptian, Syrian troops and Palestinian settlements by contaminating water supplies. Carus adds that a BW programme was later established in the Israeli Ministry of Defence, but that no further reliable information exists on its activities. Little is known about the Syrian programme, which appears to have started in 1990 and involved work on ricin toxin; likewise, beyond its establishment of a research programme in the 1960s, almost nothing is known about Egypt's involvement with BW. Carus notes that China, Iran and N. Korea are likely to have possessed their own BW programmes about which very little is known.

The following thematic sections provide some further detail about these BW programmes. In this short overview, it is impossible to be comprehensive and the examples should be regarded as indicative of particular features of biological weapons activities.

5.3 Offensive Aspects of BW Programmes

The key feature that distinguishes offensive from defensive BW programmes is the pursuit, in offensive programmes, of weapons designed to kill or incapacitate humans, crops or livestock (so-called anti-personnel, anti-crop and anti-animal). Beyond this clearly offensive pursuit, the boundaries between offensive and defensive research can blur, for example, work on pathogens could ostensibly be used to create weapons or alternatively to create novel vaccines. This section will not catalogue every microorganism researched as a potential biological weapon but instead highlights key developments and activities within offensive programmes in the latter part of the twentieth century.[17]

From work on a wide range of microorganisms, the US post-WII BW programme standardized or weaponized the following organisms[7]:

Bacillus anthracis (lethal)
Franciscella tularensis (lethal)
Bruccella suis (incpacitating)
Coxiella burnetti (incapacitating)
Yellow Fever virus (lethal)
Venezeulan equine encephalitis (incapacitating)
Botulinum toxin (lethal)
Staphylococcal enterotoxin Type B (incapacitating)
Saxitoxin (lethal)

List note: within the context of this chapter, standardisation refers to the selection of one weapon type from a range of possible choices. In a similar list, Leitenberg et al. (p. 288) omit saxitoxin and yellow fever virus.[18]

In addition, the US military standardised anti-plant agents that caused wheat stem rust and rice blast disease. Kirby argues that the US programme encompassed different ideas about what constitutes a biological weapon.[19] He notes that the United States ended WWII with some BW capability in the form of cluster munitions (where several bomblets are packed into a single missile). After the war, BW research continued to focus on the acquisition of biological bombs, now envisaged as weapons that would complement nuclear weapons. As the United States became more enamoured with nuclear weapons, interest in biological weapons declined. Within the BW programme, Kirby notes, BW was re-conceptualised, as studies from the early 1950s onwards investigated the potential of using sprayed clouds of pathogens to cover huge tracts of land (or sea) in a biological attack, this was named the Large Area

Concept. In this respect, several series of large scale, outdoor tests were carried out including Operation Saint Jo, which commenced in 1953 and involved simulated anthrax attacks in St. Louis, Minneapolis and Winnipeg, and later, in the 1960s, Project 112, consisting of around 50 separate large-scale trials of both chemical and biological agents in the Pacific.[20] A third conceptualization of BW, as a sabotage or assassination weapon on a small scale, ran alongside the battlefield concepts of biological weapons as bombs or large area sprays.

The UK BW programme mirrored the US evolution regarding what constituted a biological weapon. In addition, prior to WWII, a special advisory committee to the UK government had conceptualized BW in terms of the disruption to public health following the damage to infrastructure from a conventional bombing raid.[21] During and after WWII, the UK programme prioritised the search for a biological bomb, with research focus shifting throughout the 1950s to the Large Area Concept along with the policy drift to a defensive programme. By this point, military planners and defence scientists regarded the United Kingdom as a potentially vulnerable target for a large-scale spray attack.

The Soviet programme developed 13 officially approved pathogen and delivery systems.[22] Despite research interest in the emerging field of recombinant DNA technology, these systems were all based on traditional agents. While the Soviet BW scientists investigated a very wide range of organisms, 10 were eventually validated as suitable for weaponization:

Bacillus anthracis (lethal)
Brucella species (incapacitating)
Coxiella burnetii (Q fever) (incapacitating)
Franciscella tularensis (lethal)
Burkholderia pseudomallei (possibly) (Melioidosis) (lethal?) [sic]
Yersina pestis (lethal)
Marburg virus (lethal)
Venezeulan Equine Encephalitis virus (incapacitating)
Variola virus (lethal)
Botulinum toxin (lethal)

In terms of delivery systems, the Soviet Union again mirrored the United States, with the Soviet programme developing bomblets that could be loaded into cluster munitions and at least two wide-area spray dispersal systems for use from aeroplanes. They also tested a short-range ballistic missile although this never became part of the Soviet arsenal.

As Leitenberg and Zilinskas point out 'past state programmes have investigated many pathogens for biological weapons use, but in the end few have been found that possess the combination of characteristics that make them attractive for militaries to attempt to weaponize' (p. 287).[23] Indeed, factors such as fragility, availability of protection, incubation period and many other variables mean that the list of potential pathogens for candidate biological agents is far longer than the list that has historically been weaponised.

5.3.1 Human Exposure and Experimentation

We know of experiments that exposed humans to biological agents, either as part of the aim of the experiments or indirectly, in at least two post-war national BW research programmes: the United Kingdom and the United States. In the US programme, Operation

Whitecoat ran between 1955 and 1973, it involved volunteer Seventh Day Adventists and other conscientious objectors in trials with a wide variety of diseases, for example, Q Fever, tularemia and typhoid.[24] Beyond Operation Whitecoat scientists sprayed aerosols and injected volunteers with various agents in military settings, such as Fort Detrick at the heart of the US biological weapons programme, and also civilian settings such as Ohio State Penitentiary. Scientists also undertook outdoor BW trials which exposed human populations to non-pathogenic organisms such as *Bacillus globigii* and *Serratia marcescens*, as well as with fluorescent zinc cadmium sulphide. All were intended to simulate the behaviour of a pathogenic cloud of organisms so that scientists could monitor their potential spread over large areas. The trials took place in the 1950s and 1960s (including Operation St Jo, mentioned earlier), in rural and urban areas as far afield in the USA as Florida, San Francisco and New York, together with several other sites. As late as 1965, scientists conducted spray tests at a Greyhound bus terminal and airport in Washington DC; and a year later in the New York City subway.

The United Kingdom's involvement with human experimentation was more tangential. No volunteer studies akin to Operation Whitecoat occurred. Nonetheless, and as with the United States, outdoor field trials using pathogens, non-pathogenic organisms and zinc cadmium sulphide formed part of the UK research programme.[25] During the offensive phase of the programme, several series of tests using sprays and munitions took place at sea. In one test off the coast of Stornaway in the Hebrides, a trawler accidentally strayed into the test area just as scientists exploded a bomb containing plague bacteria. Without intervening, which would have risked compromising the secrecy of the test, scientists and naval crew tracked the trawler for several weeks. This tracking was mainly to ensure there was no outbreak of plague but in that period scientists also gained empirical proof of the otherwise theoretical calculation that the crew would not have been infected. As the research programme moved into a defensive phase in the mid-1950s and into the 1960s, scientists conducted tests of the Large Area Concept (discussed earlier), initially with clouds of zinc cadmium sulphide, sprayed from aeroplanes and later the non-pathogenic bacteria *B. globigii* and *Escherichia coli*, sprayed from naval vessels operating off the south coast of England. While humans were not the subjects of the experiments, which were designed to map the spread of the cloud, and no data was gathered about human exposure, the experimental clouds all disseminated over populated areas.

5.4 Non-state Actors

This overview has focused on State-sponsored BW programmes, but terrorist groups have also shown sporadic interest in using biological weapons. As with State programmes, this interest in acquiring biological weapons by such groups has outpaced actual use. One explanation is offered by Moodie, in a scholarly essay on the analogous case of chemical weapons terrorism.[26] He notes that terrorist groups are far from homogenous and therefore the motivation to acquire and use chemical or biological weapons may differ. Hence, a group with strong and defined political goals might avoid a weapon that is as unorthodox and unpredictable as a biological or chemical weapon. Such use might jeopardize any attempt to be regarded as credible political actors. In contrast, Moodie argues, quasi-religious or similar groups, with less tangible or more apocalyptic goals might find unorthodoxy an attractive feature of biological or chemical weapons.

Indeed, the confirmed instances in the late twentieth century of serious attempts by terrorist groups to acquire biological weapons have fallen under Moodie's second, quasi-religious, category. Carus has described the attempts by an American group of teenagers, calling themselves RISE, to acquire biological weapons in the early 1970s.[27] The RISE manifesto set out its aim, in light of the detrimental impact of humanity on the global environment, to eliminate all humans except for RISE members. Carus explains how the group managed to procure samples of pathogens, sometimes by simply writing to an organization such as a hospital. The group acquired samples of *Corynebacterium diphtheria, Neisseria meningitidis, Salmonella typhi, Shigella sonnei* and possibly *Clostridium botulinum*. While RISE planned various scenarios for use, it appears that their most advanced plan was an attack on water treatment plants around Chicago. The plot was foiled by members of the group abandoning the plan and reporting RISE to the police. Carus notes that the group was unlikely to have succeeded as it had made an extremely poor attempt at culturing their samples of microorganisms.

A more widely known, and more successful, use of biological weapons was in Oregon, United States, in 1984, by a quasi-religious cult known as the Rajneeshee cult. The cultists, followers of the guru Bhagwan Shree Rajneesh, had moved from India to Oregon in 1981 and built a sizeable commune with some 2000 resident cult members by the time of the BW attack. Wheelis and Suishima describe the increasingly fractious relationship between the cult members and local authorities, as the cult tried to expand its territory and, at the same time, fell under suspicion for violating immigration laws.[28] The solution for the cult members was to attempt to get their own members elected to the county government and form a majority. To help ensure their success, the cult travelled across the United States recruiting homeless people to its ranks, moving them to its commune and registering them to vote. A second tactic was to resort to biological weapons and try to contaminate food and drink in public places with *Salmonella enteritica,* aiming to incapacitate enough voters and prevent them from participating in the upcoming local election. As a test, glasses of salmonella-tainted water were given to two county commissioners paying a visit to the commune, making both ill. The next step, also a test for a larger-scale attack, involved cult members contaminating food at several restaurant salad bars in the city of The Dalles, Oregon. As a result, at least 751 cases of salmonella were identified by public health officials. The full attack never came as the cult fell apart because of internal conflicts, and when many of the leadership team fled to Europe, it attracted the attention of the police authorities. *Salmonella* samples were found at the commune and linked to the restaurant outbreaks, with several of the lead perpetrators eventually prosecuted.

Towards the end of the twentieth century, yet another quasi-religious group, Aum Shinrikyo, in Japan, attempted to use biological weapons and successfully carried out chemical weapons attacks.[29] Between 1990 and 1995, cult members attempted to use botulinum toxin and *Bacillus anthracis* in a variety of public settings and disseminated through a variety of routes, such as water contamination, sprays and balloons. The cult aimed to poison the entire Japanese nation in pursuit of religious purification and their own salvation. While none of the group's attempts at bioterrorism succeeded, Aum carried out attacks using a chemical weapon, sarin nerve agent. The first attack, in Matsumoto city in 1994, involved cultists spraying sarin from a specially designed truck, killing 7 people and injuring around 250. Police initially decided this had been an accident rather than a terrorist incident. The second attack was more high-profile and took place on 20th March 1995 on the Tokyo underground system. Here, the sarin left 17 in critical condition, a further 37 with serious symptoms (shortness of breath, vomiting, severe headache, muscular twitching, gastrointestinal problems) and around 1000 others with mild symptoms. Returning to Aum's failed attempts at launching BW attacks, Ben Ouagrham-Gormley argues that it was not simply lack of technical expertise, but also the authoritative, hierarchical and secretive nature of the cult that contributed to this failure.[30]

Although stretching just beyond the period covered by this overview, it is important to mention the post-19/11 anthrax attacks that occurred in the weeks following the 2001 Al Quaeda attack on the Twin Towers and Pentagon in the United States.[31] The biological weapons attack involved two batches of mail contaminated with anthrax spores, which were posted to media outlets and politicians. Five people were killed and 17 others became ill, including postal workers who had been in the vicinity of the letters at mail sorting facilities. Guillemin's interviews with postal workers from these facilities also drew attention to the invisible damage from the attack, as several surviving workers described the severe stress and anxiety they suffered, particularly on return to their work environment.[32] While the content of the letters implied that Al Quaeda was responsible for the attack, the prime suspect turned out to be a US biodefence scientist, Bruce Ivins. Ivins, most likely, was motivated by the lack of attention he felt was being paid to the BW threat; he was never charged as he committed suicide shortly before police could arrest him. Finally in this section, it is worth noting that Al Quaeda undertook biological weapons research in the 1990s, but their work terminated because of the 2001 US invasion of Afghanistan.[33]

5.5 Drivers and Inhibitors of State BW Programmes

Stepping back from the features of particular BW activities, it is possible to identify a number of factors that have acted as drivers and barriers to BW programmes.[34] The list, drawn up by Dando et al., draws together some key themes, therefore a précis and slight modification of the list is worth reproducing here:

Drivers:

- Military conflict or impending military conflict.
- Scientific discoveries that indicate the effectiveness of BW.
- Intelligence sources indicating hostile nations involvement in BW.
- Dedicated organisations for BW research, development and deployment; 'product champions' associated with such organisations who advocate for BW.
- Weapons, equipment and doctrine for BW adopted by military forces.
- Secrecy, particularly around defensive programmes, so actual intent of programmes remains opaque.
- Blurring and interaction between defensive and offensive programmes.

Barriers:

- Scientific and technical difficulties in realizing the theoretical potential of BW.
- Prevention of the formation of dedicated BW organisations.
- Moral opprobrium or general disapproval of BW.

- International arms control agreements.
- Public support for the prohibitory norm against BW embedded in international arms control agreements.
- Transparency and openness in biodefence.

Two supplementary points can be made concerning these barriers. Firstly, advocacy for the prohibition of biological weapons since 1945 (and so, in part, preceding the norm enshrined in the BWC) was evident among some organized social movements. In Europe and the United States, for example, non-governmental protest groups, such as the Women's International League for Peace and Freedom (WILPF) and the Campaign for Nuclear Disarmament (CND) at times have been highly engaged in public education about and protests against biological weapons.[35] On the international stage, the Pugwash group of scientists concerned about weapons of mass destruction have also contributed to 'soft diplomacy' and provided expertise on biological arms control and disarmament. It is difficult to assess the specific impact of these campaigning groups, but their role cannot be discounted as part of creating a general atmosphere of opposition to BW.

A second point is that a range of relatively recent work in history and sociology of science has argued that 'technical difficulties', as a barrier to bioweapons cannot be divorced from its wider social and political context.[36] Such studies have shown that often a lack of so-called tacit knowledge, the difficult to codify, everyday knowledge gained in the laboratory can act as an inhibitor to the proliferation of biological weapons. Tacit knowledge is often compared to the knowledge needed to ride a bike, where a set of written instructions cannot capture all the knowledge needed to competently become a bike rider. This knowledge is gained through social interaction between skilled scientists and engineers, rather than from written instructions. Earlier in this chapter, we saw how tacit knowledge might have played out with the RISE and Aum Shinrikyo cases, where access to resources, including samples of biological agents, did not automatically provide the tacit expertise to culture and weaponize these agents. Moreover, historical studies discussed earlier of the Soviet, United States and some smaller BW programmes have pointed to the crucial role played by the acquisition of tacit knowledge, but also to wider organizational features such as secrecy and hierarchy, that helped determine the success or failure of parts of each programme.

5.6 Conclusions

This brief historical overview demonstrates that interest in biological weapons between 1946 and 2000 was relatively widespread but use was much rarer. Moreover, as Carus has pointed out, during this period there existed only two large-scale programmes, the United States and USSR, with some smaller programmes such as France and the United Kingdom, but with many existing as small-scale operations, often focused on relatively crude sabotage or counter-insurgency weapons.[33] A related point is that the historical record indicates there were different concepts about what constitutes a biological weapon. In the 1945–2000 period, the dominant paradigms for biological weapons were bombs (or bomblets in cluster munitions), large area sprays, and sabotage weapons.

Each programme involved choices by actors such as scientists, politicians, civil servants and the military. So, we can point to more than technical factors when trying to account for the success or failure of attempts to acquire or use biological weapons. As stated at the outset, this chapter argues against technological determinism, which is the view that technology simply builds on prior technology such that advances in technology are inevitable. Determinism dismisses the important role of social or political intervention that shapes and directs any technology along different pathways. Acknowledging the broader range of factors involved in driving or inhibiting BW suggests that contemporary attempts to intervene in the control and abolition of these weapons, while challenging, are not destined to fail. Again, this is because history shows us that the development of BW science and technology is socially shaped rather than inevitable.

Author Biography

Dr. Brian Balmer is Professor of Science Policy Studies, Department of Science and Technology Studies, University College London. His research interests focus on the nature of scientific expertise, and the role of experts in science policy formation, particularly within the life sciences. His specific interests combine historical and sociological approaches and include the history of chemical and biological warfare. He has also published work on policies for biotechnology and genetics; scientific migration and the 'brain drain'; the role of volunteers in biomedical research; and science policy and the sociology of science.

References

* Denotes key readings. For edited volumes, the key reading is flagged at the first reference but refers to the whole book.

***1** Carus, S. W. (2017) A century of biological-weapons programs (1915–2015): reviewing the evidence. *The Nonproliferation Review*, 24, 129–153.
2 Leitenberg, M. (2001) Biological weapons in the twentieth century: a review and analysis. *Critical Reviews in Microbiology*, 27 (4), 267–320.
***3** Leitenberg, M. *et al* (2012) *The Soviet Biological Weapons Program: A History*, Harvard University Press, Cambridge Mass.
4 Leitenberg *et al* 2012. op. cit.
5 Kuhn, J.H. and Leitenberg, M. (2016) Chapter 4: the Soviet biological warfare program. in Lentzos, F. (Ed.) *Biological Threats in the 21st Century*, Imperial College Press, London, 79–102.
6 Filippa, L. (2016) *Biological Threats in the Twenty-First Century*, Imperial College Press, London, 43–67.
***7** Wheelis, M. *et al* (2006) *Deadly Cultures: Biological Weapons since 1945*, Harvard University Press, Cambridge Mass, 9–46.
8 Balmer, B. (2001) *Britain and Biological Warfare: Expert Advice and Science Policy, 1935–1965*, Palgrave, Basingstoke.
9 Avery, D. (2014) *Pathogens for War: Biological Weapons, Canadian Life Scientists, and North American Biodefence*, University of Toronto Press, Toronto.
10 Lepick, O. (2006) Chapter 5: the French biological weapons program. in Wheelis, M. *et al* (Eds.) *Deadly Cultures Biological Weapons since 1945*, Harvard University Press, Cambridge, 108–131.

11 Lepick. op. cit.
12 Trevan, T. (2016) Chapter 5: the Iraqi biological warfare program. in Lentzos, F. (Ed.) *Biological Threats in the 21st Century*, Imperial College Press, London, 113–129.
13 Pearson, G. (2006) Chapter 8: the Iraqi biological weapons program. in Wheelis, M *et al* (Eds.) *Deadly Cultures Biological Weapons since 1945*, Harvard University Press, Cambridge, 169–190.
14 Gould, C. and Hay, A. (2006) Chapter 9: the south African biological weapons program. in Wheelis, M. *et al* (Eds.) *Deadly Cultures Biological Weapons since 1945*, Harvard University Press, Cambridge, 191–212.
15 Cross, G. (2017) *Dirty War: Rhodesia and Chemical Biological Warfare: 1975–1980*, Helion & Company, Solihull.
16 Carus, S. W. (2017) *A Short History of Biological Warfare: From Pre-History to the 21st Century*, National Defence University Press, Washington D.C.
17 Moon, 2006, op. cit.
18 Leitenberg *et al* 2012. op. cit.
19 Kirby, R. (2007) *The Evolving Role of Biological Weapons*, Army Chemical Review, Army Chemical School, Fort Leonard Wood.
***20** Guillemin, J. (2005) *Biological Weapons: From the Invention of State Sponsored Programmes to Contemporary Bioterrorism*, Columbia University Press, New York.
21 Balmer 2001 op. cit.
22 Leitenberg *et al* 2012. op. cit.
23 Leitenberg *et al* 2012. op. cit.
24 Regis, E. (1999) *The Biology of Doom: The History of America's Secret Germ Warfare Project*, Henry Holt, New York.
25 Balmer 2001 op. cit.
26 Drell, S. D., Sofaer, A. D. and Wilson, G. D. (Eds.). (1999) *The New Terror: Facing the Threat of Biological and Chemical Weapons*, Hoover Institution Press, Stanford.
27 Carus, W. S. (2016) Chapter 7: RISE, the Rajneeshees, aum Shinrikyo and Bruce Ivins. in Lentzos, F. (Ed.) *Biological Threats in the 21st Century*, Imperial College Press, London, 171–197.
28 Wheelis, M. and Sugishima, M. (2006) Chapter 14: terrorist use of biological weapons. in Wheelis, M. *et al* (Eds.) *Deadly Cultures Biological Weapons since 1945*, Harvard University Press, Cambridge, 284–303.
29 Wheelis and Sugishima, 2006, op. cit.
30 Ouagrham-Gormley, S. B. (2012) Barriers to bioweapons: intangible obstacles to proliferation. *International Security*, 36 (4), 80–114.
31 Carus, 2016, op. cit., pp. 171–197.
32 Guillemin, J. (2011) *American Anthrax: Fear, Crime, and the Investigation of the Nation's Deadliest Bioterror Attack*, Times Books, New York.
33 Carus, 2017, op. cit.
34 Dando, M. *et al* (2006) Chapter 17: analysis and implications. in Wheelis, M. *et al* (Eds.) *Deadly Cultures Biological Weapons since 1945*, Harvard University Press, Cambridge, 355–373.
35 Balmer, B. (2020) 'Science was digging its own grave': the Women's International League for Peace and Freedom and the campaign against chemical and biological warfare. *The Nonproliferation Review*, 27, 323–341.
36 For example, Ben Ouagrham-Gormley, 2012, op. cit.

6

The Problem of Dual Use in the Twenty-first Century

Kathryn Nixdorff

Department of Microbiology and Genetics, Technical University of Darmstadt, Darmstadt, Germany

Key Points

1) Science and technology (S&T) issues have always been at the heart of the deliberations surrounding the Biological and Toxin Weapons Convention (BTWC), but with the extremely rapid, unforeseen advances that have occurred over the years since the negotiation of the Convention, biosecurity concerns have become more and more significant for the regime.
2) The evolution of the dual-use dilemma can be seen by tracing the appearance of the results of experiments in the scientific literature over more than two decades that have raised biosafety and biosecurity concerns. The COVID-19 experience has amplified concerns.
3) Such security-relevant research resulted in the formulation of criteria for experiments of particular concern. These criteria together with examples of actual research meeting these criteria in each case can be useful in tagging research of concern for life scientists in the process of carrying out research projects.
4) Dealing with dual use in order to minimise the dangers that could arise from such experiments is nonetheless challenging for the researcher. For example, oversight policies with emphasis on select agents as a key tagging criterion will result in some research of real concern not being taken into account, a case in point being the mouse-pox experiment. In addition, there has been little effort to extend the seven categories of concern defined in 2004 by the Fink Committee to meet the expansion of the threat spectrum through advances in S&T.
5) For researchers to be best equipped to deal with dual use in the twenty-first century, it is essential that they are knowledgeable of the dual-use biosecurity issues involved, which the majority of life scientists are not. Education of life scientists about biosecurity dangers that could arise from their work is imperative, and the proactive support by administrations at institutions of learning in helping to establish curricula in this regard is crucial.

Essentials of Biological Security: A Global Perspective, First Edition. Edited by Lijun Shang, Weiwen Zhang, and Malcolm Dando.

Summary

This chapter deals with the problem of dual-use biosecurity concerns involving work being carried out in the life sciences. The relevance of advances in the life sciences and technology for the Biological and Toxin Weapons Convention (BTWC) over the years is examined in the first section. The evolution of the dual-use concept is then traced by examining in detail the series of experiments reported in the scientific literature over more than two decades and analysing the debates about what should be done to minimise the biosecurity dangers that could arise from such work. Through this analysis, it can be seen that dealing with dual use is still exceedingly challenging for the life sciences researcher in the twenty-first century. It is argued that to be best equipped to deal with dual use now and in the future, it is essential that life scientists are knowledgeable about dual-use biosecurity issues, which the majority are not. Education of life scientists about biosecurity dangers that could arise from their work is, therefore, imperative. Administrations at institutions of higher learning are called upon to proactively support the establishment of dual-use biosecurity instruction in life sciences curricula.

6.1 Relationship of the Advances in Science and Technology to the BTWC

The Biological and Toxin Weapons Convention (BTWC or BWC)[1] is the treaty banning biological weapons (BWs) at the international level. In effect, it prohibits the development, production and stockpiling of microbiological or other biological and toxin agents as weapons (for more detailed information on the BTWC, see Chapter 10). The prohibitions are explicitly and comprehensively set out in Article I of the Convention in the form of a General Purpose Criterion:

> **Article I**
> 'Each State Party to this Convention undertakes never in any circumstances to develop, produce, stockpile or otherwise acquire or retain:
>
> 1) Microbial or other biological agents, or toxins whatever their origin or method of production, of types and in quantities that have no justification for prophylactic, protective or other peaceful purposes;
> 2) Weapons, equipment or means of delivery designed to use such agents or toxins for hostile purposes or in armed conflict.'

This essentially says that all undertakings with biological agents with the aim of producing a biological weapon are prohibited. At the same time, all undertakings with biological agents for 'prophylactic, protective or other peaceful purposes' are allowed. This is a clever formulation, as it categorically prohibits all BWs, while allowing progress in science and technology (S&T) to proceed. The prohibition pertains to all biological agents, 'whatever their origin or method of production'. The Convention is thus not locked into the technology of the 1970s but applies to all current or future developments. These formulations, while all-encompassing, create the situation of having to decide what intent is behind particular work with a biological agent. As the treaty has no formal verification regime that could provide adequate means of examining whether the activities of a State Party are in compliance with the provisions of the Convention or not, there is no clear roadmap that would help decide the issue. Article IV of the BTWC solely obligates the States Parties to 'take any necessary measures to prohibit and prevent the development, production, stockpiling, acquisition or retention of the agents, toxins, weapons, equipment and means of delivery specified in Article I of the Convention'. This means in effect that each State Party is obligated to carry out the process of implementation of the Convention by adopting at the national level the appropriate laws, regulations and policy directives that would allow enforcement of the treaty provisions.

With the extremely rapid developments in S&T since the negotiation of the BTWC, research activities in the life sciences and related fields with relevance to compliance provisions have become more and more significant for the Convention. While many of the advances in biotechnology have great potential benefits for health and well-being, some of these same developments also have the potential to be used to cause harm to humans, animals, plants or the environment. This is the dual-use aspect inherent in much of life science research which poses a dilemma for scientists in trying to deal responsibly with it. That is why it is imperative that national implementation measures include those that would govern modern life science research activities in the way of minimising potential biosafety and biosecurity risks associated with this work as best as possible.

6.2 Evolution of the Dual-Use Dilemma

The evolution of the dual-use dilemma in the life sciences has been a process covering some two or more decades, depending on what one might take as the actual origin of the process. For purposes of the present discussion, the focus will be set on advances in modern biotechnology that have considerably raised the potential to be misused to cause harm in both classical and novel ways. The evolution of the dual-use dilemma can be followed by tracing the appearance of the results of experiments in the scientific literature over more than two decades that have given cause for significant biosafety and biosecurity concerns.

At the time of the negotiation of the BTWC, the prohibition focussed on traditional or classical microorganisms and also toxins, that are poisonous substances produced by living beings. Genetic engineering was not yet developed to the stage where its significance for the Convention was generally recognised. With the advent of genetic engineering soon after the Convention entered into force there was suddenly the possibility that novel organisms with entirely new characteristics might be created that could prove especially problematic for BWs control. However, as discussed earlier, the General Purpose Criterion of the BTWC assured that such novel agents would fall within the scope of the agents covered in Article I prohibitions. Still, the potential danger of this technique to be misused was of concern, due in great part to several publications in the scientific literature that appeared over the years thereafter that reported on research experiments resulting in the production of potentially dangerous microorganisms with novel characteristics rendered by application of genetic engineering. The potential of this new technique for achieving better health and well-being was recognised, but at the same time, it was feared that it could motivate rogue states to develop novel types of BWs. The concept of dual use at that time meant civil versus military application. With new advances in biotechnology in the years that followed, this dual-use concept was extended to include the potential misuse of the knowledge, technologies and products gained by life science work not only for biological warfare but also for terrorist attacks or use by misguided, hostile individuals to cause harm in some way.

A seminal study by the National Research Council of the US National Academies of Science entitled *Biotechnology Research in an Age of Terrorism* published in 2004[2] examined such biosecurity-relevant developments. The committee charged with this task was the Committee on Research Standards and Practices to Prevent the Destructive Application of Biotechnology, commonly referred to as the Fink Committee named after its chairman. The charge to the committee as stated in its report was:

> '... to consider ways to minimize threats from biological warfare and bioterrorism without hindering the progress of biotechnology, which is essential for the health of the nation. This task is complicated because almost all biotechnology in service of human health can be subverted for misuse by hostile individuals or nations.'

In this sense, the given goal of the committee was:

> '... to make recommendations that achieve an appropriate balance between the pursuit of scientific advances to improve human health and welfare and national security.'

This study was the first to deal intensively with the term 'dual use' in the broader sense as applied to life science work of this advanced nature including potential misuse by other actors in addition to the military.

Before continuing the discussion of the Fink report, the problem of the definition of 'dual use' should be briefly addressed. There have been many definitions made in different contexts over the years, but it would be important to have a definition that would be useful in dealing with it in a practical way. If the definition of dual use is too broad, it runs the danger of obstructing progress in science in that researchers would be spending too much time assessing projects whose results would have only remote potential for misuse. On the other hand, if the definition is too narrow, this may cause the assessment of some projects to be overlooked that at first appear to be benign but could have serious consequences in the end.

In the context of the World Health Organisation (WHO) report entitled *Global guidance framework for the responsible use of the life sciences*[3] 'dual use' is defined as:

> 'Knowledge, information, methods, products or technologies generated by peaceful and legitimate research that *may be* appropriated for non-peaceful or harmful purposes.' (Emphasis added)

A frequent argument of scientists is that this is characteristic of most work in the life sciences and that it would be impractical to apply it to research whose results would have little consequence or little potential for misuse; that progress in science would thus be impeded. In answer to this, another rendition has emerged that attempts to narrow the definition for practical purposes and has been termed dual-use research of concern (DURC). The definition of DURC as found in the WHO global guidance framework Glossary is:

> Dual-use research of concern (DURC) describes research that is conducted for peaceful and beneficial purposes, but *could easily be misapplied to do harm with no, or only minor, modification* (Emphasis added)

The emphasis added in each case points to the fine differences between the two definitions. In the case of DURC, the results of this type of research definitely *have* the potential to *easily* be misused, whereas the results of research that is termed only 'dual use' have less certain potential for misuse. There is no need to add that these definitions are nonetheless problematic from the viewpoint of their practical application.

The Fink report went on to elaborate on several examples of research reported in scientific journals that were considered of 'dual-use' potential. These included:

> 'The Australian *ectromelia* virus (mousepox) experiment, total synthesis of the poliovirus genome and recovery of infectious virus, and the comparison of the immune response to a host defense function from *vaccinia* and smallpox ...'[4]

The description of these experiments with reference to the original work can in each case be found in the report. For the reader, two of these will be briefly summarised here, plus two others from different sources.

6.2.1 Example 1. The Mousepox Experiment (2001)

In an attempt to control a plague of rodents in Australia, researchers developed a vaccine meant to function as a contraceptive. As a vector to deliver the desired genes, they used a genetically engineered mousepox virus.[5] They outfitted the virus with a gene which encoded a protein on the surface of mouse oocytes, which was meant to trigger an antibody response in inoculated mice that would prevent fertilisation of the egg and thus act as an immune-contraceptive. The resulting antibody response in these inoculated mice was lower than desired, so to boost the response, the researchers added another gene encoding a cytokine of the immune system, interleukin-4 (IL-4), known to enhance antibody responses in general. Unexpectedly, the mice inoculated with the virus carrying the IL-4 gene died, even though they were genetically resistant to the mousepox virus infection or had been immunised against it. Apparently, overproduction of IL-4 suppressed the function of another arm of the immune system, the production of cytotoxic T cells, which are necessary to contain the viral infection. Instead of a contraceptive, the researchers created a virus with enhanced lethality.

This experiment was really the first to cause a shock reaction and quite an uproar of response from the media, the scientific and defense communities as well as policy analysts, bringing into focus the concept of dual use in the biosecurity sense. Although the mousepox virus is not a virus that normally infects humans, it was feared that this work could be a blueprint for the construction of a more virulent smallpox virus that could overwhelm the immune system even in vaccinated individuals.

6.2.2 Example 2. Synthesis of the Poliovirus Genome and Recovery of Infectious Virus (2002)

Wimmer and colleagues reported in 2002 that they had synthesised the genome of the poliovirus from oligonucleotides chemically linked together.[6] This represented the first instance of the construction of the complete, synthetic genome of a microorganism. The chemically synthesised genome was not active, but it was subsequently transformed into infective virus particles in a cellular extract containing the necessary biological components. The media presented this experiment as a recipe for terrorists to reconstruct the poliovirus or even to make any other virus as a biological weapon. Arguments to the contrary pointed out that the poliovirus is not one on the list of usual potential BWs. Also, the poliovirus is a relatively simple virus structurally, and the complete construction of more complex viruses such as the influenza virus or the smallpox virus and their transformation into infective virus particles would be much more challenging than the production of an artificial poliovirus.

As an illustration of just how fast S&T can develop, a mere three years later the reconstruction of the extinct 'Spanish Flu' influenza virus was reported.

6.2.3 Example 3. Reconstruction of the 'Spanish Flu' Influenza Virus of 1918 (2005)

The report of a study published in *Science* in 2005 on the reconstruction of the 1918 so-called 'Spanish flu' virus gave great cause for concern.[7] The virus that caused the pandemic was exceptionally virulent in that it had a particularly high death rate in relation to any other known influenza viruses and killed an estimated 20–50 million people. The genome of the influenza A virus is much larger than that of the poliovirus and is divided into eight different segments. In addition, the virus contains many proteins that are required for, amongst other things, replication of the genome. The virus itself had been lost for decades, but the researchers succeeded in synthesising all eight coded segments of the genome by applying reverse genetics, using sequences gained from archived sources. The reconstruction was then achieved by supplying a relatively harmless influenza virus with all eight encoding segments, which made the virulence of this host virus similar to that of the original 1918 virus.

According to the authors, the reasons for undertaking this study included:

> '... the beliefs that a future influenza pandemic is likely, that better understanding of the 1918 pandemic influenza virus could aid our understanding of potential novel influenza viruses, and that the research could identify targets for therapeutic development'.[8]

The biosecurity-relevant issue associated with this work is that it represented a construction plan for the production of a microorganism that would be highly dangerous for humans.

6.2.4 Example 4. Alteration of the Host Range and Increase in the Transmissibility of the H5N1 Avian Influenza Virus (2012)

The avian influenza virus designated H5N1 causes a severe disease with high mortality (60% of the infected) in birds (fowl), but is not easily transmitted to humans, and when, then usually through direct, physical contact with infected birds.

In 2012, a group of researchers in the Netherlands and another working in Japan and the United States[9] experimented with the highly pathogenic H5N1 virus, producing mutated (genetically changed) variants of the virus that could infect mammals (in this case ferrets) through airborne transfer. The researchers had no malign intent by carrying out this work. Their stated aim was to determine what mutations could alter the host range of the virus and enable it to be transmitted through the air from individual to individual, in order to be better prepared for such changes that might occur in nature. Nevertheless, they had in effect created a potentially highly dangerous virus for humans, although the effect on humans could of course not be directly tested due to ethical reasons.

This work provoked a heated debate in public and scientific circles, as to whether it should have been carried out at all, both for safety (the virus could escape the lab) and security reasons (a blueprint for those with malign intent for creating viruses with similarly dangerous characteristics). Furthermore, it is questionable, as to whether the same mutations that caused a change in host range and transmissibility in the H5N1 virus experimentally would ever occur in nature or apply to other viruses.

In general, this type of work on highly dangerous pathogens that leads to results that are of particular biosafety and biosecurity concern has been labelled gain-of-function (GOF) research, as the modified H5N1 virus in the example above gained characteristics it did not have before: an alteration in host range (from fowl to mammals) and the ability to be transmissible through the air. Consequently, the term GOF has frequently been mistaken to mean research that is always dangerous or at least very risky. It should be noted that GOF work is carried out daily by life scientists; not all GOF experiments are of DURC character, and the technology, in general, is a valuable tool for learning about the characteristics of microorganisms including what makes them pathogenic. In order to clearly differentiate work that is characteristic of basic research of benign character from that of real biosecurity concern, the object of modification and the general goal of the research should be clearly stated when referring to GOF work.

6.3 DURC Criteria with Examples in Each Case of Published Research Reports of Work That Has DURC Character

Based on reports of such studies with DURC character, the Fink Committee in its report provided criteria to:

> 'assist knowledgeable scientists, editorial boards of scientific journals, and funding agencies in weighing the potential for offensive applications against the expected benefits of an experiment in this arena.'[10]

Table 6.1 Criteria for the identification of DURC along with an example of work published in scientific Journals that illustrate each type.

Criteria for DURC: Experiments that Would:	Examples of Published Work of this DURC Type
1) Demonstrate how to render a vaccine ineffective	Mousepox Experiment (discussed in Ref. 2, pp. 25–27)
2) Confer resistance to therapeutically useful antibiotics or antiviral agents	Conferring Resistance to a Therapeutically Useful Antibiotic (Ref. 11)
3) Enhance the virulence of a pathogen or render a nonpathogen virulent	Reconstruction of the 'Spanish Flu' Influenza Virus of 1918 (discussed in Ref. 3, Annex 2, pp. 172–175); Mousepox Experiment (discussed in Ref. 2, pp. 25–27)
4) Increase transmissibility of a pathogen	Alteration of the Host Range and Increase in the Transmissibility of the H5N1 Avian Influenza Virus (Ref. 9)
5) Alter the host range of a pathogen	Alteration of the Host Range and Increase in the Transmissibility of the H5N1 Avian Influenza Virus (Ref. 9)
6) Enable the evasion of diagnostic/detection modalities	Protection of an Adenovirus Vector from Neutralising Antibody through Microencapsulation (Ref. 12)
7) Enable the weaponisation of a biological agent or toxin	Successful Gene Transfer Using a Nanoparticle Carrier as Vector (Ref. 13)

Source of DURC Criteria: From Reference 2/National Academies Press, pp. 114–115.

In this regard, the committee identified seven types of experiments that could act as a flag to help scientists in deciding whether their proposed or ongoing work has DURC character and would, therefore, require a risk–benefit assessment. A list of these DURC characteristic criteria along with an example of work published in scientific journals to illustrate each type is in Table 6.1.

The examples of DURC in Table 6.1 for criteria 1, 3, 4 and 5 have been discussed in Section 6.2 and will not be repeated here, but just to note that the mousepox experiment as well as the H5N1 avian influenza virus work each embody two of the criteria characterising DURC.

The example for criterion 2 in the table[11] represents work involving *experiments that would confer resistance to a therapeutically useful antibiotic*, in this case, tetracycline which is an antibiotic of choice for the treatment of anthrax. The study explicitly demonstrates the experimental transfer of a plasmid encoding tetracycline resistance to *Bacillus anthracis*, the bacterium causing the disease anthrax, in effect abrogating the therapeutic use of this antibiotic in fighting the disease. The researchers carried out this study in order to investigate the mechanisms of plasmid transfer among *Bacillus* species; by using the tetracycline resistance plasmid they could easily detect whether the plasmid transfer to *B. anthracis* was successful or not. The experimental method for transfer of the plasmid was certainly not high-tech, and conferring antibiotic resistance to a potential BWs agent in a relatively easy-to-do manner does indeed flag the experiment as having DURC character. However, as *B. anthracis* is susceptible to several other antibiotics in addition to tetracycline, and the bacterium can acquire tetracycline resistance naturally, this diminishes the degree of concern to a degree.

Criterion 6, *experiments that would enable the evasion of diagnostic/detection modalities*, would include changes in the agent to avoid recognition by the immune system or detection/identification by standard diagnostic antibodies. It also would include the alteration in gene sequences to avoid detection by standard molecular methods. The reference used in the table to exemplify this criterion[12] is work that describes in detail the microencapsulation of an adenovirus vector (intended for use as a vaccine) in order to avoid recognition and subsequent attack by the immune system. Adenovirus vectors have been used extensively as transport vehicles for therapeutics in both experimental and clinical settings for gene, cancer and immune-therapy. One problem encountered early on was that surface proteins on adenoviruses are strongly immunogenic, rapidly eliciting immune responses against the virus after delivery. This can limit or diminish the therapeutic efficacy of the delivery system due to immune clearance of the vector before it can transfer its cargo to the target cell. By shielding the immunogenic surface structures of the vector, it stands a better chance of surviving attack by the immune system long enough to more effectively deliver its cargo to the target site. In the example of Zeng et al., the authors encapsulated an adenovirus vector containing a foreign beta-galactose gene used for expression detection into a cationic polyethylene glycol derivative. Their results in the experimental system showed good protection for the adenovirus vector against neutralising antibody along with improved uptake of the vector into target cells followed by gene transfer designated by beta-galactose expression in the target cell. Developing viral vectors for therapeutic interventions is of course not only legitimate but also promises improved methods in the treatment of infections and complex illnesses. The research has nonetheless DURC character in that it points to ways of protecting pathogens from recognition and attack by the immune system needed to contain an infection.

Regarding DURC criterion 7, *experiments that would enable the weaponisation of a biological agent or toxin*, the authors of the work used to exemplify this citerion[13] demonstrated the successful experimental delivery of DNA encoding an enzyme (chloramphenicol acetyltransferase [CAT]) via a stable, cationic lipid nanocarrier across the mucosal barrier into the nasal epithelial cells of mice. CAT enzyme activity was measured as an indicator of the transfer, release and expression of the DNA gene cargo at the target site. One needs only to recall the great success of the nanocarrier version of the COVID-19 vaccine in delivering a protein antigen encoded in mRNA. The method of nasal instillation of the lipid nanocarriers that Kim et al. used to deliver the DNA to the target site does not exactly mimic delivery via an aerosol, which would be the method of choice for delivering a biological weapon. Nevertheless, the work shows in principle that a BWs agent encoded in DNA (for example, a protein toxin or bioregulator) incorporated into a similar type nanocarrier could feasibly be successfully delivered via inhalation. Kim et al. carried out this study with the intent of delivering cancer therapeutics, not BWs. Still, the work has obvious DURC character that would require biosecurity review.

6.4 Problems in Dealing with Dual Use: Debates About What Should Be Done

The criteria for DURC together with examples of such research are meant to aid researchers in determining whether the work they propose to carry out has DURC character. This is a means to start the reflection process in dealing responsibly with the work. This process usually means weighing benefits against potential risks involved and coming to a decision about how the project should proceed so that potential risks are minimised, or possibly whether the research should be halted altogether. (For a more thorough discussion of dual-use governance including risk management see Chapters 13 and 14).

Despite having criteria to flag potential risks, dealing with dual use in order to minimise the dangers that could arise from such experiments is nonetheless challenging for the researcher. The National Research Council's Summary of a Workshop[14] reported in 2015 discusses some of the most prominent challenges in coming to grips with DURC, particularly that termed GOF research that emerged following the publication in 2012 of the work on the H5N1 avian influenza virus. This work was a real wake-up call to the significance of the dangers associated with security-relevant research that resulted in a renewed shift of attention internationally towards potentially risky research. Above all, it provoked an often extremely heated debate that divided the life science and security communities alike on the question whether the benefits of this research outweighed the risks or vice versa, including the consideration as to whether the work should have been carried out at all. The COVID-19 pandemic has refuelled this debate in regards to performing GOF research on coronaviruses that are potential pandemic pathogens (see Chapter 15: Lessons from ePPP Research and the COVID-19 Pandemic).

This points to one of the problems with risk assessment in dual-use biosecurity-relevant work, and that is making a final decision after weighing risks versus benefits, which in the end is based on the uncertainty of judgement:

> 'We should not kid ourselves into thinking we can come up with some formula to plug in all the variables and produce something that shows that the risks outweigh the benefits or vice versa. It needs to be acknowledged that it will be difficult to quantify the equation and, in addition, if we were able to determine exact numbers, then different individuals would place different values on different variables ...'

and therefore:

> 'One of the best things to come out of the risk assessment would be to convince ourselves and the public that we considered the issues in depth and that whatever decision we made was not pulled out of thin air, but rather the result of a careful deliberative process.'[15]

In addition, oversight policies with emphasis on select agents as a key tagging criterion will result in some research of real concern not being taken into account; researchers might disregard any possibility of DURC in their work if it does not involve a pathogen. A case in point is the mousepox experiment. As the mousepox virus is not a pathogen for humans, and if work involving a pathogen is a key criterion for biosecurity review, the researcher might consider that any work with the virus has little or no potential risk and go ahead with the research without any further reflection. The same can be said for policies based on individual technologies. Since the development of genetic engineering, great strides have been achieved in cutting-edge technologies such as synthetic biology and the clustered regularly interspaced short palindromic repeats (CRISPR)/Cas method (see Chapter 7: Key Cutting-Edge Biotechnologies Today) that have expanded the concept of genetic engineering to revolutionise the technique of genome editing. However, it is not the technologies themselves that need to be governed but rather what is done with them:

> '... list-based approaches to governance in the life sciences can be limited. Owing to the speed of advancements, lists can quickly become outdated, creating gaps in the biorisk management system because new technologies and their associated risks are not listed. *Overarching frameworks with sufficient flexibility to apply to new technologies as they arise may avoid this problem.*' (Emphasis added)[16]

There has also been little effort to extend the seven categories of concern defined in 2004 by the Fink Committee to meet the expansion of the threat spectrum by advances in S&T. For example, bioregulators have expanded the agent threat spectrum beyond the classical categories of microorganisms and toxins for the BTWC or beyond the classical categories of toxic chemicals and toxins for the Chemical Weapons Convention (CWC). Bioregulators, as an expansion of the threat spectrum of biological and chemical agents, are discussed in the report of the Lemon-Relman Committee of the National Research Council.[17] Bioregulators are biochemicals which to a great extent steer the proper function of vital processes within the nervous, endocrine and immune systems such as respiration, heartbeat, body temperature, cognition, mood and immune responses. The bioregulators of relevance in these cases are neurotransmitters/neuropeptides, hormones and cytokines. For an example of work with a bioregulator that has DURC character see Ref. 3, Annex 1, Scenario 2. Neurobiology, pp. 126–130. It should also be recalled that the mousepox experiment is another case of successful delivery of a bioregulator (IL-4), in this case, encoded in a gene contained in the genome of a viral vector.

Given all the challenges involved, researchers would be best equipped to deal responsibly with dual use in the twenty-first century if they are comprehensively knowledgeable of the dual-use biosecurity issues involved, which the majority of life scientists are not. The WHO global guidance framework for responsible research sees education as fundamental for the risk management system. At the same time, this is one element of risk management that is blaringly lacking in most cases, as:

> 'A chronic and fundamental challenge in biorisk management is a widespread lack of awareness that work in the area of the life sciences could be conducted or misused in ways that result in health and security risks to the public.'

In this regard:

> 'Globally, many scientists conducting life sciences research are not trained in biosecurity, not familiar with the BWC *(12)* ... and not incentivized to devote time and resources to biorisk management.'

Indeed,

> '... any biorisk management system must include education, awareness building, and creation of a culture of individual and institutional investment in biosafety, biosecurity and oversight of dual-use research.'

Accordingly,

> *'If they are unaware of the potential for misuse and potential malicious application, stakeholders cannot accurately weigh the risks and benefits of proposed research.'* (Emphasis added)[18]

According to the WHO framework, institutions play a critical role in instituting security-relevant education: 'The lack of awareness can be reinforced by a lack of institutional incentives to attend to safety and security concerns...'. Thus, education of life scientists about biosecurity dangers that could arise from their work is imperative, and the proactive support by administrations at institutions of research and learning in helping to establish curricula, in this regard, is crucial. There is a wide variety of educational tools available that could help institutions of learning and practice to initiate biosecurity education in ways appropriate for specific cases or areas of studies or for different countries with different cultures. For a more thorough discussion of the education aspect and what to do about it, see Chapter 20: Towards an International Biological Security Education Network (IBSEN).

Author Biography

Dr. Kathryn Nixdorff is a professor emeritus in the Department of Microbiology and Genetics at the Technical University of Darmstadt, Germany. Her natural science research is focussed on the interaction of microorganisms with the immune system. Her biosecurity research is focussed on developments in science and technology and their relevance for biological arms control. From 2012 to 2014, she worked as an outside expert with the working group on biosecurity for the German Ethics Council, which produced Biosecurity – Freedom and Responsibility of Research (2014), an Opinion on Biosecurity, including recommendations for dealing with security-relevant research in the life sciences.

References

1 Schindler, D. and Toman, J. (2004) Convention on the Prohibition of Development, Production and Stockpiling of Bacteriological (Biological) and Toxin Weapons and on their Destruction. in *The Laws of Armed Conflicts*, Brill Nijhoff, 135–160.

2 National Research Council. (2004) *Biotechnology Research in an Age of Terrorism*, The National Academies Press, Washington, D.C.

3 World Health Organization. (2022) Global Guidance Framework for the Responsible Use of the Life Sciences: Mitigating Biorisks and Governing Dual-Use Research, World Health Organization, Geneva. ISBN 978-92-4-005610-7, Glossary, page xix. (electronic version).

4 National Research Council (2004) op. cit. (Page 25)
5 National Research Council (2004) op. cit. (Pages 25–27)
6 National Research Council (2004) op. cit. (Pages 27–28)
7 World Health Organization (2022) op. cit. (Annex 2, Pages 172–175)
8 World Health Organization (2022) op. cit. (Annex 2, Page 173)
9 Herfst, S. *et al* (2012) Airborne transmission of influenza A/H5N1 virus between ferrets. *Science*, 336 (6088), 1534–1541. Imai, M. *et al* (2012) Experimental adaptation of an influenza H5 HA confers respiratory droplet transmission to a reassortant H5 HA/H1N1 virus in ferrets. *Nature*, 486 (7403), 420–428.
10 National Research Council (2004) op. cit. (Page 108)
11 Green, B. D. *et al* (1989) Involvement of Tn4430 in transfer of *Bacillus anthracis* plasmids mediated by *Bacillus thuringiensis* plasmid pXO12. *Journal of Bacteriology*, 171 (1), 104–113.
12 Zeng, Q. *et al* (2012) Protection of adenovirus from neutralizing antibody by cationic PEG derivative ionically linked to adenovirus. *International Journal of Nanomedicine*, 7, 985–997.
13 Kim, T. W. *et al* (2000) *In vivo* gene transfer to the mouse nasal cavity mucosa using a stable cationic lipid emulsion. *Molecules and Cells*, 10 (2), 142–147.
14 National Research Council. (2015) *Potential Risks and Benefits of Gain-of-Function Research: Summary of a Workshop*, The National Academies Press, Washington, D.C.
15 National Research Council (2015) op. cit. (Page 19)
16 World Health Organization (2022) op. cit. (Section 4.3.1.c. Flexible frameworks, Pages 59–60).
17 National Research Council. (2006) *Globalization, Biosecurity, and the Future of the Life Sciences*, The National Academies Press, Washington, D.C.
18 World Health Organization (2022) op. cit. (Section 2.3, Page 28)

7

Key Cutting-Edge Biotechnologies Today

Xinyu Song and Weiwen Zhang

Center for Biosafety Research and Strategy, Tianjin University, Tianjin, China

Key Points

1) Briefly mention the key cutting-edge biotechnologies, such as synthetic biology and genome editing and their impact on human society.
2) Brief overview of current and potential future applications of the cutting-edge biotechnologies.
3) Highlight the potential biosafety and biosecurity risks and ethical implications of the cutting-edge biotechnologies.

Summary

Synthetic biology and genome editing are two most exciting and rapidly advancing technologies in the life sciences. Both technologies have many beneficial applications, such as creating new vaccines and treatments for diseases. The artificial designs, both in depth and scope, consistently pioneer new frontiers, achieving genome synthesis in viruses, prokaryotes, and single-cell eukaryotes. While these advancements in the life sciences are invaluable for beneficial research, they also carry the potential for misuse or intentional abuse, resulting in immeasurable harm. Increased risk of misuse and abuse of key cutting-edge biotechnologies, highlighting the dual-use nature of synthetic biology and genome editing. This chapter provides an overview of the advancement of synthetic biology and genome editing, as well as the detailed explanation of the dual-use concerns of these technologies.

Essentials of Biological Security: A Global Perspective, First Edition. Edited by Lijun Shang, Weiwen Zhang, and Malcolm Dando.

7.1 Introduction

Rapid development of life sciences and the accelerating globalization process have benefited human society significantly. In 2016, synthetic biology's estimated market stake was US$3.9 billion worldwide, and it is expected to grow at a compound growth rate of just under 25%. By 2026, the market for synthetic biology could grow to $US14 billion. In addition, private-sector synthetic biology companies raised a record sum of US$3.8 billion in funding, in 2018, and the first two quarters of 2019 suggested that funding will keep that record pace for this year. There is also substantial investment in public financing. In the United States, a reported US$200 million was invested in synthetic biology research in 2014, while multiple national investments, including priorities set by the Chinese Academy of Sciences (CAS) and the national 973 and 863 Programme, have awarded just under US$80 million of annual funding for synthetic biology research in China. The National Key Research and Development Program of China also committed approximately $320 million to synthetic biology research for 2018–2022. It is expected that in the next 5–10 years, synthetic biology and genome editing technologies will have a significant impact on various areas, including national security, population and health, agriculture, industry, environment and biosecurity.[1] As technological barriers lower, experimental procedures become more convenient, and associated costs decrease, misuse and abuse of dual-use biotechnologies present significant challenges to the biosafety and national security defense systems of countries worldwide. This chapter provides an overview of the advancement of synthetic biology and genome editing, as well as the detailed explanation of the dual-use concerns raised by these technologies.

7.2 Development and Application of Synthetic Biology

7.2.1 Landmark Achievement in Synthetic Biology

Although synthetic biology has only been around for two decades, it has already become one of the most hyped research topics this century. As early as 2002, Cello et al. first synthesised poliovirus by assembling positive- and negative-strand polar oligonucleotides, demonstrating the potential for synthesising infectious agents *in vitro*. Early synthetic biology research focused on the design of the genetic toggle switch and repressilator. As more genetic parts and building blocks were characterised and assembled, scientists were able to construct a variety of devices, such as switches, memory elements, oscillators, pulse generators, digital logic gates, filters and communication modules.

In the second decade of the twenty-first century, from 2010 to 2020, synthetic biology blossomed and delivered many new technologies and landmark achievements. In 2010, Gibson et al. created new *Mycoplasma mycoides* cells with synthetic chromosomes, which were assembled using digitised genome sequence information. As the 'Human Genome Project-Write' was launched, life science scientists began to explore the synthesis of more complex eukaryotic genomes. In 2016, Nielsen et al. reported Cello, a remarkable end-to-end computer-aided design system for logic circuit construction in *E. coli*. In 2017, a team from four countries completed the de novo design and total synthesis of five yeast chromosomes (Nos. 2, 5, 6, 10 and 12) through international cooperation, which led to breakthroughs in the synthesis of eukaryotic genomes. It is interesting that synthetic yeasts with engineered chromosomes display high consistency with natural yeast strains. All these advances demonstrate that synthetic biology has the potential to revolutionise many areas of science and industry, including medicine, chemical industry, artificial food and sustainable agriculture.

7.2.2 Opportunities for Medical Application

In the field of medicine, synthetic biology can be used to develop new treatments and therapies for various diseases, for example, engineering cells to produce specific proteins or molecules that can be used to treat diseases such as cancer or diabetes. The recently approved chimeric antigen receptor (CAR) T-cell therapy tisagenlecleucel has a high per-patient cost, which could exceed $1 million due to the costs of managing treatment-related adverse events and subsequent procedures. The success of CAR T-cell therapy in the treatment of blood cancers demonstrates the potential of using genetically engineered cells as therapeutic agents.

The use of L-homoserine as a pharmaceutical intermediate has good market prospects, but the large-scale application of L-homoserine has been limited due to issues such as production intensity and economy. However, recent research has made progress in improving the efficiency and yield of L-homoserine production through systematic metabolic engineering and fermentation process improvement. Mu et al. engineered a fermentation pathway with an overall balance of reducing power and reused the CO_2 released in the reducing power supply pathway. As a result of these improvements, they achieved a fermentation level for homoserine that exceeds 84 g/L, with a conversion rate of 50%, which represents a significant improvement in the production of L-homoserine and has good economic potential. These techniques can be applied to other compounds and may pave the way for more efficient and cost-effective production of a wide range of pharmaceuticals and other high-value chemicals.

In the field of basic medicine, researchers from MIT's synthetic biology group developed a technology that uses implanted bacteria in stretchable, flexible hydrogels to detect specific substances. The implanted bacteria glow when they come into contact with certain chemical molecules, providing a way to detect toxic substances, pathogens and allergens. This technology may have applications in the detection of hazardous materials on protective gloves and, in initial, medical diagnosis of human skin.

7.2.3 Benefits to Agricultural Development

Synthetic biology has potential to greatly benefit agriculture and address some of the challenges facing the world, such as increasing crop yields, improving pest and disease resistance and reducing the environmental impact of farming practices.[2] The construction of artificial pathways utilising synthetic biology offers the possibility of producing valuable compounds in a sustainable and efficient manner. Cai et al. constructed the artificial starch anabolic pathway (ASAP) through computational pathway design, modular assembly and substitution and protein engineering of three bottleneck-associated enzymes and obtained a chemoenzymatic system with spatial and temporal segregation. In this system, the conversion from CO_2 to starch reached a rate of 22 nanomoles of CO_2 per minute per milligram of total catalyst, an ~8.5-fold higher rate than starch synthesis in maize.

Insecticides derived from natural sources, such as pyrethrins extracted from pyrethrum (*Tanacetum cinerariifolium*), have been used in agriculture for many years. However, the synthesis process for these compounds can be complicated and expensive. Synthetic biology offers the possibility of producing these compounds more efficiently and at a lower cost through the engineering of microorganisms to produce key intermediates and enzymes. Additionally, synthetic biology can be used to engineer crops with increased pest and disease resistance, reducing the need for harmful pesticides and improving crop yields through the engineering of traits such as pathogen recognition and immune responses in crops.

7.2.4 Changing the Future of Foods

With increasing environmental pollution, climate change and population growth, it is becoming challenging to keep the food supply safe, nutritious and sustainable.[3] Kumar et al. predicted that by 2050, the production of animal products would double, from 229 billion kilograms in 2000 for 6.0 billion people to 465 billion kilograms for 9.1 billion people. One proposed solution is synthetic meat, a new type of food that requires cells to be extracted from living animals and grown in a laboratory environment without the need for raising and slaughtering animals. The large-scale production and application of synthetic meat will help solve many environmental problems associated with livestock production systems. It is estimated that synthetic meat could reduce land use by 99%, water use by 96% and energy consumption by 45%. In addition, due to the strict control of cultured meat production and the limited human–animal interaction, cultured meat safety could be better than that of conventional livestock production.

Other food-related applications of synthetic biology include the production of artificial honey, which involves mimicking the process of honey production in bees using microbial fermentation in a laboratory setting. Honey bees are just one species out of a staggering 20,000; therefore, growing honey bee colonies may harm wild and native bee populations as the demand for honey increases. Using synthetic biology, the MeliBio company replicated real honey down to the molecular level without the need to harm the planet's essential pollinators, and the artificial honey completely restored its texture and taste.

Aside from food for humans, it is worth mentioning the dazzling news 'Industrial Carbon Monoxide Synthesis of Protein' produced by the Feed Research Institute of the Chinese Academy of Agricultural Sciences. This technology for synthesising *Clostridium* ethanol protein already has an industrial production capacity of 10,000 tons, and it can be used as feed for pigs, cattle, sheep and chickens on farms.

7.2.5 Creation of Sustainable Energy

Sustainable energy refers to energy sources that are renewable, meaning they can be replenished naturally over time and do not contribute to environmental pollution or the depletion of natural resources.[4] Examples of sustainable/renewable energy sources include solar energy, wind energy, geothermal energy, hydropower and bioenergy, all of which are considered more environmentally friendly than fossil fuels, such as oil and coal, which contribute to air pollution, climate change and depletion of natural resources.

In recent years, there has been a growing focus on sustainable energy as a way to reduce reliance on fossil fuels and move towards a more environmentally sustainable future. For example, new 'cell factories' capable of generating energy from traditional and non-traditional forms of feedstock have been created by synthetic biology.[5] A new generation of synthetic bioenergy, including cellulosic ethanol, higher alcohols, aliphatic hydrocarbons, biogas, biohydrogen and bioelectricity, has gradually been developed. However, the production cost makes it difficult to compete with fossil energy, and large-scale promotion has not yet been fully launched. At present, the total production of biofuels in the world is close to the equivalent of 100 million tons of standard oil, i.e. 86.72 million tons of fuel ethanol produced from first-generation grain and sucrose, equivalent to 56.35 million tons of standard oil (accounting for 59%); 38 million tons of first-generation biodiesel 10,000 tons, equivalent to 33 million tons of standard oil (accounting for 35%); and approximately 5.52 million tons of second-generation biomass diesel, equivalent to 5.7 million tons of standard oil (accounting for 6%).

In addition to producing sustainable energy, there is also a growing focus on improving the efficiency and sustainability of energy storage and distribution systems. This focus includes research into new battery technologies, as well as the development of smart grid systems that can more efficiently manage and distribute energy from renewable sources. The combination of modern nanomaterials, chemistry, electronic circuits and other fields with synthetic biology provides new insight into the development of biobatteries. Recently, Yin et al. proposed a promising strategy to fabricate high-strength and high-performance composite solid electrolytes with bacterial cellulose scaffolds through a mild and green biosynthesis process.

7.2.6 Approaches for New Materials

Synthetic biology offers new approaches to producing materials with unique properties, including artificial spider silk, which is one of the strongest biomaterials available in nature.[6] Its mechanical properties make it a good candidate for applications in various fields ranging from protective armour to bandages for wound dressing to coatings for medical implants. A large amount of research has been directed towards harnessing the spectacular potential of spider silks and using them for different applications. Traditional fermentation engineering and chemical engineering may not be sufficient to analyse and reduce components of ultrahigh-molecular-weight proteins, such as spidroin; therefore, the interdisciplinary approach of synthetic biology is an ideal tool to study these spider silk proteins and work towards the engineering and production of synthetic spider silk.

A challenge is that the conditions for living cells are not conducive to material processing and require continuous water and nutrients. Embedding living cells into materials can provide added functionality, such as sensing, energy production and physical movement, making materials multifunctional. It seems that the 3D printer developed by Gozalez et al. is able to mix material and cell streams to create 3D objects. The use of *Bacillus subtilis* spores within the material and their subsequent germination on the exterior surface could lead to some interesting applications. Additionally, the material was shown to be resilient to a variety of extreme stresses, which could make it useful for a range of applications.

7.3 Development and Application of Genome Editing

7.3.1 Landmark Progress in Genome Editing

Genome editing, also known as gene editing, is a group of technologies used to engineer specific modification(s) within target genes and has proceeded through three generations of development. Traditional genome editing tools, including ZFNs and TALENs, are defined as first- and second-generation technologies, respectively.[7] The shortcomings of these traditional genome editing technologies, such as low target recognition rate, high cost, high off-target probability and complex structure, motivated the development of the third-generation genome editing technology, the CRISPR/Cas system, which utilises a guide RNA (gRNA) to target the nuclease Cas9 to specific DNA sequences, enabling precise and efficient genome editing with reduced off-target effects.

With the continuous deepening of research on CRISPR systems, certain defects and limitations have also been exposed, such as serious off-target effects. In September 2015, Feng Zhang's research group in MIT reported a new type 2 CRISPR effector: Cpf1, a tracrRNA-independent endonuclease mediated by a single RNA. Cas9 simultaneously cuts both strands of DNA molecules at the same position, forming blunt ends, while Cpf1 cuts and forms two strands of different lengths, which are called sticky ends. At the same time, Cpf1 can recognise thymine (T)-rich protospacer adjacent motif sequences, which can expand the editing range of CRISPR.

Many known genetic diseases are caused by point mutations; however, current methods to correct point mutations are inefficient or cause defects such as random deletions or insertions. In April 2016, David Liu's research group in Harvard University reported a new approach to genome editing technology, 'base editing', that enables the direct, irreversible conversion of one target DNA base into another in a programmable manner without requiring dsDNA backbone cleavage or a donor template. They also developed second- and third-generation base editing technology to further improve the efficiency of base editing. On this basis, Chen Jia's research group at Shanghai Tech University and collaborators jointly developed an enhanced base editor. They constructed a transformer base editor system that induces efficient editing using only background levels of genome-wide and transcriptome-wide off-target mutations.

Considering that genome editing at the DNA level causes permanent changes in the genome, Zhang Feng's research group further reported that a CRISPR-based approach targeting RNA allows the generation of transient up- or downregulated changes with greater specificity and functionality than existing RNA interference techniques. In June 2016, Zhang Feng's research group discovered an effector, C2c2 (now called cas13a), from *Leptotrichia shahii* with the ability to be programmed to cleave specific RNA sequences in bacterial cells, which makes it an alternative molecular biology toolbox for genome editing.

7.3.2 Potential in Curing Diseases

Treating disease is one of the greatest expectations for CRISPR technology. In December 2013, Wu et al. published a study using CRISPR/Cas9 to treat cataracts in a mouse model with cataracts caused by base deletions. They demonstrated that the CRISPR–Cas9 system can be used to cure a genetic disease in mice by directly correcting the genetic defect through gene editing. In another study conducted during the same period, Schwank et al. corrected the cystic fibrosis transmembrane conductor receptor locus by homologous recombination in cultured intestinal stem cells of cystic fibrosis patients using CRISPR/Cas9 technology. This study provides a proof of concept for gene correction by homologous recombination in primary adult stem cells derived from patients with a single-gene hereditary defect.

Since then, CRISPR technology has been applied in a wide range of disease treatments, including, but not limited to, genetic disorders, cancer and infectious diseases. The first *in vivo* gene editing trial using CRISPR/Cas9 was conducted in China to treat lung cancer. The researchers used CRISPR/Cas9 to modify T cells extracted from the patient's blood to improve their ability to fight cancer cells, which were then infused back into the patient. In another study published by Young et al., CRISPR/Cas9 was used to treat Duchenne muscular dystrophy in a mouse model. The researchers used CRISPR/Cas9 to correct the mutation in the dystrophin gene that causes the disease, which resulted in improved muscle function in the treated mice. In 2017, the US Food and Drug Administration (FDA) approved the first CRISPR-based therapy, voretigene neparvovec (Luxturna). The treatment involves injecting a modified virus carrying a functional copy of the RPE65 gene directly into the patient's eye, which has been shown to improve vision in clinical trials.

More recently, Foy et al. generated patient-specific T cells for personalised cancer treatment utilising CRISPR genome editing technology. Their study demonstrated the manufacture of neoantigen-specific T-cell reporter (neoTCR)-engineered T cells at clinical grade, the safety of infusing up to three gene-edited neoTCR T cell products, and the ability of the transgenic T cells to traffic to the tumours of patients. In another study, Li et al. used CRISPR/Cas9 technology to correct a mutation in the MYBPC3 gene, which causes hypertrophic cardiomyopathy, a genetic disorder that affects heart function. They demonstrated the feasibility and safety of using CRISPR/Cas9-mediated genome editing in correcting a pathogenic gene mutation in a human embryo.

These studies suggest the potential of CRISPR genome editing in treating a variety of genetic diseases, including cancer and heart disease, and highlight the importance of continued research and development in this field to realise its full potential in clinical applications.

7.3.3 Supporting Sustainable Agriculture

Modern agriculture is facing a global weed problem that reduces crop yield; thus, the development of herbicide-tolerant crops is an excellent choice to control weeds and hence produce higher crop yields. In recent years, emerging genome editing technologies have provided a new methodology for crop improvement through the precise manipulation of endogenous genes in plant genomes. The development of transgene-free herbicide-tolerant plants by targeted genome editing is currently the most suitable alternative to traditional genetic engineering approaches. Moreover, in many countries, genome-edited plants created via CRISPR-based genome editing technologies have been excluded from genetically modified organism (GMO) regulation. Nevertheless, unintended off-target edits can arise that might confer risks when present in gene-edited food crops.[8] Through an extensive literature review, Sturme et al. gathered information on CRISPR–Cas off-target edits in plants and concluded that the frequencies of off-target mutations present in CRISPR/Cas9-edited plants are lower than those of conventionally bred plants.

In addition to herbicide-tolerant crops, CRISPR technology has also been used to create crops with desirable traits, such as disease resistance, higher yield or increased nutritional components. For example, Zafar et al. used CRISPR/Cas9 to engineer rice plants with enhanced resistance to bacterial blight, a devastating disease that causes significant yield loss in rice crops. Similarly, using CRISPR/Cas9, many negative regulatory genes, such as the gibberellin-regulated gene GASR7 that controls grain length in rice, and the RING-type E3 ligase GW2 that controls rice grain weight, have been knocked out to improve wheat yields and quality. Many nutrient elements, such as carotenoids, γ-aminobutyric acid, iron and zinc, are effective against inflammation, cancer and oxidation. CRISPR/Cas9-mediated genome editing has been applied in carotenoid biofortification in rice, tomato and banana. Li et al. used a multiplex CRISPR/Cas9 method to delete SlGABA-Ts and SlSSADH, which resulted in GABA levels increasing by approximately 20-fold but with accompanying high penalties in tomato fruit size and yield.

7.4 Main Biosafety and Biosecurity Concerns Associated with Key Cutting-Edge Biotechnologies

7.4.1 The Increasing Accessibility of Biotechnology Tools and Techniques Exacerbates Safety and Security Risks

The combination of cutting-edge biotechnologies with the unique traits of infectious diseases could lead to advances in the diagnosis, treatment and prevention of infectious diseases. However, the declining cost of biotechnology tools and equipment has made them more accessible to individuals and organisations with malicious intent. Gene synthesis cost per base pair has dropped from approximately $10 per base pair to approximately $0.10 per base pair over the past 10 years.[9] The current cost of sequencing, a whole human genome is approximately $1000 dollars, less than 1% of the original cost 10 to 15 years ago.[9] The cost of DNA sequencing has drastically decreased, making it easier for individuals to access genetic information and potentially use it for harmful purposes such as bioterrorism. The development of low-cost, portable genome editing tools such as CRISPR-Cas9 also raises concerns about their potential misuse. Additionally, the increasing availability of online information and tutorials for biotechnology techniques has made it easier for nonexperts to gain access to and manipulate biological materials, potentially creating a risk for biosecurity breaches.

7.4.2 Emerging/Re-emerging Infectious Diseases Aggravate the Misuse and Abuse Risk of Cutting-Edge Biotechnologies

Emerging or re-emerging infectious diseases pose a significant biosafety risk, as they can break through traditional physical boundaries and rapidly spread globally. With the increasing interconnectedness of the world and the ease of international travel, a disease outbreak in one part of the world can quickly become a global epidemic. Moreover, climate change, environmental degradation and population growth can create conditions that favour the emergence of new diseases or the resurgence of previously controlled diseases. Despite significant efforts at the global, regional, national and community levels, in the era of rapid development of dual-use biotechnology, the world is no safer now from infectious diseases than it was 20 years ago.[10] For instance, the COVID-19 pandemic has demonstrated how a highly infectious disease can rapidly spread across the world and cause significant public health and economic impacts. Effective vaccine development using biotechnologies has been critical in shortening processes that previously took years to only months; however, certain kinds of biotechnologies have the potential to enable pathogen engineering, raising the risk for biological events, including pandemics of the largest scale, with the potential to destabilise society.

7.4.3 Integration and Innovation in the Field of Cutting-Edge Technologies Aggravate Safety and Security Risk

As different fields of cutting-edge, science and technology (S&T) continue to integrate and innovate, there is a growing concern that biological safety hazards could be exacerbated. For instance, advances in synthetic biology have enabled the creation of synthetic organisms, which may pose risks if accidentally released or intentionally misused. Similarly, the convergence of biotechnology with other fields such as nanotechnology, artificial intelligence (AI) and robotics may introduce new and unforeseen risks. If the genetic information of pathogenic bacteria is misused and abused by terrorists and biohackers, precise genome weapons targeting specific human groups or specific races could be developed, and the resynthesis of new types of organisms that are more toxic and infectious threats is even possible. In August 2017, researchers at the University of Washington in the United States converted computer commands into DNA sequencing data by implanting malicious codes in gene particles and obtained full control of the computer when the computer processed the sequencing data, successfully creating the world's first exploit gene attack on computer software. Relevant researchers have demonstrated the use of manipulated blood or saliva samples to hack into computer equipment and steal police methods in research institutions.

7.5 Conclusions

Although cutting-edge biotechnologies, such as synthetic biology and genome editing, have the potential to bring significant benefits to human health and agriculture, among other areas, there are also potential risks and ethical concerns that need to be addressed carefully. One of the main biosafety concerns associated with synthetic biology is the potential for the accidental or intentional release of synthetic organisms into the environment, which could have unpredictable and potentially harmful ecological effects. There is a possibility for synthetic organisms to outcompete or disrupt natural populations, leading to ecological imbalances. The main biosafety concern of genome editing is the potential for off-target effects, where unintended changes are made to the genome, resulting in unexpected mutations or changes to the organism's phenotype. There are also biosecurity concerns about the deliberate misuse of cutting-edge biotechnologies, such as the creation of bioweapons or the enhancement of human traits, which could have significant ethical and social implications. Overall, it is important to carefully consider the potential biosafety and biosecurity risks, and ethical implications of cutting-edge biotechnologies and to develop appropriate regulations and guidelines to ensure their safe and responsible use.

Author Biography

Dr. Xinyu Song is an associate professor at the Center for Biosafety Research and Strategy at Tianjin University, China. Xinyu's research focuses on the application of cutting-edge biotechnologies to address the issues of the sustainable development. She is also dedicated to the development of biosafety/biosecurity strategies to avoid the misuse of the biotechnologies. Xinyu is the awardee of the Young Scientist Fund from the National Natural Science Foundation of China (NSFC). She was invited as a core member to attend the China–United States Track 2 Dialogue on Biological Security and the Biological Weapons Convention Meetings of Experts.

Dr. Weiwen Zhang is Baiyang chair professor of Tianjin University, director for Laboratory of Synthetic Microbiology, and Center for Biosafety Research and Strategy (CBRS) at Tianjin University of China. Dr. Zhang has broad research experience in microbial synthetic biology and has authored more than 250 peer-reviewed scientific papers. Dr. Zhang is currently chief scientist for the National Key R&D Research Program of China – Synthetic Biology programme, and chief investigator for the Key Strategic Project of the Chinese Association for Science and Technology on dual-use biotechnology governance. Dr. Zhang is also the founding director of Center for Biosafety Research and Strategy (CBRS) and has served in a number of scientific advisory boards on biosecurity, biosafety, food science and technology and so on for multiple ministries and agencies in China.

References

1 Wintle, B. C. *et al* (2017) A transatlantic perspective on 20 emerging issues in biological engineering. *eLife*, 6, e30247.
2 Sargent, D. *et al* (2022) Synthetic biology and opportunities within agricultural crops. *Journal of Sustainable Agriculture and Environment*, 1 (2), 89–107.
3 Lv, X. *et al* (2021) Synthetic biology for future food: research progress and future directions. *Future Foods*, 3, 100025.
4 Moriarty, P. *et al* (2012) What is the global potential for renewable energy? *Renewable and Sustainable Energy Reviews*, 16 (1), 244–252.
5 Furtado, A. *et al* (2014) Modifying plants for biofuel and biomaterial production. *Plant Biotechnology Journal*, 12 (9), 1246–1258.
6 Poddar, H. *et al* (2020) Towards engineering and production of artificial spider silk using tools of synthetic biology. *Engineering Biology*, 4 (1), 1–6.

7 Gaj, T. *et al* (2016) Genome-editing technologies: principles and applications. *Cold Spring Harbor Perspectives in Biology*, 8, 12.
8 Sturme, M. H. J. *et al* (2022) Occurrence and nature of off-target modifications by crispr-cas genome editing in plants. *ACS Agricultural Science & Technology*, 2 (2), 192–201.
9 Wetterstrand, K.A. (2014). *DNA Sequencing Costs: Data from the NHGRI genome sequencing program (GSP)* http://www.genome.gov/sequencingcostsdata.
10 National Academies of Sciences. (2017) Medicine, Infectious Diseases, Pandemic Influenza, and Antimicrobial Resistance: Global Health Security Is National Security. in *Global Health and the Future Role of the United States*, National Academies Press (US).

8

Convergence of Science and Technology

Ralf Trapp

Independent Consultant, Chessenaz, France

Key Points

1) Convergence of science and technology is accelerating advances in the life sciences and drastically shortening the time from discovery to practical application.
2) Whilst convergence is expected to result in a wide range of beneficial applications, it also increases the latent misuse potential inherent in the life sciences.
3) It is important to adapt national and international regulatory frameworks, including arms control measures, to mitigate against these emerging risks.
4) Given the pace and complexity of these advances, however, government action alone will not suffice; it needs to be complemented by the adoption of a culture of responsibility by professionals working in the life sciences.

Summary

Convergence of science and technology is an approach to problem-solving that cuts across disciplinary boundaries. It has become a key feature of the life sciences in the twenty-first century. Examples include synthetic biology, mathematical modelling and the use of engineering approaches in biology. Digitalisation, including artificial intelligence (AI), enables the design of materials with new properties some of which mimic desired biological functions. Automation is linking up design, synthesis, experimental screening and process development into a loop to speed up, fine-tune and optimise the development of new materials. Cloud manufacturing and data exchanges are dramatically changing the way in which goods and services are being provided. These developments promise significant and wide-ranging benefits for society. But, they also increase the latent misuse potential of the life sciences, given their ability to automate key steps in biological weapons development and manufacturing, their potential to create new types of biological weapons and an increased exposure to cyberattacks. Given the dual-use

Essentials of Biological Security: A Global Perspective, First Edition. Edited by Lijun Shang, Weiwen Zhang, and Malcolm Dando.

nature of the technologies involved and the rapid pace of their advancement and dissemination, regulatory action and adaptations in the operations of arms control regimes alone will not suffice to mitigate against these risks. Government action needs to be complemented by the actions of other stakeholders, including an adoption of a culture of responsibility by all life science professionals.

8.1 Introduction

Science and technology progress along two complementary trajectories: specialisation (investigations focusing on specific phenomena and processes using ever-more sophisticated tools and concepts developed within a scientific discipline to deepen insights into the workings of nature and society) and convergence (bringing together knowledge, concepts and techniques from multiple disciplines to develop new and/or more complex theoretical concepts and investigative approaches).

The US National Academies characterised convergence as an approach to problem-solving that cuts across disciplinary boundaries. It integrates knowledge, tools and ways of thinking from a wide range of scientific and engineering disciplines, mathematics and computational sciences and beyond, stimulating innovation from basic science discovery to translational applications.[1]

At the turn of the twenty-first century, convergence has become a characteristic feature of the life sciences. Concepts and methods from engineering disciplines and mathematics are being applied to biology and combined with vastly enhanced computational power and an ever-increasing capacity of the Internet to enable the creation and use of huge data storages as well as of remote collaborations. This has changed the way biology is done, and it has manifested itself at the interface between chemistry and biology.

This chapter describes what convergence in the life sciences means, and how it affects practical applications of the life sciences in society, including with regard to dual-use risks. It then briefly discusses mitigation strategies to address the dual-use risks associated with convergence.

8.2 Convergence of Science and Technology in the Life Sciences

Convergence in the life sciences is contributing to better understand, mimic, manipulate and eventually design complex biological systems. It is expected, amongst others, to provide new and more effective medical treatments such as precision medicine, to expand the chemical and biochemical space available for the development of new materials and biochemical agents, to revolutionise industrial processes (e.g. biomanufacturing; additive manufacturing (AM); development, testing and manufacturing in the cloud; and application of AI) and more.

A key example of convergence in the life sciences is synthetic biology. It has been described as the application of engineering-based modelling and building techniques to modify existing organisms and microbes or construct new ones from scratch.[2] Based on engineering approaches to design, efforts are being made to develop standardised biological building blocks ('bio bricks') that can be used to construct biological systems with desired functionality. This may lead to practical applications such as regulatory circuits, data storage systems, biologically functional organelles and much more. The ambition of the field is to move biology from a descriptive to a prescriptive science to allow the design of desired functionality from basic principles.

Engineering approaches and principles are increasingly deployed in the life sciences in combination with other enabling technologies. Examples include vastly increased computing power and interconnectivity (for example, distributed computation to model 3D protein folding), automation of experimentation (for example, combinatorial libraries and high throughput synthesis and screening) and the creation of large data repositories (for example, DNA and omics data banks).

Still accelerating in pace, the field is expected to make significant contributions in medicine (new vaccines, delivery of therapeutics and treatments), energy (biofuels), environmental remediation, food production and general industrial manufacturing of goods (detergents, adhesives and perfumes).

A central aspect of conversion is the digitalisation of the life sciences. Machine learning (ML) and AI are being deployed to vastly expand the chemical space of potentially useful molecules whilst at the same time enabling molecular design to target desired properties and functionality (for example, new lead molecules for drug design, new drugs to overcome antibiotics resistance and new types of vaccines). Digitalisation is enabling new ways of designing materials with desired properties and functionality (for example, novel materials used in regenerative medicine, 'smart' textiles incorporating biosensors linked up with devices that deliver medicines in precise dosage when required).

Automation of manufacturing and robotics implemented in clouds, technologies such as AM, biomanufacturing and biofoundries, are beginning to dramatically change the way goods and services are being provided. This is expected to lead into more user-involved, distributed and digitalised ways of manufacturing and trade (also called industry 4.0, i.e. the integration of automation and data exchange into manufacturing), including a shift from transferring materials and products to transmitting information.

Genetic engineering is being deployed to develop crops with increased resistance to pests, reduced water consumption and higher yields. These methods are being combined with locally distributed or remote sensors and surveillance platforms (e.g. drones) to detect early stages of disease or drought and to enable automated and targeted delivery of remedial measures.

Furthermore, convergence facilitates scientific collaborations. This is enabled and driven by digitalisation, increasing interconnectivity and rapid dissemination and sharing of new methods, concepts and data (for example, the rapid global dissemination of the gene editing tool clustered regularly interspaced short palindromic repeats associated protein [CRISPR–Cas]). Such collaborations occur not merely across disciplinary boundaries but also across physical borders, resulting for example in the creation of virtual laboratories distributed between different countries and facilities that work together on a project basis. Despite existing cultural, institutional and political roadblocks, convergence can enable regional and global frameworks for beneficial international collaborations and partnerships in the life sciences.

At the same time, investments into a bioengineered world are seen by many States as of strategic importance. Countries are spending vast amounts of money in this field, and expectations are high for making an impact in such areas as public health, including pandemic preparedness, food and energy production, managing the effects of global warming and more. These strategic investments are helping scientists to push further the frontier of their research. They provide significant resources in addition to what industry and private funders already provide.

8.3 Convergence and Arms Control and Security

Advances in science and technology can affect the operation of the two cornerstone disarmament treaties in the field of chemistry and biology – the Biological Weapons Convention (BWC) and the Chemical Weapons Convention (CWC). The CWC prohibits the development, acquisition, possession, transfer and use of poisons as weapons; the BWC does the same primarily with regard to weaponising disease. Both treaties are almost universally adhered to by States, and both compel their States Parties to adopt measures to domesticate these prohibitions and apply them to their subjects. For more see Chapter 10.

The provisions of both treaties are anchored in the sciences and technologies that enable the use of poison and disease as weapons. Advances in science and technology may alter the scientific and industrial environment within which the treaties function; new discoveries might even challenge the very scope of the prohibitions.

To safeguard the integrity of the prohibitions under the two Conventions against the discovery or design of new types of chemical or biological agents, both treaties rely on a so-called general-purpose criterion.[3] This legal construct aims at making the prohibitions future-proof. The norms are designed to cover any discovery or development of new types of chemical or biological weapons, or of new methods for their development, production or use. It is, however, prudent to monitor and review advances in the sciences to detect developments that could become 'game changers' and may have the potential of undermining the foundations of the prohibitions.

But even without challenging the fundamentals of the regimes, scientific and technological advances might call for adaptations of practical and administrative measures employed to implement treaty requirements. Advances in science and technology and in the organisation and operations of the related industries affect the implementation environment of the two treaties. This may result in the need to adapt certain national implementation measures or to create additional mechanisms to manage the risks associated with these advances and to prevent misuse.

This is why both treaties require regular monitoring of scientific and technological advances and the evaluation of their impact on the functioning of the treaty mechanisms. As part of such a regular review, developments at the intersection of chemistry and biology were studied by the scientific advisory board (SAB) of the Organisation for the Prohibition of Chemical Weapons (OPCW) already in 2003.[4] The SAB was able to draw on findings of an international conference held by the International Union of Pure and Applied Chemistry (IUPAC) that had reviewed how these advances in science and technology affected the CWC implementation.[5] In its report to the First CWC review conference, the SAB highlighted advances in molecular biology (such as genomics and proteomics) and other developments at the intersection of chemistry and biology. Neither IUPAC nor the SAB used the term 'convergence' at the time. The SAB took up these issues again in its report to the Second CWC review conference in 2008, prompting the OPCW Director-General to conclude that the convergence between chemistry and biology needed further reflection, as it clearly impacted on the scientific basis of the CWC. In 2011, the Director-General established a temporary working group of the SAB to further explore the issue. Its report contained a range of recommendations, including a recommendation to continue the conversation on convergence of chemistry and biology in some form.[6]

Mirroring these efforts in the CWC context, discussions on convergence were also taken up in the context of the BWC. Reviews prepared by several State Parties of the BWC and international science unions as well as the BWC implementation support unit drew attention to the rapid pace of advancement in the life sciences. In addition to general reviews of advances in science and technology relevant to the BWC (a standing agenda item for the intersessional BWC process), BWC review conferences and annual meetings included specific scientific topics in the intersessional work programmes, amongst them: enabling technologies (2012), infectious disease and toxin surveillance, detection, diagnostics and mitigation (2013), understanding pathogenicity, virulence and toxicology and immunology (2014), advances in production, dispersal and delivery (2015), biological risk assessment and management (2018–2020), voluntary model code of conduct for biological scientists and all relevant personnel, biosecurity education (2018–2020) and genome editing (2018). At the 9th review conference of the BWC in 2022, BWC Member States decided to establish a working group on the strengthening of the Convention, and tasked that group, amongst others, to prepare recommendations on how to establish a mechanism to review and assess scientific and technological developments relevant to the Convention and to provide States Parties with relevant advice.[7]

An initial question raised in these conversations was whether convergence could create uncertainties about the application of the overlapping treaty requirements. Both treaties cover toxic chemicals of biological origin such as toxins and bioregulators. Examples are ricin and saxitoxin, two naturally occurring toxins that are specifically listed in Schedule 1 of the CWC, but that would also fall under the provisions of Article I of the BWC if their quantities cannot be justified for prophylactic, protective or other peaceful purposes. However, the regimes established under the two treaties differ significantly: the management of the implementation of the BWC is left to the States Parties themselves, supported by a small implementation support unit set up in the United Nations Office for Disarmament Affairs (UN ODA) which supports consultative mechanisms such as intersessional work, expert and diplomatic meetings and review conferences; with regard to suspected uses of biological and toxin weapons, there is also the UN Secretary-General's Mechanism to investigate such reports – this complements the diplomatic mechanisms foreseen under the BWC. The CWC, on the other hand, has set up a complex institutionalised system involving at the international level, amongst others, expert and diplomatic meetings, international verification measures implemented by the Technical Secretariat, technical assistance to States Parties, a programme to foster international cooperation and assistance, and mechanism for fact-finding and compliance assurance.

As convergence became more manifest in the life sciences, and (more importantly) in the different domains of practical application of science and technology (industry, agriculture, public and animal health, environmental management, the fight against climate change, etc.), the question of how the two treaty regimes would interact at the intersection of chemistry and biology became a practical issue. This was reflected also in how the arms control community initially described convergence – somewhat simplistically – as 'chemistry making biology, biology making chemicals'.

It was recognised, however, that the overlap between chemistry and biology was not itself a new issue. The disciplinary overlap had existed all along, but convergence as experienced today had certain practical implications which could affect the implementation systems of the two treaties. This included:

- An acceleration of the pace of discovery in the life sciences and a compression of the time required from scientific discovery to practical application,
- A change in the drivers for these developments with players in industry becoming more prominent; these new actors complemented government funding streams by resources and incentives emanating from the markets and from within industry and the research community,
- Changing patterns of scientific collaboration and wider diffusion of scientific, technical and industrial capabilities,
- A changing industrial landscape, in terms of technology, organisation within the industry and trade relations.

These effects can put existing treaty implementation systems under pressure, posing questions about whether and how national implementation systems and international verification may have to be adapted. The convergence trends also call for a closer interaction and communications between the communities involved, respectively, in the implementation of the CWC and the BWC.

8.4 Technologies of Particular Relevance for Possible Misuse of Biology for Nefarious Purposes

Convergence in science and technology (understood not simply as convergence between chemistry and biology but in its broader sense as described above) can contribute to a growing misuse potential inherent in the life sciences. It increases the accessibility of relevant knowledge, materials, and equipment that could be misused for nefarious purposes, including by non-State actors. It may also create new incentive for State actors to exploit these advances for military purposes. This is why advances in the life sciences need to be monitored to identify trends and discoveries that could become 'game changers'.

Efforts to monitor emerging technologies and evaluate their potential impact on the regime governing chemical and biological (CB) weapons arms control are being done in different frameworks. In addition to the reviews undertaken under the two Conventions (the OPCW SAB and the science and technology (S&T) reviews prepared by the implementation support unit (ISU), as well as the outcomes of expert meetings organised by the two treaty systems), reviews are also undertaken by certain countries that have taken a particular interest in the matter. Examples include:

- The Spiez Convergence conference series – a workshop series organised by Spiez Laboratory that brings together experts from CBW arms control, life science research, other scientific and technology disciplines and industry to assess the impact of advances

in science and technology at the intersection of chemistry and biology on CBW arms control – has over the years analysed advances in a range of scientific and technological disciplines in an attempt to clarify possible risks and benefits and identify potential game changers for CBW arms control.[8] In recent meetings, Spiez Convergence paid particular attention to ML and AI. The conference series also recognised the significance of deploying several emerging technologies in a combined fashion (such as the combination of AI with synthetic biology, automation and robotics). These developments lead to a shift in relevance from sensitive hardware (equipment/laboratories) and materials to data and people, which could challenge some of the assumptions underlying export controls as well as verification.
- National working papers prepared for BWC review conferences and annual expert and diplomatic meetings. A recent example was a UK national working paper[9] that analysed developments in science and technology with particular emphasis on BWC Article VII and experiences from the COVID-19 pandemic. It presented lessons-learned for a more effective response to future disease outbreaks, the enhancement of international cooperation and assistance and the application of scientific discoveries for disease prevention and future pandemic preparedness, regulatory and ethical issues and for tools to assist investigations of suspected uses of biological weapons.

One field of research that has attracted much attention for its inherent misuse potential is synthetic biology. Its techniques and concepts could be deployed to create dangerous pathogens, invasive organisms or other disruptive biological agents. Using a comprehensive analytical framework which assessed the maturity of the technologies in question, their usability as weapons, the requirements for the actors concerned and the mitigation potential available, the US National Academies ranked ways of deploying synthetic biology for weapons purposes into five different degrees of concern as shown in Figure 8.1:[10]

Genome editing in particular has attracted attention of the security community. In 2016, the then US Director of National Security James R. Clapper included genome editing amongst the major threats in the area of weapons of mass destruction, given its broad distribution, low cost, accelerated pace of development and misuse potential.

More recently, the focus has widened beyond synthetic biology and genome editing and now includes reference to AI. The Director of National Intelligence Avril Haines testified before the US Senate Select Committee on Intelligence that '[N]new technologies – particularly in the fields of AI and biotechnologies – are being developed and proliferating faster than companies and governments are able to shape norms governing their use, protect against privacy challenges associated with them and prevent dangerous outcomes that they can trigger.'

Advances in three specific emerging technologies could facilitate the development or production of new biological weapons and their delivery systems: AM, AI and robotics.[11] This conclusion is based on a number of factors: the ability of these technologies to automate key steps in biological weapons development and manufacturing, their potential to create new types of biological weapons uses and an increased exposure of digitalised biological data and operating parameters to cyberattacks. There are also difficulties to control these technologies given their dual-use nature, their digitalisation, and the fact that they are mainly being developed by the civilian/private sector.

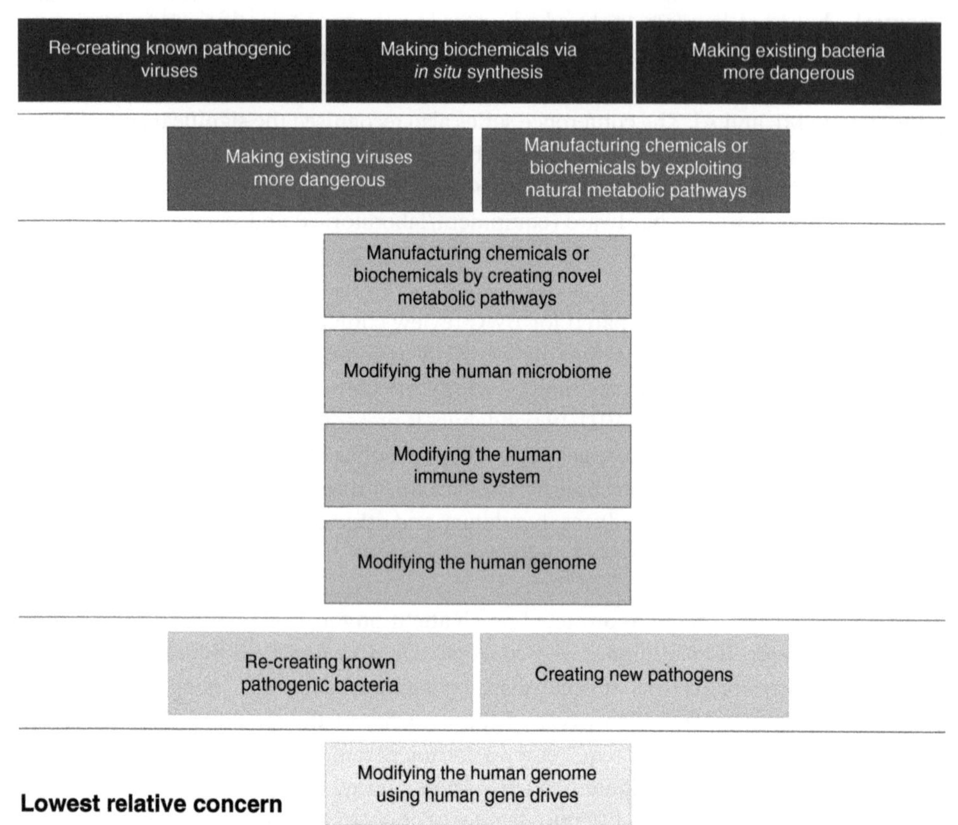

Figure 8.1 Relative ranking of concerns related to synthetic biology-enabled capabilities, according to US National Academies of Sciences, Engineering, Medicine (2018)/National Academies Press.

Spiez Convergence concluded in 2022 that ML and AI are close to becoming game changers that could profoundly affect the regimes prohibiting CB weapons. It is important to realise that it is not simply the advancement of any individual technology that acts as a game changer, however important it may be. For example, AI and ML are providing tools to dramatically expand the space of conceivable biologically active chemical compounds, which could be exploited to develop new types of chemical or biological weapons. But what really makes these advances highly relevant is the integration of AI/ML with other technologies, thereby linking design, synthesis, experimental screening and process development for eventual manufacturing into a loop to speed up, fine-tune and optimise the development of new materials (molecules, composites, synthetic structures that mimic biological systems/functionality, etc.).

At the same time, the threats posed by these developments ought to be assessed in context, taking account of motivations and capabilities of the relevant actors. The impact of these technologies on the engineering of biological weapons and their delivery systems should not be exaggerated. The expertise required to exploit these technologies for the purpose of developing and producing biological weapons remains significant and continues to pose a barrier to most actors.[12]

For many years, following the terrorist attacks of 11 September (2001), much attention has been paid to the possibility of terrorists attempting to use chemical and biological weapons. In this context, many non-proliferation and biosecurity efforts were designed and implemented around threat perceptions associated with non-State actors. Recent development – the use of chemical weapons in Syria and a number of assassination attempts by State actors using highly toxic chemicals – have given rise to rethinking this orientation. The report of Spiez Convergence 2022 observed that 'the (common) perception about a limited military utility of CB weapons may no longer be shared by all. Concerns of the arms control community about utilising scientific advances for chemical or biological weapons may have to focus on states and less on terrorists or lone actors. This recognition is important with regards to the large strategic investment into a bioeconomy. The results will inherently contain also some misuse potential. That all states will resist the temptation is in today's geopolitical climate not obvious and a challenge for the future'.[13]

8.5 Mitigation of the Evolving Misuse Potential Resulting from Convergence

Advances in science and technology have consistently been included by BWC and CWC meeting reports and in statements and reports by governments, NGOs, science organisations and observers as amongst the trends that may create both benefits and new risks to the norm against CB weapons. Although motivated by creating deeper insights into nature and aimed at rendering beneficial effects for society in a wide range of domains, these advances also contribute to a steady increase of the latent misuse potential of the life sciences.

Convergence in the life sciences can accelerate the expansion of this latent misuse potential. The integration into the life sciences of knowledge, tools and concepts from other scientific and engineering disciplines, mathematics and computational sciences does not simply speed up the incremental acquisition of new data and knowledge. It can also result in 'non-linear' progress, unexpected discoveries (serendipity) and the removal of existing roadblocks.

Monitoring advances in science and technology and responding to emerging challenges that emanate from these developments for the scope as well as the effective functioning of the BWC and the CWC therefore are key aspects of maintaining and strengthening the norm against biological and toxin weapons. This was recognised, amongst others, by including the item of 'measures on scientific and technological developments relevant to the Convention' amongst the tasks given to the newly established working group on the strengthening of the BWC set up by the 9th BWC review conference.

States see their investment into the advances in the life sciences as strategic investments. In certain areas such as brain research, bio-foundries or AI, certain States are spending vast amounts of money – based on expectations that these may bring about future solution to critical societal problems but also to enhance defences against external threats. Big powers such as the United States or China clearly aim at achieving technological dominance in critical areas. Others such as many European countries and the European Union (EU) aim at more technological and economic independence. As biotechnology is becoming better, cheaper, more versatile and widely distributed, the question may arise whether certain States may be tempted to take advantage of these advances in the form of new military means or means of suppression and control. Will the international norms be strong enough to guard against such attempts to break out of the regimes that are meant to prohibit and prevent the misuse of the life sciences, including chemical and biological warfare?

Strengthening the international norms against chemical and biological warfare, therefore, remains an important task. The resilience of the norms set out in the CWC and BWC will depend on how governments respond to changes in the threat landscape and deal with new challenges – be they scientific discoveries, new research and industrial methods and processes, changes in the security landscape, or newly emerging actors relevant to the norm. This will also include the manner in which States deal with compliance concerns, and how they address uncertainties in good faith to resolve compliance concerns or respond to demonstrated cases of non-compliance.

There is, at this moment in history, a growing reason for concern. In 2022, the report on Spiez Convergence observed that '[m]any discussions in the arms control communities are stuck in the past and no longer reflect the emerging misuse potential. The same applies to the verification system of the CWC and the compliance assurance mechanism of the BWC, which were designed for past weapons programmes. It is, therefore, important to also strengthen non-treaty mechanisms by embedding dual-use ethics with scientists, they are the first line of defence'.[13] Nongovernmental organisations (NGOs) and civil society need to monitor and evaluate these trends closely and keep reminding governments of their responsibilities and commitments with regard to protecting the norm against CB warfare.

But, it is not States and governments alone that may pose concerns, or that will need to take mitigation action to manage these risks. The norm against the misuse of biology manifests itself at different levels and involves a wider range of actors than governments alone. Understood as a set of behavioural standards with regard to how the advances in the life sciences are brought to use in society, the norm has legal, political, social, professional and ethical dimensions. Its strength and resilience do not merely depend on the quality of the law, the effectiveness of the structures and institutional mechanisms that apply and enforce it and the willingness and ability of the parties of the Conventions to ensure compliance. It is also conditioned by other factors and actors, including by perception of and actions taken by professional communities, civil society, the media, industry and others.

The advances in the life sciences are driven today not only by government funding and investment but also by private investors and industry. Motivations come from high expectations that the biological industrial revolution will help developing sustainable solutions to pressing problems such as managing global warming, ensuring food security, delivering public and animal health and helping to protect the environment.

Furthermore, the advances in the life sciences have also attracted the interest of communities such as a growing do-it-yourself (DIY) biotech community. The activities by these communities have been identified by several authors as a potential risk factor given the ease of availability of genome editing techniques and access to information, equipment, material and learn techniques – creating an inadvertent misuse potential as well as a potential problem for public safety.

None of these actors appear to be motivated by nefarious purposes. However, it is well established that certain terrorist groups (the Aum Shinrikyo in Japan or more recently the Islamic State (IS) and some groups associated with it) have shown an interest in deploying biological or chemical agents. The growing (albeit latent) misuse potential of the life sciences may be something that such organisations may decide to tap into, or that lone actors with the right skill set and knowledge might decide to exploit for their purposes. States, too, may feel tempted to encourage or conduct targeted small-scale/clandestine operations such as assassinations using CB agents that are highly potent, and difficult to detect and/or distinguish from naturally occurring disease factors.

It is essential to respond to such evolving threats by adapting traditional arms control and security instruments, including international treaties and national laws. But adapting the law is a slow and solemn process. The life sciences and related technologies and their applications in society, by contrast, are advancing at an ever-increasing pace. Regulatory and institutional responses, therefore, will not suffice to properly manage the risks associated with these new scientific discoveries and technological advances. They need to be complemented by actions taken by stakeholders other than governments, and in particular by actors in the professional communities that drive these advances.

A first step towards such a multidimensional approach to protecting and strengthening the norm against the misuse of the life sciences is to create awareness and understanding of the risk potential in these communities. This is the first critical step towards developing effective mechanisms that are capable of detecting and preventing/responding to negligence or deliberate misuse attempts. It is important therefore to widen and deepen awareness about dual-use risks.

Given the pace, complexity, multiplicity of actors and variety of motivations, States and the international actors implementing arms control in the CBW domain (the UN, the OPCW and others) cannot provide all the answers. Strategies and practical measures to prevent misuse must include efforts to widen and deepen awareness about dual-use risks in the communities that work in or intersect with the life sciences, to develop mitigation strategies against misuse and strengthen a culture of responsible conduct (for example, the Tianjin Biosecurity Guidelines or the OPCW's The Hague Code). A key challenge for education and outreach in this respect is the multidisciplinary nature of convergence and, consequently, the diversity of the life science community. This is why conversations and activities across different communities will be critical to address these arms control and security concerns.

Author Biography

Dr. Ralf Trapp is a chemist and toxicologist by training. Between 1983 and 1992, he participated as a technical adviser in the negotiations of the Chemical Weapons Convention. Thereafter, he worked for 13 years at the OPCW, dealing with issues of verification, international cooperation, government relations, strategic planning, and science advice. Since leaving the OPCW in 2006, he has been providing consulting services on chemical and biological security and arms control to, amongst others, the United Nations, the OPCW, EU institutions, and several academic and research organisations. He has published extensively on chemical and biological weapons arms control.

References

*1 National Research Council *et al* (2014) *Convergence: Facilitating Transdisciplinary Integration of Life Sciences, Physical Sciences, Engineering, and beyond*, National Academies Press.

2 Trump, B. D. *et al* (2021) Biosecurity for Synthetic Biology and Emerging Biotechnologies: Critical Challenges for Governance.

*3 For an explanation in the context of the CWC see R. Trapp (2022) "The chemical weapons convention", chapter 20 of Eric Myjer and Thilo Marauhn (Eds.), Research Handbook on International Arms Control, E Elgar Publishing, 280–284.

4 OPCW Document RC-1/DG.2 (2003). Available at https://www.opcw.org/sites/default/files/documents/CSP/RC-1/en/RC-1_DG.2-EN.pdf.

5 Parshall, G. W. *et al* (2002) Impact of scientific developments on the chemical weapons convention (IUPAC technical report). *Pure and Applied Chemistry*, 74, 2323–2352.

6 OPCW document SAB/REP/1/14 (2014), *Recommendation 13*. Available at https://www.opcw.org/sites/default/files/documents/SAB/en/TWG_Scientific_Advsiory_Group_Final_Report.pdf.

*7 BWC document BWC/CONG.IX/9, *II. Decisions and Recommendations, E. Review of Scientific and Technological Developments Relevant to the Convention, Para. 19*. Available at https://unodaweb-meetings.unoda.org/public/2022-12/2022-1221%20BWC_CONF_IX_9%20adv%20vers.pdf.

*8 Spiez Laboratory and ETH Zürich, *Spiez Convergence, Reports of Meetings Held in October 2014, September 2016, September 2018, September 2021 and September 2022*. A Next Meeting in This Series is Scheduled for September 2024. Meeting reports are available at https://www.spiezlab.admin.ch/en/home/meta/refconvergence.html.

9 United Kingdom Working Paper BWC/CONF.IX/WP.9. Available at https://documents-dds-ny.un.org/doc/UNDOC/GEN/G22/576/29/PDF/G2257629.pdf?OpenElement.

*10 National Academies of Sciences, Engineering, and Medicine. (2018) *Biodefense in the Age of Synthetic Biology*, The National Academies Press, Washington, DC.

11 Brockmann, K. *et al* (2019) *BIO PLUS X: Arms Control and the Convergence of Biology and Emerging Technologies*, SIPRI.

12 National Research Council *et al* (2014) *Convergence: Facilitating Transdisciplinary Integration of Life Sciences, Physical Sciences, Engineering, and beyond*, National Academies Press.

13 Spiez Convergence (2022), op. cit. 8, p. 11.

9

Role of the Life Science Community in Strengthening the Web of Prevention for Biosafety and Biosecurity

Tatyana Novossiolova

Law Program-Center for the Study of Democracy, Sofia, Bulgaria

Key Points

1) The effective management of biological risks requires the implementation of both biosafety and biosecurity policies, regulations, controls and measures.
2) Awareness of biological threats in the life sciences is an important element of the process of ensuring health security.
3) Life science stakeholders can play a key role in driving innovation in biosafety and biosecurity risk management and promoting the ethos of science for peace.

Summary

This chapter focuses on the ways in which the life science community contributes to the process of upholding the norm against biological weapons. It uses the concept of a 'web of prevention' to elucidate the interconnectedness between health security and biological disarmament to ensure that life sciences and related fields are used only for peaceful purposes and to benefit humanity and the environment. The chapter begins by discussing the importance of a holistic approach to the management of biological risks regardless of their origins, whether natural, resulting from accidents or deliberate acts. It then outlines a set of indicative elements that are essential for addressing the risk of deliberate biological events. The chapter draws upon practical examples and initiatives to demonstrate possible ways in which life science stakeholders can help enhance the resilience of the existing biosafety and biosecurity regimes and ensure that biological agents, toxins and related information are not misused.

Essentials of Biological Security: A Global Perspective, First Edition. Edited by Lijun Shang, Weiwen Zhang, and Malcolm Dando.

9.1 Introduction

Disease outbreaks – whether naturally occurring, or resulting from the unintentional or deliberate release of pathogens or toxins – can have significant consequences in terms of human, economic, societal, environmental and material costs and strain a state's capacity to counter threats to public, animal or plant health.[1] The scale of life loss and socio-economic devastation that communities have suffered during the COVID-19 pandemic is instructive. Overcrowded hospitals and deserted cities became an everyday reality, as the disease raged across the world. Social distancing that was vital for curbing the spread of the infection heightened the risk of social isolation instilling a widespread perception of helplessness and desperation. Social anxieties and recurring misconceptions that were, at least in part, amplified by the deliberate dissemination of misleading narratives and far-fetched conspiracy theories permeated (and at times, hijacked) policy and scientific debates, as demands for quick solutions shaped the overall context of the response to the pandemic.

COVID-19 has turned into a painful reminder that pathogens recognise no borders and that, in a densely globalised international environment, air travel and trade can facilitate the spread of a disease even before its nature and source have been established. Four years after the start of the pandemic, the question of the origins of SARS-CoV-2 is yet to receive a definitive answer. In 2021, a global study convened by the World Health Organisation (WHO) concluded that an introduction through an intermediate host was a very likely pathway of emergence of the virus.[2] Unresolved concerns that the infection could have spread, as a result of a laboratory accident have polarised scientific and policy communities and offered fodder to malign actors seeking to sow doubt and confusion.[3]

To prevent and respond to biological events regardless of their origins, states require robust national capacities for disease surveillance and detection and preparedness in case of a disease outbreak. International coordination and cross-sectoral collaboration are key instruments for developing and sustaining such capacities. Establishing multidisciplinary teams which can tap diverse areas of expertise (e.g. health, agriculture, research, industry, law enforcement, security, academia, etc.) and facilitating dialogue and collaborative action amongst professionals from various fields is crucial for enhancing cross-sectoral interoperability and fostering sustainable partnerships amongst the different types of emergency responders.

Whereas the delivery of healthcare and disease containment are states' utmost priorities during a biological event, conducting a timely investigation to establish the cause of the disease, can provide authorities with important insights, especially if there is suspicion that the event has occurred as a result of an accident, or a deliberate action. Equally, if the event is natural, data collected during the investigation can be instrumental for countering disinformation campaigns and preventing malign actors from misusing the crisis to pursue self-serving objectives.

This chapter focuses on the role of the life science community in managing the risk of biological events regardless of their origins. This includes the risk of misuse of life sciences to threaten and/or harm human, animal or plant health. The chapter uses the concept of a 'web of prevention' to elucidate the need for continuous interaction between the international regime that seeks to prevent the unintentional/accidental release of pathogens and toxins, including naturally occurring diseases (i.e. biosafety), and the international regime that seeks to prevent the deliberate release and misuse of pathogens and toxins (i.e. biosecurity).[4] It first discusses the importance of a holistic approach to the management of biological risks regardless of their origins, whether natural, resulting from accidents or deliberate acts. Secondly, the chapter outlines a set of indicative prerequisites that are essential for addressing the risk of deliberate biological events. And thirdly, it highlights possible ways in which life science stakeholders can help enhance the resilience of the existing biosafety and biosecurity regimes and ensure that biological agents, toxins and related information are not misused.

9.2 Integrating Biosafety with Biosecurity: The Web of Prevention as a Model Concept

The international prohibition of the use of disease as a weapon of war dates back at least to the early twentieth century, but it was not until the 1970s that the development, stockpiling, production and retention of biological and toxin weapons were outlawed altogether. Evidence that states which had ratified (Union of Soviet Socialist Republics [USSR], South Africa) or signed (Iraq) the 1975 Biological and Toxin Weapons Convention (BTWC) remained eager to operate clandestine offensive biological weapons programmes, the changed international security landscape after the end of the Cold War, and the rapid progress of life sciences over the past two decades are key factors that have shaped the context of biological disarmament. Approaching this context requires a lens that is broad enough to capture and evaluate the underpinning multifaceted dynamics. The concept 'web of prevention' provides such a lens and makes it possible to appreciate the interconnected nature of the different international regulatory and guiding frameworks relevant to the prevention and countering of the misuse of life sciences, and the importance of their coherent implementation through appropriate lines of action.

The concept 'web of prevention' in relation to biological disarmament originated in the work of Pearson in the early 1990s and reflected the growing appreciation of the need for effective protection against the entire chemical and biological weapons (CBW) spectrum that ranges from classical chemical weapons (e.g. chlorine, nerve agents, etc.), through industrial or pharmaceutical chemicals, to 'mid-spectrum' agents such as bioregulators, peptides and toxins, to genetically modified and traditional biological weapons (e.g. viruses, bacteria, whether tailored/modified or not).[5] Initially defined as a web of deterrence or a web of reassurance, the web of prevention comprises several complementary elements:

- Effective protective and preparedness measures, and response plans that reduce the utility of biological weapons.
- Effective international and national prohibition regime reinforces the norm that all biological weapons are comprehensively outlawed.
- Broad international and national controls on the handling, storage, use and transfer of biological agents and toxins to increase the difficulties of acquiring biological warfare agents and/or the necessary technology and equipment for their production.
- A political commitment to react vigorously and with determination to any use of biological weapons or threat, thereof, with a range of national and international responses to ensure that those responsible are held to account.

The idea of a web of prevention provides a useful conceptual tool for understanding the need for coherent and effective implementation of the different international regulatory and guiding instruments that underpin the process of ensuring that life sciences and related fields are used only for peaceful ends. Depending on their purpose, these instruments fall in two broad categories: (1) biosafety and (2) biosecurity.[6] International biosafety instruments address the risk of the unintentional (accidental) release of biological agents and toxins, including naturally occurring diseases. International biosecurity instruments address the risk of the deliberate release of biological agents and toxins which includes the use of biological weapons. The international biosafety regime comprises three main elements:

- Human, animal and plant health security and food safety.
- Preservation of biodiversity and responsible management of living modified organisms (LMOs).
- Safe handling, shipment, transfer and transport of biological agents and toxins.

The international biosecurity regime comprises four main elements:

- Prohibition of biological and toxin weapons.
- International counter-terrorism.
- Export controls.
- Secure storage, handling and transport of biological agents, toxins and related technology and equipment and management of dual-use life science research.

The international biosafety and biosecurity regimes are mutually reinforcing for managing the risk of biological events regardless of their origins. To illustrate their complementary role, the next section outlines a set of indicative prerequisites that are essential for effective protection against deliberate disease and safeguarding life sciences against misuse.

9.3 Addressing the Threat of Deliberate Biological Events and Life Science Misuse

United Nations Security Council (UNSC) Resolution 1540 which was adopted unanimously in 2004 decided that all States shall take and enforce effective measures to ensure the physical protection, including security of transport, transfer and shipment of biological agents and toxins, as well as to combat the illicit trafficking in such materials through appropriate legislation such as export controls.[7] Subsequent Security Council's Resolutions have reiterated these obligations and called upon states 'to take into account the developments on the evolving nature of risk of proliferation and rapid advances in science and technology in their implementation of resolution 1540'.[8] UNSC Resolution 1540 is a key element of the international biological disarmament and non-proliferation architecture, insofar as it is binding on all states regardless of whether they participate in relevant international agreements, e.g. the BTWC. As such, it provides a regulatory framework for identifying emerging biosecurity concerns and developing appropriate policies, standards, measures and controls to mitigate the risk of misuse.

The biological threat spectrum comprises different forms of hostile activities which makes it particularly daunting (Figure 9.1).

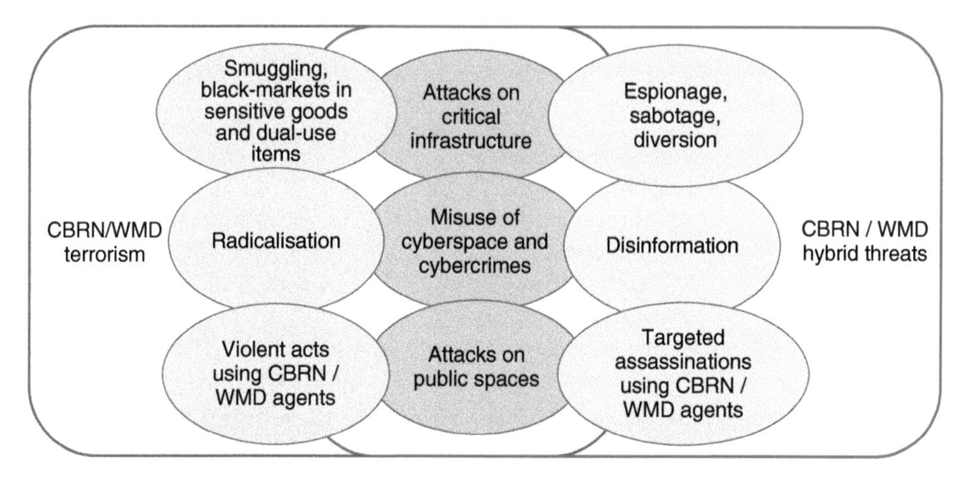

Figure 9.1 Biological threat spectrum. *Source:* The author. Based on Figure 2 in Novossiolova and Martellini.[9]

Firstly, biological threats vary depending on their source, i.e. a state or non-state actor, and underlying motivation. Secondly, as such threats may involve the release of biological agents or toxins, their effects can be far-reaching and difficult to contain. And thirdly, since biological events can have significant consequences, even the threat of deliberate release of biological agents or toxins can result in mass panic. An operational web of prevention for biosafety and biosecurity is essential for countering biological threats. The rest of this chapter outlines a set of indicative prerequisites for addressing the risk of deliberate disease and life science misuse.[10]

9.3.1 Multi-layered Framework for Response to Deliberate Biological Events

A biological attack could take different forms, and it is possible that the origin of the disease may not be immediately evident. The availability of robust and reliable healthcare infrastructure and services is of utmost importance to tackle the immediate effects of a disease outbreak. To assist states with the implementation of the 2005 International Health Regulations (IHRs) and measure their progress in developing capacity to prevent, detect and respond to biological events that threaten public health, the WHO has launched a voluntary evaluation process that uses a metrics system of 56 indicators covering 19 technical areas – the Joint External Evaluations (JEE) Tool (Table 9.1).[11]

The JEE Tool is grounded in a one-health approach that takes into account the interconnected nature of human and animal health and the health of natural ecosystems. It acknowledges the relevance of a strong health-security interface for addressing suspicious biological events, as well as the importance of equity considerations regarding vulnerable populations at all stages of disease prevention and preparedness. The JEE Tool is intended to support states in their efforts to strengthen national systems for health security through

Table 9.1 Overview of the Joint External Evaluations (JEE) Tool.

JEE Tool: Technical Areas	
Prevent	P1. Legal instruments P2. Financing P3. IHR coordination, national IHR focal point functions and advocacy P4. Antimicrobial resistance P5. Zoonotic disease P6. Food safety P7. Biosafety and biosecurity P8. Immunisation
Detect	D1. National laboratory system D2. Surveillance D3. Human resources
Respond	R1. Health emergency management R2. Linking public health and security authorities R3. Health services provision R4. Infection prevention and control R5. Risk communication and community engagement
IHR-related hazards	PoE: Points of entry and border health CE: Chemical events RE: Radiation emergencies

Source: The author. Adapted from World Health Organisation.[11]

the identification of needs and matching opportunities for improvement. It further seeks to enhance awareness amongst relevant national authorities and stakeholders of the mechanisms for reporting biological events of international concern (e.g. IHR requirements, World Animal Health Information System [WAHIS]) and requesting assistance (e.g. Global Outbreak Alert and Response Network [GOARN]). The BTWC contains specific provisions for assistance in case a State Party is exposed to the use of biological weapons and the WHO administers an early warning, alert and response system (EWARS) which aims to improve the detection of disease outbreaks in emergency settings, including conflict situations.[12]

Under the United Nations Secretary-General's Mechanism (UNSGM) for investigation of alleged use of CBW, any UN Member State can trigger an internationally mandated investigation if it is suspected that biological or toxin weapons have been used.[13] Member States can nominate expert consultants, qualified experts and analytical laboratories for the UNSGM roster to take part in investigation activities, such as fact-finding missions or sample analysis. The UNSGM is separate from the BTWC but complements the effective implementation of its provisions and objectives.

To support national-level capacity building for the investigation of suspicious biological events, the World Organisation for Animal Health (WOAH) has published organisational and operational guidelines.[14] These guidelines are primarily directed at veterinary services and stress the need for intersectoral relationships and interregional cooperation in the

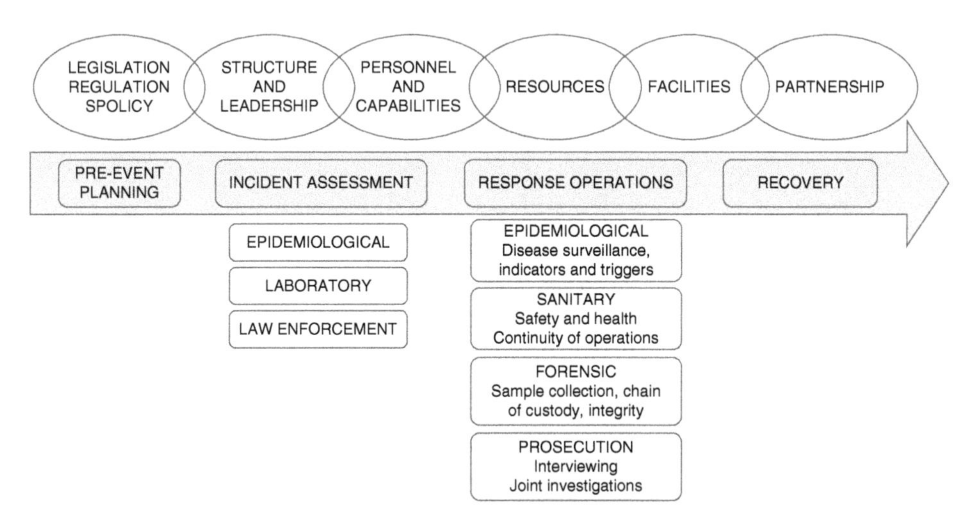

Figure 9.2 WOAH organisational and operational guidelines for investigation of suspicious biological events. *Source:* The author. Adapted from Based on World Organisation for Animal Health.[14]

management of biological threats. Building partnerships is a core element of the proposed organisational setup for the development and implementation of response strategies. This includes engaging with stakeholders across (1) government departments such as 'health, law enforcement, interior, border and customs, environment, trade/commerce, foreign affairs, defence, finance, crisis management, civil protection and agriculture or rural affairs'; (2) industry such as 'producers, transport, processing, distributors and retail'; (3) civil society and general public; (4) bordering countries; and (5) regional/international organisations. The guidelines outline an indicative process for organising investigation and response activities based on pre-event planning (Figure 9.2). They further suggest three groups of indicators for assessing suspicious biothreat events: (1) epidemiological (e.g. eradicated disease; novel/emerging disease; disease that is exotic to the country/ geographic region); (2) laboratory (e.g. theft, security breach); and (3) law enforcement (e.g. intelligence/credible threats; suspicious behaviour).

Law enforcement at the national level is a critical element in the process of preventing and countering the misuse of biological agents and toxins. The United Nations Inter-Regional Crime and Justice Research Institute (UNICRI)'s *A Prosecutor's Guide to Chemical and Biological Crimes* provides a consolidated overview of the evolving threat landscape in the area of biological (and chemical) security, key regulatory instruments and the procedural aspects of the process of investigating and building a case for prosecuting biological crimes.[15] This Guide discusses the life cycle of biological crimes (Figure 9.3) highlighting the relevance of national legislation that applies to the 'following elements as potential or actual crimes:

- Encouraging or assisting a crime, incitement, attempting to commit an offence and conspiracy to commit it.
- Planning a criminal act.

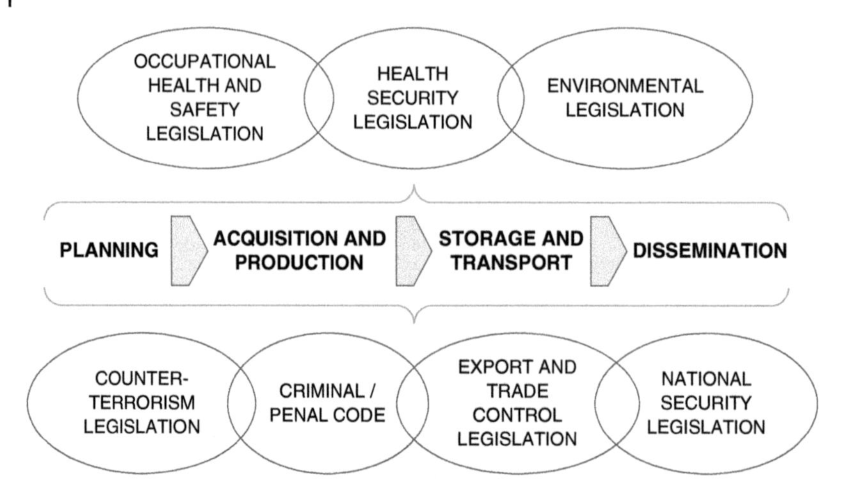

Figure 9.3 Indicative legal frameworks applicable to the biological crime lifecycle. *Source:* The author. Adapted from United Nations Inter-Regional Crime and Justice Research Institute.[15]

- Breach of regulatory or industry civil and criminal penalties.
- Deliberately contaminated soil, water or air ways with chemical or biological material that may cause harm.
- Finance operations linked to the deliberate use of chemical or biological materials/agents.
- Illegally acquired security-sensitive chemicals or biological materials.
- Inappropriate storage and use of restricted chemicals or security sensitive biological agents outside an approved holding facility'.

9.3.2 An Integrated Approach for Biological Risk Management in Life Science Research and Innovation

Research and innovation in life sciences are crucial for advancing health and protecting the environment. To prevent the accidental release of or exposure to biological agents and toxins and safeguard life science activities against misuse and diversion requires a shared commitment to safety and security amongst public and private stakeholders – biological laboratories, research institutes, companies and businesses – that are on the forefronts of driving biotechnology innovation.

WHO's *Laboratory Biosafety Manual* provides guidance for those who work with biological agents or in facilities where personnel may be exposed to potentially infectious substances that present a hazard to human health.[16] This manual discusses biosafety and biosecurity aspects of laboratory policy and practice including review, assessment and management of dual-use life science research and emerging technologies. Laboratory biosecurity aims to ensure that biological agents and toxins, as well as related equipment and information are not stolen, misused or diverted for illicit purposes. Assessing biosecurity risks may require input from different stakeholders such as law enforcement, intelligence, defence and civil protection. Risk mitigation measures seek to guarantee the comprehensive protection of laboratory assets and typically include inventory, information, access control, personnel reliability, transport security and emergency planning.

Dual-use life science research (of concern) is research conducted for peaceful and beneficial purposes that has the potential to provide knowledge, information, products or technologies which could be directly misapplied to pose a threat to health or the environment (see Chapter 6).[17] Due to the evolving technology and increasing digitisation, including the advent of more sophisticated artificial intelligence-enabled tools, the span of life science advances that raise dual-use concerns is constantly expanding and not constrained by disciplinary boundaries. Developments in genetics, microbiology, neuroscience and pharmacology that hold tremendous promise for societal betterment could also cause grave harm if misused and deployed for malicious ends. Addressing the dual-use potential of life science research and innovation requires risk mitigation strategies that complement but extend beyond laboratory biosecurity measures. Such strategies should apply along the entire cycle of research: from its conception and funding to publication and data/product dissemination. Examples include institutional requirements for dual-use risk assessment and continued review of identified dual-use research projects; evaluation criteria for funding dual-use research; compliance with national export and trade control regulations; mechanisms for supplier/customer screening; editorial policies for dual-use risk management; dissemination and risk communication plans that promote the responsible use of research results.

Cybersecurity is a cross-cutting element of laboratory biosecurity and the management of dual-use life science research. Its primary goal is to guarantee the reliability and resilience of electronic systems in case of sabotage or other forms of interference and to ensure information security, including personal and other sensitive data that is stored and managed electronically. Digital services and automated processes are an integral part of life science research and innovation and safeguarding these against cyberattacks is essential for business continuity. Cyberthreats can take different forms ranging from data theft through service disruption to infrastructural damage. The emergent field of cyber-bio-security examines vulnerabilities at the interfaces of life science, cyber, cyber-physical, supply chain and infrastructure systems to counter unwanted surveillance, intrusions and malicious and harmful activities.[18] Addressing such vulnerabilities has both a technical and human dimension which means that advanced cybersecurity systems at the institutional level should be coupled with efforts to enhance the situational awareness and vigilance of staff to identify, detect and tackle cyber threats.

9.3.3 Biosecurity Risk Communication and Public Engagement

Biological events regardless of whether they affect human, animal or plant health can cause lasting damage to communities and their livelihoods. Community engagement is a core element of an effective response strategy, as it can help contain the spread of disease, facilitate the distribution of medical counter-measures and therapeutics and mitigate the psychological impact of disease outbreaks to create an enabling environment for recovery. Access to timely and reliable information about the biological event and how it is addressed by competent authorities is of paramount importance to alleviate societal anxieties and enhance public awareness of the steps and measures that individuals can take to protect themselves and their closed ones. Risk communication strategies that life science stakeholders adopt should acknowledge gaps and contradictory findings in scientific data with a view to encouraging and facilitating transparent and inclusive debate that benefits the public interest. They should also acknowledge the wide range of information sources and tools that are available to the public to get updates and track the evolving situation including the media and online social networks.

Modern information and communication technologies allow sharing news and story items across countries and continents within seconds. During a crisis situation, this trend can have negative manifestations and result in an overwhelming pool of conflicting information that creates confusion and makes it difficult to discern fact from fiction. The phenomenon of infodemic that was observed during the COVID-19 pandemic is instructive. An infodemic is too much information including false or misleading information in digital and physical environments during a disease outbreak.[19] If left unchecked, an infodemic can hinder the implementation of disease control and prevention measures, erode trust in healthcare authorities and infrastructure and trigger social unrest and risk-taking behaviours. Depending on the type of narratives that make up the infodemic, additional effects can include exacerbating social divisions, polarising opinions and marginalising vulnerable groups within communities. In case of a man-made biological event, whether accidental or deliberate, an infodemic can interfere with the investigation operations and twist public perceptions of the gravity of the threat. *WHO Competency Framework: Building a response workforce to manage infodemics* aims to promote capacity building for infodemic management at health institutions and organisations across five main areas:

- 'Strengthening the scanning, review and verification of evidence and information.
- Strengthening the interpretation and explanation of what is known, fact-checking statements and addressing mis/disinformation.
- Strengthening the amplification of messages and actions from trusted actors to individuals and communities that need the information.
- Strengthening the analysis of infodemics, including the analysis of information flows, monitoring the acceptance of public health interventions and an analysis of factors affecting infodemics and behaviours at individual and population levels.
- Strengthening systems for infodemic management in health emergencies'.[20]

In the context of biosecurity, disinformation encompasses the production and spread of intentionally misleading or deceptive information about biological threats that can potentially cause serious political, financial and physical harm to governments, international organisations, the scientific community, academia, industry and the population at large.[21] Countering disinformation requires understanding of the purpose and design of disinformation operations. Disinformation is a power-projection tool for advancing one's agenda through deceit and social manipulation. Historically, it has been used by state and non-state actors alike. A recent example of a state-sponsored disinformation campaign of relevance to biosecurity concerns Russia's allegations that Ukrainian biomedical laboratories are developing biological weapons which should be assessed against the backdrop of similar disinformation activities that the Kremlin has carried out over the recent years.[22] Whereas such disinformation operations are intended to serve political purposes, their effects can be far-reaching, not least because they devalue the principles on which international biomedical cooperation rests and undermine public trust in the goals and outcomes of science.

9.4 Implications for the Governance of Biotechnology in the Twenty-first Century

A robust and effective web of prevention for biosafety and biosecurity is the result of the full and effective implementation of pertinent international standard-setting initiatives, international and national regulatory and guiding frameworks, institutional policies and procedures, codes of conduct and other binding and non-binding instruments and measures that seek to prevent and address biological events regardless of whether these occur naturally, or due to an accident or intentional act. In other words, fostering and maintaining a web of prevention for biosafety and biosecurity requires the active engagement of life science stakeholders across disciplines (e.g. microbiology, genetics, systems biology, bioinformatics, medicine, pharmacology, neuroscience, etc.), sectors (e.g. government, academia, industry, citizen science and 'do-it-yourself' [DIY] biology) and professional domains (e.g. practising researchers, academics, research institutions, private companies, learned societies and trade associations, funding bodies, science publishers, science administrators and policymakers). There are, at least, three ways in which life science stakeholders contribute to the sustainability of biosafety and biosecurity governance arrangements:

- by promoting responsible conduct of life sciences.
- by driving biosafety and biosecurity innovation.
- by reinforcing the norm against biological weapons.

Responsible conduct of science entails respect for the values and fundamental principles of scientific research, such as rigour, honesty, reliability of results, openness and accountability, but it also entails an appreciation of the role and place of scientific research in society.[23] It other words, life scientists have duties both to their peers and to society as a whole. Biosafety and biosecurity professional and institutional standards, practices and routines are reflections of these duties. For example, observing appropriate laboratory biosafety procedures in the course of research is essential for preventing the unintentional exposure to biological agents or toxins of staff, people outside the facility or the environment. Equally, following laboratory biosecurity rules and guidelines helps ensure that research materials, equipment or data are not misused to cause harm and, thus mitigates the risk of reputational damage and reinforces public trust in research and the institutions and infrastructure that supports it. As life science stakeholders integrate biosafety and biosecurity requirements within their professional practice, this creates a normative expectation of what constitutes responsible behaviour in life sciences. Such a normative expectation is key to sensitising the next generation of life scientists to the importance of the web of prevention for biosafety and biosecurity.

The socio-political, economic and technological context of life science research and innovation changes which inevitably gives rise to novel biosafety and biosecurity challenges. Identifying opportunities to address such challenges requires expert assessment and foresight, in order to ensure that the proposed solutions are fit-for-purpose, viable, cost-effective and sustainable. As the ones on the frontlines of research, life science stakeholders can provide important insights into the process of strengthening biosafety and biosecurity arrangements and offer technical advice across sectors.

Life scientists and their institutions, professional organisations and networks play a vital role in promoting the ethos of science for peace which is a fundamental element of the international biological and chemical prohibition regime. The Hague Ethical Guidelines and the Tianjin Guidelines for Codes of Conduct for Scientists are both examples of standard-setting initiatives developed by scientists for scientists specifically to raise awareness of the different ways in which scientific communities help uphold the norms against biological and chemical weapons.[24]

Author Biography

Dr Tatyana Novossiolova is Senior Analyst at the Center for the Study of Democracy, Bulgaria combining research, training and project-management experience. Her work covers the cross-cut between international law and security with a focus on policy sustainability. She has examined the global governance of chemical, biological, radiological and nuclear (CBRN) security. In 2021, Tatyana served as an international consultant to the Organisation for Security and Cooperation in Europe (OSCE) Project Coordinator in Ukraine on the development of an advanced training programme on biosafety and biosecurity for local laboratory professionals. She is experienced with the application of active learning methods and Information and communication technology (ICT) for training and has published widely in academic and policy circles.

References

1 On this point, see Novossiolova, T. (2017) Comparing responses to, accidental, and deliberate biological events. *OIE Scientific and Technical Review*, Special issue by Beckham, T. (ed.) Biological Threat Reduction, 36, 2. https://doc.woah.org/dyn/portal/index.xhtml?page=alo&aloId=34818.

2 World Health Organisation. (2021) *WHO-Convened Global Study of Origins of SARS-CoV-2: China Part*, Joint WHO-China study. 14 January – 10 February 2021. 30 March. https://www.who.int/publications/i/item/who-convened-global-study-of-origins-of-sars-cov-2-china-part.

3 Cohen, J. (2023) A new pandemic origin report is stirring controversy. *Science*. 21 March. https://www.science.org/content/article/covid-origin-report-controversy; Mallapaty, S. (2023) WHO abandons plans for crucial second phase of COVID-origins investigation. *Nature*, 3 March. https://www.nature.com/articles/d41586-023-00283-y; Lenharo, M. and Wolf, L. (2023) US COVID-origins hearing renews debate over lab-leak hypothesis. *Nature*, 9 March, https://www.nature.com/articles/d41586-023-00701-1.

4 Novossiolova, T. *et al* (2021) The vital importance of a web of prevention for effective biosafety and biosecurity in the twenty-first century. *One Health Outlook*, 3, 17. https://doi.org/10.1186/s42522-021-00949-4.

5 Pearson, G. S. (2015) The idea of a web of prevention. in Whitby, S. *et al* (Eds.) *Preventing Biological Threats: What Can You Do*, University of Bradford, UK. https://bradscholars.brad.ac.uk/handle/10454/7821.

6 Novossiolova, T. *et al* (2021) The vital importance of a web of prevention for effective biosafety and biosecurity in the twenty-first century. *One Health Outlook*, 3, 17. https://doi.org/10.1186/s42522-021-00949-4.

7 United Nations Security Council Resolution 1540. (2004) https://www.un.org/en/sc/1540/resolutions-committee-reports-and-SC-briefings/security-council-resolutions.shtml.

8 United Nations Security Council Resolution 2663. (2022) https://www.un.org/en/sc/1540/resolutions-committee-reports-and-SC-briefings/security-council-resolutions.shtml.

9 Novossiolova, T. and Martellini, M. (2021) Effective and comprehensive CBRN security risk management in the 21st century. EU Non-Proliferation and Disarmament Paper Series, 75. https://www.nonproliferation.eu//activities/online-publishing/non-proliferation-papers/.

10 This Section of the Chapter Is Based on the *MASC-CBRN Catalogue of Best Practices for Stakeholder Engagement*, https://masc-cbrn.eu/catalogue-on-best-pracices, 2021.

11 World Health Organisation. (2022) *Joint External Evaluation Tool*, third ed. https://www.who.int/publications/i/item/9789240051980.

12 World Health Organisation. (2023) *Early Warning, Alert and Response System (EWARS).* https://www.who.int/emergencies/surveillance/early-warning-alert-and-response-system-ewars.

13 *United Nations Secretary-General's Mechanism for Investigation of Alleged Use of Chemical and Biological Weapons (UNSGM).* (2023) https://disarmament.unoda.org/wmd/secretary-general-mechanism.

14 World Organisation for Animal Health. (2018) *Guidelines for Investigation of Suspicious Biological Events.* https://www.woah.org/en/document/guidelines_investigation_suspicious_biological_events.

15 United Nations Inter-Regional Crime and Justice Research Institute. (2022) *A Prosecutor's Guide to Chemical and Biological Crimes.* https://unicri.it/Publication/Prosecutor-Guide-Chemical-Biological-Crimes.

16 World Health Organisation. (2020) *Laboratory Biosafety Manual*, fourth ed. https://www.who.int/publications/i/item/9789240011311.

17 World Health Organisation. (2022) *Global Guidance Framework for the Responsible Use of the Life Sciences: Mitigating Biorisks and Governing Dual-Use Research.* https://www.who.int/publications/i/item/9789240056107.

18 Murch, R. *et al* (2018) Cyberbiosecurity: an emerging new discipline to help safeguard the bioeconomy. *Frontiers in Bioengineering and Biotechnology*, 6, 39. https://doi.org/10.3389/fbioe.2018.00039; Richardson, L. et al (2019) Cyberbiosecurity: a call for cooperation in a new threat landscape. *Frontiers in Bioengineering and Biotechnology*, 7, 99. https://doi.org/10.3389/fbioe.2019.00099.

19 World Health Organisation. (2023) *Infodemic.* https://www.who.int/health-topics/infodemic#tab=tab_1

20 World Health Organisation. (2021) *WHO Competency Framework: Building a Response Workforce to Manage Infodemics.* https://www.who.int/publications/i/item/9789240035287.

21 United Nations Inter-Regional Crime and Justice Research Institute. (2023) *Handbook to Combat CBRN Disinformation.* https://unicri.it/Publication/Hanbook-to-combat-disinformation.

22 United Nations. (2022) *United Nations Not Aware of any Biological Weapons Programmes in Ukraine*, Senior Disarmament Affairs Official Tells Security Council. *Press release,* 27 October. https://press.un.org/en/2022/sc15084.doc.htm; Leitenberg, M. (2020) False allegations of biological-weapons use from Putin's Russia. *The Nonproliferation Review*, 27, 4–6. https://doi.org/10.1080/10736700.2021.1964755.

23 Interacademy Partnership. (2016) *Doing Global Science: A Guide to Responsible Conduct in the Global Research Enterprise*, Princeton University Press, Princeton, NJ. https://www.interacademies.org/publication/doing-global-science-guide-responsible-conduct-global-research-enterprise.

24 Organisation for the Prohibition of Chemical Weapons. (2015) *The Hague Ethical Guidelines*. https://www.opcw.org/hague-ethical-guidelines; Johns Hopkins Center for Health Security, Tianjin University, and Interacademy Partnership (2021). *Tianjin Biosecurity Guidelines for Codes of Conduct for Scientists*. https://www.interacademies.org/publication/tianjin-biosecurity-guidelines-codes-conduct-scientists.

10

The 1925 Geneva Protocol and the BTWC

Jez Littlewood

Independent Consultant, Edmonton, AB, Canada

Key Points

1) Preventing the use of biological and toxin weapons occurs through a wider anti-biological weapons regime that consists of a range of agreements and mechanisms implemented by States, non-State actors in industry and civil society and attitudes over the risk of dual use science and technology.
2) The Biological and Toxin Weapons Convention (BTWC) and the Geneva Protocol are core elements of this wider regime. They are the legal basis for a complete prohibition on the development, production, stockpiling, acquisition and use of biological and toxin weapons.
3) The BTWC has evolved in an incremental manner. There are over 140 additional understandings about the Convention and how it should be implemented. These understandings are not legally binding but do represent a road map to the what and the how of biological disarmament.
4) Differences about verification of the BTWC have plagued the Convention since it entered into force in 1975. A new working group established by the Ninth Review Conference of the BTWC in 2022 is tasked with identifying and developing ways to strengthen the Convention, including through verification. Even if this working group is successful, more detailed work will be required after 2027. Legally binding verification measures are at least 10 years away from adoption and implementation by BTWC States Parties.

Essentials of Biological Security: A Global Perspective, First Edition. Edited by Lijun Shang, Weiwen Zhang, and Malcolm Dando.

Summary

The Biological and Toxin Weapons Convention (BTWC) and the Geneva Protocol are important parts of a wider regime, but preventing biological warfare is not completely dependent on either legal agreement alone or in combination. The laws exist within a wider normative constraint against the use of poisons and disease as a weapon and within a deeper anti-biological and anti-chemical weapons regime that consists of many different types of agreement. The BTWC has evolved slowly and in an incremental manner during its nearly 50-year life to date through agreements at the five-yearly review conferences. These agreements are the key to understanding how the Convention should be implemented by its States Parties because they capture the 'what' and the 'how' of implementation that all States Parties agree upon. There are over 140 additional understandings, as they are known. At the Ninth Review Conference of the BTWC in 2022, States Parties agreed to a series of meetings to address challenges facing the Convention (e.g. peaceful cooperation in the life sciences, a mechanism to review scientific and technological developments, compliance and verification). The recommendations from these meetings will be considered by States Parties and further work will be required to develop, adopt and implement any additional measures. This means any legally binding mechanisms to strengthen the Convention will not be developed until after 2027 and implementation will be after 2030 at best. Preventing the use of biological and toxin weapons will continue to rely on States, civil society, intergovernmental organisations and the public and private sector for the foreseeable future.

10.1 Introduction

At the heart of this chapter is an argument that preventing biological warfare requires the strengthening of the much wider and broader anti-biological weapons (BWs) regime that already exists today. The Biological and Toxin Weapons Convention (BTWC) and the Geneva Protocol are important parts of that wider regime, but preventing biological warfare is not completely dependent on either legal agreement alone or in combination.

The BTWC entered into force nearly 50 years ago in 1975. As of late-2023, 185 states have signed and ratified, or acceded, to the Convention and are States Parties. Four States have signed but not ratified the BTWC. These signatory States are Egypt, Haiti, Somalia and the Syrian Arab Republic. A further eight States are non-signatory States: Chad, Comoros, Djibouti, Eritrea, Israel, Kiribati, Federated States of Micronesia and Tuvalu.

The Geneva Protocol was signed in 1925 and is approaching its centennial. It covers the use of chemical and biological weapons (CBWs) and is considered customary international law, binding all States whether they have signed and ratified it or not.

In practice, this means that any use by a State of biological or toxin weapons against another State is a violation of customary international law. In addition, any development, production, stockpiling or acquisition of BWs or means of delivery designed to use such weapons is a violation of the BTWC for 185 States Parties. Any use of a toxin weapon would also violate the Chemical Weapons Convention (CWC). These clear legal prohibitions mean the Geneva Protocol and the BTWC are the two key foundations to prevent the use, development, production, stockpiling and acquisition of biological and toxin weapons. They are not, however, the only laws, mechanisms, agreements, initiatives or practices in place to prevent such use or the development, production or acquisition of these weapons.

The laws exist within a wider normative constraint against the use of poisons and disease as a weapon and within a deeper anti-biological and anti-chemical weapons regime that consists of many different types of agreement (e.g. other treaties such as the CWC, obligations placed on all States by United Nations (UN) Security Council resolutions, like-minded export control regimes, multilateral initiatives against proliferation of weapons and global public health agreements).

As a consequence of the wider normative constraints, deeper regime and foundational and comprehensive legal prohibitions, we live in a world where no State admits to having or having any interest in BWs, and few States are suspected of having an interest in BWs and the historical record of known use of such weapons is thin: use is unusual and rarely has any major impact on a conflict.[1]

None of the above means there are no risks or threats around BWs. Nor do the norm, regime and prohibitions guarantee such weapons will remain out of the arsenals or plans of any hostile actor. They do, however, mean that fears about BWs or known and potential risks from social, political or scientific and technological developments can be overblown or detached from the wider context in which weapons development occurs.

10.2 The Origins and Evolution of the 1925 Geneva Protocol and the BTWC

There is a long history of normative, cultural and legal constraints against the use of poisons in warfare and conflict (see Chapter 2), but the contemporary basis of the prohibition on the use of biological, chemical and toxin weapons rests on the 1925 Geneva Protocol.

At least since the early seventeenth century, the use of poison or poisoned weapons has been condemned in different forms of law. In the nineteenth and early twentieth centuries, this can be seen in the 1874 Brussels Declaration, the 1899 Hague Peace Conference and 1907 Hague Conference. Greater specificity emerged after World War I in the 1919 Versailles Treaty in Article 171 with 'asphyxiating, poisonous or other gases and all analogous liquids, materials or devices' and in the 1922 Washington Treaty and the Preamble to the 1925 Geneva Protocol.[1]

The Geneva Protocol emerged in the aftermath of World War I where use of chemical weapons such as chlorine and blister agents such as mustard gas was widespread. Chemical weapons use and their impact were fresh in the minds of diplomats and civil society. In the aftermath, States sought to prevent the future use of chemical weapons. This led to the *Protocol for the prohibition of the use in war of asphyxiating, poisonous or other gases and bacteriological methods of warfare* in June 1925. More often referred to the Geneva Protocol, it entered into force on 8 February 1928, and France serves as the depositary government.[2]

The inclusion of 'bacteriological' within the Geneva Protocol is now taken for granted, but the inclusion of bacteriological was a very late addition to the text of the Geneva Protocol. Unlike the widespread experience with chemical weapons, BW, in contrast, were limited to small-scale sabotage operations, and it remains unclear if the use of such weapons was even detected during the war. The potential of disease as a weapon was recognised and their inclusion in the Geneva Protocol addressed less of an extant contemporary threat and more of an awareness of interest in BWs among a range of actors, including non-State actors.[3]

The prohibition on use of chemical and bacteriological (biological) weapons in war between 'high contracting parties' inherent to the Geneva Protocol is now considered customary international law. As such, the use by any state of chemical or BWs is a violation of established law whether or not that state is a party to the Geneva Protocol.

However, its prohibition on use left certain loopholes related to development, production and stockpiling of such weapons. In addition, the reservations attached to the Geneva Protocol and interpretation of its text were such that it was understood as a no-first-use agreement. The first use of CBWs was recognised as a violation of the Geneva Protocol by most States – if the State-to-State adversaries were both parties to the Geneva Protocol – but retaliation if attacked with such weapons was deemed to be permitted. These interpretations meant that chemical weapons were used between 1925 and 1945 in Morocco, Abyssinia and Manchuria.[1]

After 1945, the advent of nuclear weapons overshadowed concerns about CBWs. Limited use of chemical weapons is known to have occurred by Egyptian forces in Yemen and both herbicides and riot control agents were used by the United States in Vietnam.[1] By the mid-1960s concerns about the proliferation of nuclear weapons resulted in the Nuclear Non-Proliferation Treaty (NPT) being opened for signature in 1968. Soon after completion of negotiation attention turned to CBWs in the UN-based Eighteen Nation Disarmament Committee (ENDC) which was the forerunner to the contemporary Conference of Disarmament (CD). A proposal from the United Kingdom in 1968 to separate consideration of a new treaty on CBWs to one focused only on BWs, eventually gained sufficient support to proceed to negotiations over 1969–1971 in Geneva. The outcome was the 1972 *Convention on the Prohibition of the Development, Production and Stockpiling of Bacteriological (Biological) and Toxin Weapons and on Their Destruction* which was opened for signature on 10 April 1972 and entered into force on 26 March 1975. This Convention, the BTWC as it is more commonly known, is approaching its 50th year of entry into force in 2025 as the Geneva Protocol approaches its centennial of being open for signature.

As a recent guide to implementing the Convention notes, the BTWC 'effectively prohibits the development, production, acquisition, stockpiling and use of biological and toxin weapons.'[4] The central prohibition around biological and toxin weapons is established in Article I of the BTWC. Under Article I, a State Party 'undertakes never in any circumstances to develop, produce, stockpile or otherwise acquire or retain: (1) microbial or other biological agents, or toxins whatever their origin or method of production, of types and in quantities that have no justification for prophylactic, protective or other peaceful purposes. (2) Weapons, equipment or means of delivery designed to use such agents or toxins for hostile purposes or in armed conflict.'

The prohibition, and the negotiators of the Convention, took the general purpose for which a biological agent or toxin was intended as the criterion of control. There is no list of prohibited agents or toxins. This means the prohibition covers all biological agents or toxins. Agents or toxins may be used legitimately for prophylactic, protective or other peaceful purposes. This General-Purpose Criterion, as it is known, is simple, effective and futureproof and ensures the scope of the BTWC remains relevant to any and all scientific and technological developments. It also means implementation arrangements must be reviewed on a regular basis at the national and international levels. This function is supported by the review conferences of the BTWC which occur at roughly five-year intervals.

10.3 The Review Conferences of the BTWC and Their Outcomes: 1980–2022

The BTWC has evolved slowly and in an incremental manner during its nearly 50-year life to date through agreements at the five-yearly review conferences. These agreements are the key to understanding how the Convention should be implemented by its States Parties because they capture the 'what' and the 'how' of implementation that all States Parties agree upon. There are over 140 additional understandings, as they are known. While they are not legally binding and fully adopted and implemented by all States Parties, they are important politically binding undertakings and a guide to how to implement the obligations of the BTWC.[5]

The Convention has evolved in three phases. A fourth phase is beginning in 2023. Phase one covers the period from entry into force, in 1975, through to the end of the Cold War in 1991. This period is marked by the first three review conferences in 1980, 1986 and 1991. The second phase covers the 1992–2002 period when the fourth and fifth review conferences occurred in 1996 and 2001–2002, respectively: the review conferences were not the major focus of work in this period. The decade of work between 1992 and 2002 was marked by the pursuit of agreement on strengthening the Convention through a multilaterally negotiated legally binding agreement (the BTWC Protocol). Phase three began after the collapse of the negotiations on the BTWC Protocol in the summer of 2001 and took shape in late 2002 when the Fifth Review Conference resumed. This phase was marked by a series of intersessional work programmes. It covers the period 2002–2022 during which four intersessional work programmes and four review conferences occurred. The review conferences were the sixth (2006), seventh (2011), eighth (2016) and ninth (2022) with work programmes in the years 2002–2005, 2007–2010, 2012–2015 and 2018–2020.

Over time, this incremental evolution via a series of politically binding agreements and understandings about implementation of the BTWC has kept the Convention relevant to most challenges around biosecurity and the prevention of biological warfare, but these agreements and understandings lack legal force. Implementation is dependent on action by each State. Some view the absence of a formal, legally binding verification regime for the BTWC as a major weakness. The political differences over verification remain, but at the Ninth Review Conference in 2022 States Parties established a new working group to try and break the stalemate over how to strengthen the Convention.

At the Ninth Review Conference of the BTWC States Parties agreed to a series of meetings to address challenges facing the Convention (e.g. peaceful cooperation in the life sciences, a mechanism to review scientific and technological developments, compliance and verification). This can be considered the beginning of the fourth phase of the Convention's evolution because it ends the intersessional work with a mandate to discuss and promote common understandings and effective action that occurred between 2002 and 2020 and marks the beginning of a new phase of activity.

The task of the new Working Group on Strengthening the Convention is to 'identify, examine and develop specific and effective measures, including possible verification measures, and to make recommendations to strengthen and institutionalize the Convention in all its aspects.'[6] These recommendations are to be submitted to States Parties for consideration and further action. In addition, the measures must be formulated and designed in a manner that their implementation supports international cooperation, scientific research and economic and technological development, while avoiding any negative impacts on those areas.

The seven areas of activity of the working group are:

- International cooperation and assistance under Article X;
- Scientific and technological developments relevant to the Convention;
- Confidence-building and transparency;
- Compliance and verification;
- National implementation of the Convention;
- Assistance, response and preparedness under Article VII; and
- Organisational, institutional and financial arrangements.

In addition to the above, the working group must make recommendations on two mechanisms: one to facilitate and support the full implementation of international cooperation and assistance under Article X and one to review and assess scientific and technological developments relevant to the Convention and provide States Parties with relevant advice.

The challenge States Parties set themselves is immense. It is made more arduous by the fact that they allocated only 60 days to this task, spread over 15 days each year. An indicative timetable of work was agreed in March 2023. Based on past experience, with the Verification Experts Group (VEREX) in 1992 and 1993, the Ad Hoc Group Protocol negotiations between 1995 and 2001, the intersessional work programmes and the current geopolitical challenges that are affecting the BTWC (e.g. the Ukraine conflict, deteriorating US–China relations and the failure of the international community to respond effectively to the pandemic in a cooperative manner), a successful outcome for the working group is far from guaranteed.

Success or failure will be relative. Even if the working group reaches consensus agreement on a range of feasible and effective measures to strengthen the Convention, those recommendations are only for consideration of States Parties. Further action will be decided in 2027 at the Tenth Review Conference. At the other end of the spectrum, if the working group fails to reach consensus on any recommendations that does not mean the BTWC cannot be strengthened, or future efforts are unlikely. If this major failure occurs in 2027, it is important to note States Parties have been here before: in 2001 when the BTWC Protocol negotiations collapsed, and the next phase of work was agreed only in late 2002. In a similar manner, success for the working group will most likely resemble the successful outcome of the VEREX discussions which led, via a Special Conference decision, to the formation of the Ad Hoc Group and negotiations on the Protocol.

The important issue here is that neither success nor failure of the working group determines the success or failure of the required future efforts beyond 2027. Even if the working group reaches a consensus outcome that does identify, examine and develop specific and effective measures, including possible verification measures, and makes recommendations to strengthen and institutionalise the Convention in all its aspects, that work will occur between the tenth and eleventh review conferences, i.e. 2027–2032 and at best any agreed legally binding measures might enter into force between 2030 and 2035.

10.4 Biological Disarmament as It Is: Strengths and Weakness of the BTWC and the Geneva Protocol in the Twenty-first Century

It is not uncommon to hear claims or statements that the BTWC is a weak agreement. If assessed in isolation, the BTWC does look weaker than many of its arms control counterparts. Unlike the Nuclear Non-Proliferation Treaty (NPT) and the CWC it lacks a formal structure of verification and an implementing organisation. The CWC is the most often cited comparison and its implementation organisation, the Organisation for the Prohibition of Chemical Weapons (OPCW) together with its robust routine verification provisions to oversee the verified destruction of over 70,000 tonnes of chemical weapons is a model that many seek to emulate for the future of biological disarmament. The NPT is bifurcated. The International Atomic Energy Agency (IAEA) is tasked with verification of the declared peaceful uses of nuclear technology, while the States Parties to the NPT address the more political aspects of achieving nuclear disarmament through agreements at review conferences. Other agreements also support nuclear arms control: the Comprehensive Test Ban Treaty (CTBT), the Treaty on the Prohibition of Nuclear Weapons (TPNW), export control regimes such as the Zangger Committee and bilateral agreements between the United States and Russia.

Neither the chemical weapons disarmament regime nor the nuclear non-proliferation and arms control regimes, rely upon a single, legally binding agreement to keep such weapons from being used.

The challenge to the BTWC is weak argument is that it fails to take into account two issues. Firstly, the BTWC is the legal basis for biological disarmament, but preventing the development, production, stockpiling, acquisition and use of biological and toxin weapons occurs not by a single treaty – the BTWC – but by a number of treaties and other instruments and mechanisms. Secondly, when assessing the strengths and weaknesses of biological disarmament based on the regime as a whole, the regime is far from weak.

The key issue to understand with regard to biological disarmament is that BTWC is part of a deeper and wider anti-BWs regime. That regime consists of the Geneva Protocol and the BTWC at the core – prohibiting the use, development, production, stockpiling and acquisition of such weapons – and supporting mechanisms that range from universal obligations such as UN Security Council Resolution 1540 (2004), the role of the United Nations Secretary General's Mechanism (UNSGM) for investigating alleged use of CBWs, the work of Interpol's chemical, biological, radiological and nuclear sub-directorate, the Global Health Security Initiative (GHSI) and Global Health Security Agenda (GHSA), the Global Partnership Against Weapons of Mass Destruction, export control regimes such as the Australia Group and more regional bodies such as the Global Emerging Pathogens Treatment Consortium that is very active in Africa. In addition to these, the World Health Organisation, World Organisation for Animal Health and the Food and Agriculture Organisation as well as regional entities that work with these bodies all have some role and activity in biosafety and biosecurity. Beyond State, regional or intergovernmental organisations civil society organisations are also active in strengthening biological disarmament. All these core and supporting mechanisms are themselves embedded within a normative framework that has its origins well before the twentieth century and still exists today.

Pearson refers to this deeper and wider anti-BWs regime as a web of prevention (see Chapter 9). As he remarked in 2016, this 'web of prevention is thus an integrated and comprehensive approach in which all elements are complementary and reinforce each other, to create an effective counter to the threat of biological weapons, whether posed by states, non-state actors or other entities.'[7]

The BTWC has never stood alone as the legal or normative bulwark against BWs. None of this is to argue that the BTWC does not need strengthening. The Convention is in need of sustained work by its States Parties, industry and civil society, but the range of mechanisms and structures in place mean that the Convention, the regime and the norm are not weak.

10.5 The BTWC Beyond 50 and the Geneva Protocol Beyond 100: Can They Prevent Biological Warfare?

Preventing biological warfare will depend on how the norm, regime and Convention develop over the next decade. Considerable faith is being placed in the new BTWC working group. The outcome of the working group cannot be predicted, but it does raise the issue of what kind of future for the BTWC? Five possible trajectories for the Convention can be envisaged.[8]

The status quo is the continuation of incrementalism based on agreements between States Parties over what good practice for implementation looks like: the what and the how of implementation in a changing world. Nothing here is legally binding and even some of the politically binding understandings, notably the confidence-building measures, are considered 'voluntary' by some States Parties. An additional aspect of the status quo approach is that elements outside of the BTWC have been adapted or developed to bolster biological disarmament. This is most notable with regard to the Geneva Protocol and the evolution and strengthening of the UNSGM for investigations of alleged use, but also evident from UNSCR 1540 (2004) and the emergence of scientific and technological review mechanisms that offer advice to States Parties at review conferences. For the Convention, the status quo is sufficient to stop it from becoming meaningless. The BTWC is law and its States Parties have unambiguous obligations not to develop, produce, stockpile, acquire or support acquisition by others of biological or toxin weapons. While the status quo may be sufficient to prevent the Convention from drifting into meaninglessness, it is not enough to prevent it from becoming less relevant to decisions around biology.

The second trajectory is convergence where the BTWC and the CWC work more closely together. This is not a merger of the Conventions or the adoption of CWC verification procedures for the BTWC, since that is legally difficult and technically pointless because verification provisions for chemical weapons will not work for BWs except in the space where the Conventions overlap, i.e. toxins. Convergence is a process that begins with a recognition of both differences and similarities between the two conventions in terms of law, mandate, membership and operation and an identification of common requirements, such as physical infrastructure, administrative services such as translation and interpretation and information technology and communications requirements. It might then seek to expand in other areas (e.g. training of inspectors for investigation procedures, chemical–biological safety and security training) and evolve over time where two separate disarmament regimes share as many common services as possible for the benefit of both. Co-location is a prerequisite for this approach to work. In practice, this means making The Hague the seat of CBW expertise in the same way that Vienna is the seat of nuclear expertise.

A third evolutionary trajectory is mimicry. This is where the BTWC adopts a legally binding verification system similar to that envisaged in the 1990s under the Protocol. It is the model many wish to emulate from the CWC in an effort to give the BTWC the type of organisation, infrastructure and financial resources that the CWC possesses to achieve the objective of chemical disarmament. For most States Parties this would be a single negotiation approach involving working on all issues of concern until everything is agreed. Others might push for a modular approach that would involve seeking agreement on issues in a discrete manner (e.g. assistance under Article VII or investigation procedures under Article VI) to produce not a Protocol but a series of protocols or other types of agreement or understandings.

A fourth trajectory is innovation. In practice, this is a mix of the status quo, convergence and mimicry in both form and function. Innovation could embrace some legally binding agreements, some politically binding undertakings and other voluntary arrangements. It may not involve all States Parties agreeing to all elements of any future strengthening and may permit opting in or out of any new mechanisms as a way forward. Innovation treats the BTWC as a foundation for strengthening rather than the only place where strengthening biological disarmament can occur. Where it is possible to enhance implementation of the BTWC via agreements made in the Convention that route will be adopted. In areas, where agreement is not possible, other avenues will be explored and different approaches adopted. Innovation views the norm, the existing wider regime and the Convention as a whole, and this approach involves further enhancements to what is already in place for the web of deterrence.

A final trajectory is erosion. This is where the BTWC and the Geneva Protocol are significantly weakened, break down or wither away. Erosion may occur through neglect or deliberate action. For some, the failure to adopt legally binding verification measures would be considered a form of neglect whereas for others the failure to adapt implementation processes and the organisational and institutional structures of both the BTWC and biological disarmament as a whole might be erosion. The use of biological or toxin weapons on a regular basis by a range of actors coupled with inaction in the face of such use by the international community would be a key indicator of erosion of both the BTWC and the Geneva Protocol. Scientific and technological developments are often considered as another indicator of erosion. Commentary is replete with claims that this or that scientific development is not covered by the General Purpose Criterion of the BTWC. Such commentary is nearly always wrong since the General Purpose Criterion covers everything. Scientific and technological developments that fall under Article I and the BTWC are covered. It is not the scope that is deficient, but awareness of issues and adoption of implementation practices to address potential risks that are absent.

Which of these trajectories the BTWC evolves along in the next 5, 10 or 15 years is yet to be determined. The trajectories are not mutually exclusive, and the future may unfold with a continuation of the status quo approach of minimal incrementalism, innovation that embraces convergence or mimicry. Erosion is, at this time, the least likely outcome but cannot be dismissed or ignored as a possibility. Whatever the future holds, any increase in the use of BWs will, as Carus observed, 'be because some past constraint has disappeared. Although technological and scientific advances might facilitate that trend, it is most likely to result from fundamental changes in attitudes towards the use of disease as a weapon.'[9]

10.6 Conclusion

Biological disarmament is a process. While it may be achieved at a certain point in time, scientists, diplomats, government officials or citizens of any State cannot afford to ignore the continued requirement to ensure they maintain a constant effort to prevent BWs use. This means attending to biosafety and biosecurity issues that emerge worldwide as well as efforts to ensure the BTWC and the Geneva Protocol remain at least functional in their current state. Improvements to the BTWC, strengthening, should occur, but in the event that it does not a wider regime and normative constraints remain in place. This is not an ideal outcome, but any actor that wishes to use BWs will have to transgress the norm, overcome numerous social and technological barriers to BWs that exist in the web of deterrence and violate two legal agreements.

Author Biography

Jez Littlewood is a policy analyst in Alberta working on strategic policy and policy management issues related to social services for government. Prior to that, he worked for the United Nations under contract as a member of the BTWC Ad Hoc Group secretariat between 1998 and 2001, on the Fifth Review Conference of the BTWC in 2001 and 2002, and under secondment to the UK Foreign and Commonwealth Office between 2005 and 2007 where he worked on the BTWC for the UK government. He holds a Ph.D. from Bradford University in the United Kingdom (awarded 2001) and spent five years in postdoctoral research at the University of Southampton (2002–2006) and 11 years at Carleton University in Ottawa at the Norman Paterson School of International Affairs (NPSIA) between 2007 and 2018.

References

1 World Health Organization. (2004) *Public Health Response to Biological and Chemical Weapon*, WHO, Geneva.
2 Spelling, A. *et al* (2015) Where did the biological weapons convention come from? *Indicative Timeline and Key Events*, 1925–1975.
3 Carus, W. S. (2017) *A Short History of Biological Warfare: From Pre-history to the 21st Century*, Government Printing Office, United States, 12–27.

4 United Nations. (2023) *Guide to Implementing the Biological Weapons Convention.*, United Nations, Geneva, 16.
5 Implementation Support Unit. (2022) *Additional Understandings and Agreements Reached by Previous Review Conferences Relating to each Article of the Convention. BWC/CONF.IX/PC/5*, United Nations, Geneva.
6 United Nations. (2022) *Final Document of the Ninth Review Conference, BWC/CONF.IX/PC/5*, United Nations, Geneva.
7 Whitby, S. M. *et al* (2015) *Preventing Biological Threats: What You Can Do*. (See, in particular, Graham S. Pearson, *The Idea of a Web of Prevention*, pp. 136–159
8 Crowley, M. *et al* (2018) *Preventing Chemical Weapons: Arms Control and Disarmament as the Sciences Converge*, Royal Society of Chemistry, London.
9 Carus, op. cit, page 45.

11

Constraining the Weaponisation of Pathogens and Toxic Chemicals Through International Human Rights Law and International Humanitarian Law

Michael Crowley

Bradford University, Bradford, UK

Key Points

1) In addition to the Biological and Toxin Weapons Convention (BTWC) and Chemical Weapons Convention (CWC), appropriate application of international humanitarian law (IHL) and international human rights law (IHRL) potentially provide further measures to prevent and respond to malign application of the rapidly advancing chemical and life sciences and to constrain the weaponisation of pathogens and toxic chemicals.
2) Relevant IHL constraints imposed by the Four Geneva Conventions and Additional Protocols include those forbidding superfluous injury or unnecessary suffering, indiscriminate weapons and deliberate attacks on civilians; and those protecting prisoners of war or others considered *hors de combat* from, for example, forced interrogations, torture or human experimentation. Of further relevance is the obligation, under Article 36 of Additional Protocol 1, to conduct legal reviews of 'new weapons'.
3) IHRL has broad applicability across the 'use of force' spectrum from law enforcement to armed conflict. State obligations to protect the rights to life, and freedom from torture and other ill-treatment, together with the attendant restrictions on the use of force, are of particular relevance in preventing weaponisation and other malign use of pathogens and toxic chemicals.
4) States must give full consideration to IHL and IHRL: firstly because of the direct obligations that arise from such law which either prohibit or severely restrict weaponisation of pathogens and toxic chemicals; and secondly because such relevant international law should inform implementation of the BTWC and CWC, where treaty language is ambiguous and interpretation is currently contested.

Essentials of Biological Security: A Global Perspective, First Edition. Edited by Lijun Shang, Weiwen Zhang, and Malcolm Dando.

Summary

Given constraints upon the Biological and Toxin Weapons Convention (BTWC) and Chemical Weapons Convention (CWC) to effectively respond to malign application of rapidly advancing chemical and life sciences, this chapter explores additional protection that can be found in appropriate application of international humanitarian law (IHL) and international human rights law (IHRL). The chapter explores existing relevant IHL constraints imposed by the four Geneva Conventions and Additional Protocols: inter alia forbidding superfluous injury or unnecessary suffering, indiscriminate weapons and deliberate attacks on civilians; and protecting prisoners of war or others considered *hors de combat* from, for example, forced interrogations, torture or human experimentation. The obligation, under Article 36 of Additional Protocol 1, to conduct legal reviews of 'new weapons' is also explored. The chapter examines IHRL, highlighting its breadth of applicability across the use of force spectrum from law enforcement to armed conflict. State obligations to protect the rights to life, and freedom from torture and other ill-treatment, are explored and applied to preventing weaponisation and other malign use of pathogens and toxic chemicals. The chapter concludes by arguing that States must give full consideration to IHL and IHRL: firstly because of direct obligations that arise from such law which either prohibit or severely restrict weaponisation; and secondly because relevant international law should inform implementation of the BTWC and CWC, where treaty language is ambiguous and interpretation is currently contested.

11.1 Introduction

The Biological and Toxin Weapons Convention (BTWC) and the Chemical Weapons Convention (CWC) were developed decades ago to prevent and address critical global-level threats primarily from the large-scale military development, stockpiling, transfer and use of weapons employing pathogens and toxic chemicals, respectively, with both Conventions covering mid-spectrum agents, including toxins (i.e. toxic chemicals of biological origin and chemically synthesised analogues and derivatives), bioregulators and certain types of central nervous system (CNS)-acting chemicals and riot control agents. Both treaties also incorporated obligations to dismantle, destroy or divert to peaceful purposes any existing biological and toxin (BTWC) and chemical (including toxin) (CWC) agents, equipment, weapons and means of delivery. While (largely) succeeding in these tasks, the two Conventions and associated regulatory regimes appear less suited to responding effectively and flexibly to the full range of contemporary threats arising from the potential malign application of the rapidly advancing life and chemical sciences, particularly when such application may occur outside of armed conflict.

While recognising the continuing central importance of the BTWC and the CWC, this chapter presents an overview of potentially applicable additional legal constraints – under international humanitarian law (IHL) and international human rights law (IHRL) – that could also be employed by States to prevent and address the weaponisation of pathogens and toxic chemicals whether derived from natural sources or that are synthesised by humans. As will be seen, both IHL and IHRL are of particular (and arguably growing) relevance, given the potential broadening of malign application of the chemical and life sciences, including and beyond the battlefield, to encompass novel forms of manipulation, punishment and repression of individuals, groups or entire populations.

At present the norm prohibiting development or use of biological weapons (BWs) appears to be strongly adhered to by States. However, the norm prohibiting development and use of chemical weapons appears to be under attack on multiple fronts. Notable examples include confirmed use of chemical weapons including nerve agents and industrial toxic chemicals by Syria against its own citizens; use of toxic chemicals for targeted assassination of individuals in Malaysia, Russia and the United Kingdom; development of CNS-acting chemical agents for law enforcement by Russia and China; and widespread misuse of riot control agents in law enforcement with unconfirmed reports of their use by certain States, including Russia, in armed conflict. Consequently, while potential application of IHL and IHRL to BWs will be noted, as appropriate, this will not be the focus of this chapter. Instead the chapter will concentrate on constraining development and use of weapons employing toxic chemicals (including toxins and other mid-spectrum agents) for military and law enforcement purposes.

11.2 International Humanitarian Law

11.2.1 Introduction

IHL is the body of law that applies during situations of armed conflict with the aim of protecting civilians and others who are no longer participating in hostilities, and regulating the conduct of such hostilities. Amongst its provisions are those regulating the means of conflict (including the weapons employed) and also the methods of warfare (how such weapons are employed). IHL is applicable to international armed conflicts, that is conflicts that arise between two or more States, even if a state of war is not recognised by one of them. It also covers all cases of partial or total occupation of the territory of a State, even if the said occupation meets with no armed resistance. In addition, IHL is applicable to armed conflicts of a non-international character, which can be considered as conflicts between the armed forces of a State and organised armed groups (or between two or more such groups) which are under responsible command, control territory and carry out sustained and concerted military operations. An armed conflict is not considered to include a situation of internal disturbances and tensions, such as riots, isolated and sporadic acts of violence and other acts of a similar nature. The determination of whether a specific situation should be considered, as an internal armed conflict or as internal disturbance requiring law enforcement can be highly contentious. As well as regulating the activities of States in armed conflicts, IHL is applicable to armed non-State actors that meet the requisite criteria.

11.2.2 Over-arching IHL Obligations Constraining Weaponisation of Toxic Chemicals and Pathogens

In addition to the weapons-specific agreements prohibiting development, possession or use of biological and chemical weapons (BCW) (notably the BTWC and CWC), there are a number of generally applicable IHL treaties – that can potentially be applied to further and more broadly constrain weaponisation of pathogens and toxic chemicals (including toxins). Of particular relevance are the four Geneva Conventions, which principally safeguard a range of 'protected persons' in the power of the opposing party, including wounded combatants, prisoners of war, medical personnel and civilians[1]; and two Additional Protocols[2] which address certain aspects of the conduct of armed conflict. Obligations deriving from these treaties or from related customary international[3] law include:

11.2.2.1 The Prohibition of Deliberate Attacks on Civilians, the Prohibition of Indiscriminate Weapons and of Attacks That Do Not Discriminate Between Civilians and Military Objectives

The above prohibitions are considered fundamental to IHL and are covered by both treaty law (e.g. Additional Protocol I) and customary IHL. These prohibitions are applicable to both international and non-international armed conflict. The International Court of Justice has stated that the principle of distinction was one of the 'cardinal principles' of IHL and one of the 'intransgressible principles of international customary law'[4]. The application of such prohibitions to the use of industrial toxic chemicals as weapons in Syria has been highlighted in 2017 by the Independent International Commission of Inquiry on the Syrian Arab Republic, which stated that: 'The use of weapons in densely populated areas which are by nature indiscriminate and whose effects cannot be limited as required by international humanitarian law is prohibited. As the dispersal pattern of gas found in chlorine bombs cannot be controlled, their use throughout residential areas in eastern Aleppo city amounts to the war crime of indiscriminate attacks in a civilian populated area'.[5]

11.2.2.2 The Prohibition of the Employment of Means and Methods of Warfare of a Nature to Cause Superfluous Injury or Unnecessary Suffering (SIRUS)

The SIRUS prohibition has been enunciated in a number of IHL treaties and is considered to be part of customary IHL. A review of military manuals by the International Committee of the Red Cross (ICRC) has shown that a number of States such as Australia, France and Germany consider that the use of chemical weapons can cause unnecessary suffering.[6] However, there is no international consensus regarding an objective means of determining what constitutes 'superfluous injury or unnecessary suffering', nor on the criteria which can be used to judge whether specific weapons potentially breach the SIRUS prohibition.

In 1997, the ICRC's SIRUS Project proposed that what constituted superfluous injury and unnecessary suffering should:

> '[B]e determined by design-dependent, foreseeable effects of weapons when they are used against human beings and cause[d]:
>
> - specific disease, specific abnormal physiological state, specific abnormal psychological state, specific and permanent disability or specific disfigurement; or
> - field mortality of more than 25% or hospital mortality of more than 5%; or
> - Grade 3 wounds as measured by the Red Cross wound classification; or,
> - effects for which there is no well-recognized and proven treatment.'[7]

Criteria 1, 2 and 4, in particular, appear to be of potential applicability for States considering the (il)legality of the weaponisation of pathogens and/or toxic chemicals.

11.2.2.3 The Protection of Persons Considered *Hors de Combat*

Common Article 3 of the Geneva Conventions stated that 'Persons taking no active part in the hostilities, including members of armed forces who have laid down their arms and those placed *hors de combat* by sickness, wounds, detention, or any other cause, shall in all circumstances be treated humanely'. In addition, certain actions against such persons are prohibited, including '(a) violence to life and person, in particular murder of all kinds, mutilation, cruel treatment and torture; (b) taking of hostages; (c) outrages upon personal dignity, in particular humiliating and degrading treatment . . .' There is also a positive obligation to ensure that 'the wounded and sick shall be collected and cared for'.[8] These obligations are also deemed to be part of customary international humanitarian law.

11.2.2.4 Requirement to Respect and Ensure Respect of International Humanitarian Law

Under Common Article 1 to the four Geneva Conventions, and Article 1 of Additional Protocol I, 'The High Contracting Parties undertake to respect and to ensure respect' for the relevant treaties 'in all circumstances'.[9] The UK Royal Society has argued that the potential malign application of advancing chemical and life sciences to facilitate development of weapons employing chemicals that affect the brain could potentially impair or degrade the cognitive ability of enemy combatants (or indeed your own soldiers), with the consequent dangers of their committing serious breaches of IHL. 'Degrading the cognitive abilities of an adversary such that they are unable to distinguish between military targets and civilians, which often require a high degree of concentration, will undermine this requirement'. This is because such cognitive impairment could easily result in an unintended attack on one's own civilians or other persons or places specifically protected by law. Furthermore, the Royal Society has contended that 'Such attacks could not be prosecuted because the perpetrators will have been rendered mentally incapable of being responsible for the offences.'[10]

11.2.2.5 Prohibition of Methods or Means of Warfare Intended to Cause Widespread, Long-term and Severe Damage to the Natural Environment; Prohibition on the Deliberate Destruction of the Natural Environment as a Form of Weapon

Additional Protocol I prohibits the use of 'methods or means of warfare which are intended, or may be expected to cause, widespread, long-term and severe damage to the natural environment'.[11] State practice has established this rule as a norm of customary international law applicable in international, and arguably also in non-international armed conflicts. However certain States, notably the United States, are 'persistent objectors' and others would dispute the customary status of the rule.

In addition, there is extensive State practice prohibiting the deliberate modification of the natural environment as a form of weapon. Under the Convention on the Prohibition of Military or Any Other Hostile Use of Environmental Modification Techniques (ENMOD), States Parties will not 'engage in military or any other hostile use of environmental modification techniques having widespread, long-lasting or severe effects as the means of destruction, damage or injury to any other State Party'. Nor will they assist, encourage or induce others to engage in activities.[12] The difference between this provision and the one in Additional Protocol I is that the latter prohibits attacks on the environment as such, regardless of the means used, whereas the ENMOD Convention refers to the deliberate manipulation of natural processes that could produce phenomena such as hurricanes, tidal waves or changes in climate. According to the ICRC while it is unclear whether the provisions in the ENMOD Convention are now customary, there is sufficiently widespread, representative and uniform practice to conclude that the destruction of the natural environment may not be used as a weapon.[13]

Both these rules are potentially applicable to the development and use of weapons employing pathogens and/or toxic chemicals, including those targeting plants as well as animals, if they were intended to be employed on a sufficient scale. Indeed, the Final Declaration of the Second Review Conference of the Parties to the ENMOD Convention reaffirmed that the military and any other hostile use of herbicides as an environmental modification technique was a prohibited method of warfare 'if such a use of herbicides upsets the ecological balance of a region, thus causing widespread, long-lasting or severe effects as the means of destruction, damage or injury to another State Party'.[14]

11.2.2.6 Obligations to Review 'New' Weapons Under International Humanitarian Law

Under Article 36 of Additional Protocol I to the Geneva Conventions, all High Contracting Parties that are engaged 'in the study, development, acquisition or adoption of a new weapon, means or method of warfare' are 'under an obligation to determine whether its employment would, in some or all circumstances, be prohibited by this Protocol or by any other rule of international law applicable to the High Contracting Party'. As of 1 June 2023, there were 174 States party to Additional Protocol I, which are explicitly bound by this obligation. In addition, the *ICRC Guide* has stated that this requirement is 'arguably' one that applies to all States, regardless of whether or not they are party to Additional Protocol I.[15] Lawland has argued that: '[E]stablishing national mechanisms to review the legality of new weapons is *especially relevant and urgent in view of emerging new weapons technologies such as* directed energy, *[chemical] incapacitants, behaviour change agents*, acoustics and nanotechnology, to name but a few' (emphasis added).[16] However, despite the widespread recognition of States' obligations to conduct legal reviews of new weapons, in 2016 Giacca, Legal Advisor at the ICRC, stated that only 'between 15 and 20 States are known to conduct their own legal reviews'; in contrast 'many others rely on manufacturer information and on the reviews conducted by others.'[17]

11.3 International Human Rights Law

11.3.1 Introduction

Although IHRL does not specifically address the use of discrete arms or security equipment, it is certainly of great relevance to the employment of all such weapons, including those that employ toxic chemicals including toxins and other mid-spectrum agents, as it regulates the use of force by law enforcement officials and other agents of the State. [IHRL would also potentially be applicable to (and prohibit) the use of BWs by all such actors.] An important strength of IHRL is its applicability in a broad range of circumstances where such weapons might be considered including domestic policing operations, non-international conflicts (whether or not the State recognised it as such), and those aspects of an international conflict occurring in territory under a State's jurisdiction. The importance of this breadth of coverage has been highlighted by Hampson, who has noted that '[S]tates frequently refuse to characterize an internal armed conflict as such, preferring to call it criminal or terrorist activity. In such a situation, they can hardly challenge the applicability of human rights law.'[18] While several human rights norms may be applicable to the regulation of weapons employing toxic chemicals (or pathogens), the rights to life and to freedom from torture and other ill-treatment together with associated obligations on the restraint of force, are the most relevant.[19]

11.3.2 Protection of the Right to Life and Restrictions on the Use of Force

The 'inherent' right to life is enshrined in many international, and regional, human rights instruments. The UN Human Rights Committee, the body that monitors the implementation of the International Covenant on Civil and Political Rights (ICCPR), has stated that the right to life is 'the supreme right from which no derogation is permitted even in time of public emergency which threatens the life of the nation' The Committee has further stated that States Parties to the ICCPR should take measures not only to prevent and punish deprivation of life by criminal acts 'but also to prevent arbitrary killing by their own security forces'. The Committee has argued that the right to life, in the case of the ICCPR at least, is a right 'which should not be interpreted narrowly'.[20]

Guidance to States on their attendant obligations to restrain and govern the use of force in law enforcement is provided by the 1990 UN Basic Principles on the Use of Force and Firearms by Law Enforcement Officials (UNBP) and the 1979 UN Code of Conduct for Law Enforcement Officials (UNCoC).[21] These two instruments specify that the use of force must be proportionate, lawful, accountable and necessary. Under Principle 5 of the UNBP, law enforcement officials are required to 'exercise restraint in such use and act in proportion to the seriousness of the offence and the legitimate objective to be achieved; minimize damage and injury, and respect and preserve human life; ensure that assistance and medical aid are rendered to any injured or affected persons at the earliest possible moment.' Furthermore, under Principle 2 of the UNBP 'Governments and law enforcement agencies should develop a range of means as broad as possible and equip law enforcement officials with various types of weapons and ammunition that would allow for a differentiated use of force and firearms. These should include the development of non-lethal incapacitating weapons for use in appropriate situations, with a view to increasingly restraining the application of means capable of causing death or injury to persons.'[22]

It is clear that the use of weapons employing pathogenic organisms, such as 'classic' BWs (e.g. anthrax, brucellosis, plague and tularaemia), a range of toxic chemicals, such as 'classic' chemical warfare agents (e.g. mustard gas, sarin and VX), industrial toxic chemicals (e.g. ammonia and chlorine) or toxins (e.g. aflatoxin, botulinum toxin, ricin and SEB), that have caused widespread injury and death would be a serious violation of IHRL, notably contravening State obligations to protect life and restrict use of force. In contrast, certain so-called 'non-lethal' or 'less lethal' weapons employing a restricted range of toxic chemicals (in limited quantities) and purportedly intended to incapacitate but not cause permanent harm, have been employed by law enforcement personnel. Of most importance is the widespread employment (and frequent misuse) of riot control agents and by contrast the isolated and highly contentious employment of CNS-acting chemical agent weapons.

11.3.2.1 Application to Riot Control Agents (RCAs)

RCAs[23] are widely employed by law enforcement officials throughout the world for activities such as the dispersal of assemblies posing an imminent threat of serious injury. When used in accordance with manufacturers' instructions and in a lawful, proportionate and discriminate manner, in line with international human rights standards, RCAs can provide an important alternative to other applications of force more likely to result in injury or death, notably firearms. However, regional and UN human rights monitoring bodies and international non-governmental human rights organisations have regularly expressed concern regarding reports of the employment of RCAs as part of the indiscriminate, excessive or lethal use of force by law enforcement officials, particularly in crowd control situations. A specific recurring concern has been the employment of RCAs in excessive quantities or in confined spaces, where the targeted persons cannot disperse and where the toxic properties of the agents can lead to serious injury or death, particularly to vulnerable individuals.

RCAs are also employed as a means to subdue prisoners and maintain order in correctional centres, prisons, police stations and other places of detention. Human rights bodies have raised concerns about the appropriateness of such application. For example, the European Committee for the Prevention of Torture and Inhuman or Degrading Treatment or Punishment (CPT) has stated that tear gas is a 'potentially dangerous substance' and that 'only exceptional circumstances can justify [its] use ... inside a place of detention – but never in a confined space such as a cell – for control purposes, and such exceptional use should be surrounded by appropriate safeguards'.[24]

11.3.2.1.1 Considerations Regarding Means of Delivery and Dispersal of RCAs A range of projectiles and delivery systems which have a narrow dispersal area and emit a limited quantity of agent (e.g. hand-thrown or weapons-launched RCA canisters and grenades or hand-held RCA spray disseminators) are widely employed by law enforcement officials, for example, in public order situations. If such devices have been properly tested and trailed, their use should not raise undue concerns as long, as it is in strict accordance with the relevant human rights standards, specifically, UNBP, UNCoC and UN Standard Minimum Rules for the Treatment of Prisoners,[25] and in conformity with, national deployment guidelines.

In contrast to the foregoing, a range of delivery mechanisms have been developed for crowd control and dispersal that deliver far larger amounts of RCAs over wider areas than could be delivered by hand-held sprays and the like. A number of these devices raise questions concerning proportionality and/or about the feasibility of their discriminate use, with the consequent danger of affecting bystanders, in potential contravention of human rights law and standards. In addition, a further range of 'wide area' RCA means of delivery have been developed and promoted which are inherently unacceptable for employment in law enforcement activities, contravening relevant use of force standards (as well as being in apparent breach of the CWC). These include munitions containing RCAs which have military utility, such as cluster munitions, aerial bombs, mortar rounds and artillery shells.

11.3.2.2 Application to CNS-Acting Chemical Agent Weapons

Proponents of CNS-acting chemical agent weapons[26] have long advocated their development and use in certain law enforcement scenarios, notably large-scale hostage situations, where there is a need to rapidly and completely incapacitate individuals or a group without causing death or permanent disability. However, the legitimacy of such weapons in any form of law enforcement is highly contested. The ICRC has argued against their development and use, highlighting the grave dangers of their employment in practice and the extremely limited circumstances under which the use of such 'potentially lethal force' should even be considered, given States obligations under IHRL.[27] Furthermore, even if such conditions are met, the obligations upon States under IHRL to protect the right to life still apply, with contingent severe constraints upon employment of these weapons as well as the requirement to take appropriate remedial measures.[28]

11.3.3 Prohibition Against Torture and Other Cruel, Inhuman or Degrading Treatment or Punishment

The prohibition against torture and other cruel, inhuman or degrading treatment or punishment (CIDTP) is recognised in a wide range of international and regional human rights agreements and is a customary norm, applicable at all times and in all circumstances, including in armed conflict. It is one of the few human rights for which no derogation has been permitted. Torture is defined under Article 1 of the UN Convention against Torture as:

> 'Any act by which severe pain or suffering, whether physical or mental, is intentionally inflicted on a person for such purposes as obtaining ... information or a confession, punishing him. . . or intimidating or coercing him or a third person, or for any reason based on discrimination of any kind, when such pain or suffering is inflicted by or at the instigation of or with the consent or acquiescence of a public official.'[29]

Elements of CIDTP have been defined in other relevant legal texts. For example, 'degrading treatment' has been defined by the European Commission of Human Rights as treatment or punishment that 'grossly humiliates the victim before others or drives the detainee to act against his/her will or conscience'.[30]

11.3.3.1 Application to Psychoactive (CNS)-Acting Chemical Agents

In its report exploring the potential use and misuse of neuroscience, the UK Royal Society highlighted the European Commission of Human Rights' definition of degrading treatment and considered it to be of 'particular importance in considering the potential applications of neuroscience that could, for example, manipulate behaviour or thought processes'. Consequently, the Royal Society stated that 'the use of potential militarised agents including noradrenaline antagonists such as propranolol to cause selective memory loss, cholecystokinin B agonists to cause panic attacks, and substance P agonists to induce depression could all be considered violations of the prohibition against degrading treatment'.[31]

Fidler has stated that: 'non-consensual, non-therapeutic use of any chemical or biochemical against detained individuals would constitute degrading treatment and could, constitute cruel or inhumane treatment and perhaps even torture'.[32]

In addition to the prohibitions against torture and CIDTP, further important potential constraints upon the non-consensual application of psychotropic drugs to detainees relate to obligations to ensure respect for the detainee's right to freedom of opinion. For example, Article 19 of the ICCPR has declared that 'everyone shall have the right to hold opinions without interference'.[33] In his legal commentary to the Convention, Nowak stated that this provision consequently 'obligates the States Parties to refrain from any interference with freedom of opinion (by indoctrination, "brainwashing", influencing the conscious or unconscious mind with psychoactive drugs or other means of manipulation) and to prevent private parties from doing so'.[34]

11.3.3.2 Application to Riot Control Agents

Despite the absolute prohibition on torture and CIDTP, the misuse of RCAs for such purposes by law enforcement officials has been reported by regional and UN human rights bodies and international human rights non-governmental organisations. Presenting the results of a 2003 study on the trade in security equipment that could be used for torture and CIDTP, the UN Special Rapporteur on Torture stated that:

> '[T]he allegations of torture that he has received from all regions of the world have involved instruments such as [inter alia] ... chemical control substances (e.g. tear gas and pepper spray). While some of the cases have involved the use of equipment which is inherently cruel, inhuman or degrading, and would per se breach the prohibition of torture, the vast majority [including RCAs] have involved the misuse of those instruments, legitimate in appropriate circumstances, to inflict torture or other forms of ill-treatment...'[35]

In certain cases, RCAs have reportedly been employed as a means of inflicting 'collective punishment' upon groups of individuals or crowds. Other cases of concern involved the use of hand-held irritant sprays against individual prisoners and detainees in a targeted fashion.

11.3.4 Obligations to Review and Monitor the Use of 'Less Lethal' Weapons

To fulfil their obligations under IHRL and to ensure the responsible use of force by law enforcement officials, States are encouraged to implement review mechanisms to ensure that so-called 'less lethal' weapons – including RCAs and related means of delivery – developed or otherwise acquired are consistent with such obligations. UNBP Principle 3 states that 'development and deployment of "nonlethal" incapacitating weapons should be carefully evaluated in order to minimize the risk of endangering uninvolved persons, and the use of such weapons should be carefully controlled.'[36] However, despite the recommendations of human rights bodies and the continuing widespread well-documented misuse of certain 'less lethal' weapons (including RCA and associated means of delivery) in human rights violations (sometimes resulting in serious injury or death), there are currently no internationally accepted procedures for evaluating new 'less lethal' weapons, for ensuring adequate training in their use or effectively controlling and monitoring their subsequent employment.

11.4 Conclusions

While the BTWC and the CWC and associated customary law are the *lex specialis* prohibiting States from developing, stockpiling and using BCW, there are a number of additional international agreements and relevant international law further constraining the weaponisation of pathogens and toxic chemicals (including toxins and other mid-spectrum agents). Consequently, States should give full and careful consideration to the application of such agreements and international law; firstly because of the direct obligations that arise from such agreements and laws which may either prohibit or severely restrict development, stockpiling, transfer or use of weapons employing pathogens and toxic chemicals (including toxins and other mid-spectrum agents), and secondly because relevant international law should inform the implementation of those areas of the BTWC and CWC where treaty language is ambiguous and interpretation is currently contested.

In terms of armed conflict, the BTWC and CWC prohibitions upon development and use of BCW are further underpinned by associated obligations derived from IHL treaties and customary international law. Indeed, the use of weaponised pathogens or toxic chemicals in armed conflict would potentially breach relevant IHL prohibitions (such as those forbidding SIRUS, indiscriminate weapons and deliberate attacks on civilians). In addition, IHL also provides protection to prisoners of war or others considered *hors de combat* which would, for example, prohibit the use of toxic chemicals against them in forced interrogations, torture or human experimentation. Furthermore, IHL provides added protection for the civil population, for example, in effect prohibiting large-scale use of toxic chemicals (such as herbicides) which have widespread, long-lasting and severe effects on the environment. However, the potential utility of IHL to the regulation of toxic chemicals (and prohibition of BWs) is curtailed due to limitations in investigation and enforcement procedures, and extremely low levels of State implementation of Article 36 legal reviews of new weapons. Furthermore, IHL is only applicable to situations of armed conflict. While much relevant IHL would extend to non-international armed conflicts, there may well be disagreements as to whether a particular situation is indeed a non-international armed conflict, with the relevant State instead claiming to be involved in law enforcement activities against criminals or terrorist organisations. However, even in such disputed situations, State action would be constrained by a second body of law, IHRL.

IHRL covers the full 'use of force' spectrum from law enforcement activities through to armed conflict, including 'grey areas' such as counter-terrorist, counter-insurgency and military operations outside of armed conflict where the use of weapons employing certain toxic chemicals have been proposed. While several human rights norms may be applicable to the regulation of such weapons, the rights to life; and to freedom from torture and other cruel, inhuman or degrading treatment or punishment; together with attendant obligations on the restraint of force, are the most relevant. An important aspect of human rights law is that there are a number of international and regional mechanisms to monitor adherence to relevant treaties. In addition, certain treaties provide for the possibility of individual petitions which can result in legally binding judgements for those States party to the relevant instrument. However, it must be recognised that these monitoring and enforcement mechanisms have limited preventative value (though they may have some deterrent effect), as they would only be initiated after a potential misuse of such weapons has occurred. Furthermore, there are currently no internationally accepted procedures, under IHRL, for evaluating new weapons such as those employing toxic chemicals, or for monitoring their subsequent use at a national level.

Author Biography

Dr. Michael Crowley is Honorary Visiting Senior Research Fellow at Bradford University and also Research Associate with the Omega Research Foundation. He has worked for over thirty years on arms control, security and human rights, including as Executive Director of the Verification Research, Training and Information Centre (VERTIC) and has acted as Chairperson of the Bio-Weapons Prevention Project. His recent work employs a Holistic Arms Control methodology to examine threats from continuing revolutionary advances and convergence of the chemical, life and associated sciences; and consequently, to develop realistic routes for States and the scientific community to address these threats while preserving the benefits of such advances for all.

References

1 Geneva Conventions (I)–(IV) (1949) *Adopted by Diplomatic Conference for the Establishment of International Conventions for the Protection of Victims of War.* Geneva, 12 August 1949.
2 ICRC (1977) Protocol Additional to the Geneva Conventions of 12th August 1949, and Relating to the Protection of Victims of International Armed Conflicts [Protocol I], (1977); Protocol Additional to the Geneva Conventions of 12th August 1949, and Relating to the Protection of Victims of Non-International Armed Conflicts [Protocol II] (1977).
3 ICRC (2005) Customary international law which can be considered as a general practice accepted as law, is binding on all States whether or not they are parties to the relevant treaties. For an authoritative analysis of relevant customary IHL see: Henckaerts, J. and Doswald-Beck, L. (2005) Customary International Humanitarian Law Volume 1: Rules, Cambridge University Press, Cambridge.
4 International Court of Justice (1996) *Nuclear Weapons Case, Advisory Opinion.* The Hague: ICJ.
5 United Nations General Assembly, Human Rights Council (2017) *Report of the Independent International Commission of Inquiry on the Syrian Arab Republic*, A/HRC/34/64, 2 February 2017, p.31.
6 Henckaerts, J. and Doswald-Beck, L. (2005) *op.cit.*, p. 244
7 Coupland, R. (ed.) (1997) *The SIRUS Project, Towards a Determination of Which Weapons Cause "Superfluous Injury or Unnecessary Suffering"*, 23. Geneva: ICRC*.
8 Geneva Conventions (I)-(IV) (1949) *op.cit.*, Common Articles 3.1 & 3.2.
9 Geneva Conventions (I)-(IV) (1949) *op.cit.*, Common Article 1; Additional Protocol I (1977) *op.cit.*, Article 1.
10 Royal Society, Science Policy Centre (2012) *Brain Waves Module 3, Neuroscience, Conflict and Security*, 20. London: Royal Society.
11 Additional Protocol I (1977) *op.cit.*, Article 35(3).
12 ICRC (1977) *Convention on the Prohibition of Military or Any Other Hostile Use of Environmental Modification Techniques* (1977), Articles 1 & 2.
13 Henckaerts, J. and Doswald-Beck, L. (2005) *op.cit.*, Rule 45, p.156.
14 ENMOD Convention (1992) Second Review Conference of the Parties to the ENMOD Convention (1992) *Final Declaration*, paragraph 633.

15 ICRC (2006) *A Guide to the Legal Review of New Weapons, Means and Methods of Warfare, Measures to Implement Article 36 of Additional Protocol I of 1977*, 4. Geneva: ICRC.

16 Lawland, K. (2006) Reviewing the legality of new weapons, means and methods of warfare. *International Review of the Red Cross* 88 (864): 926.

17 ICRC (2023) *Legal review of new weapons: scope of the obligation and best practices.* Humanitarian Law and Policy. http://blogs.icrc.org/law-and-policy/2016/10/06/legal-review-new-weapons (accessed 9 August 2023).

18 Hampson, F. (2007) International law and the regulation of weapons. in Pearson, A., Chevrier, M., and Wheelis M. (Eds.) *Incapacitating Biochemical Weapons*, Lexington Books, 244–245.

19 Palgrave Macmillan (2016) Other potentially applicable rights include those to liberty and security; to freedom of opinion, expression, association and assembly; to human dignity; and to health. For further discussion see: Crowley, M. (2016) Chemical Control: Regulation of Incapacitating Chemical Agent Weapons, Riot Control Agents and their Means of Delivery, Palgrave Macmillan UK, Basingstoke, chapter 8.

20 Human Rights Committee, General Comment No. 6: The right to life, 30 April 1982, paragraphs 1 & 3.

21 United Nations (1990) *Basic Principles on the Use of Force and Firearms by Law Enforcement Officials*, adopted by the Eighth United Nations Congress on the Prevention of Crime and the Treatment of Offenders, Havana, Cuba, 7 September 1990; United Nations (1979), *Code of Conduct for Law Enforcement Officials*, adopted by United Nations General Assembly Resolution 34/169, 17 December 1979.

22 United Nations (1990) *op.cit.*, Principles 2 &5 (a)–(c).

23 PAVA (n.d.) Riot control agents are defined under the CWC "as any chemical not listed in a schedule which can produce sensory irritation or disabling physical effects rapidly in humans and which disappear within a short time following termination or exposure". As well as synthetic chemicals such as CS, CN and CR, they include certain biological substances notably the capsaicinoids and the related synthesised analogue, PAVA.

24 CPT (2009) *Report to the Portuguese Government on the visit to Portugal by the CPT.* CPT/Inf (2009) 13, 19 March 2009, paragraph 92.

25 UN (1977) Standard Minimum Rules for the Treatment of Prisoners, approved by the Economic and Social Council by its Resolutions 663 C (XXIV) of 31 July 1957 and 2076 (LXII) of 13 May 1977.

26 Central nervous system (CNS) acting chemical agents [also referred to incapacitants or incapacitating chemical agents] are a diverse range of biological substances and human synthesised chemicals that, when used as weapons, are supposedly intended to cause a long but non-permanent incapacitation in those individuals or groups targeted. However, inappropriate doses of these toxic chemicals can cause very serious, long-lasting health effects and can be fatal to those affected.

27 ICRC (2012) *Toxic Chemicals as Weapons for Law Enforcement: A Threat to Life and International Law?* 3. Geneva: ICRC.

28 Fidler, D. (2007) *op.cit.*, Incapacitating and Biochemical Weapons and Law Enforcement under the CWC. in Pearson, A., Chevrier, M., and Wheelis, M. (Eds.) 171–194.

29 UN (1984) *Convention against Torture and Other Cruel, Inhuman or Degrading Treatment or Punishment, adopted by UNGA Resolution 39/46, 10 December 1984*, Article 1(1). It should be noted that this definition is narrower than most relevant international and regional treaties, which normally require only that a State agent intentionally inflicts severe pain or suffering, without the need to show instigation, consent or acquiescence.

30 European Commission of Human Rights, Greek case, as cited in Henckaerts, J. and Doswald-Beck, L. (2005) *op.cit.*, Rule 90.

31 Royal Society, Science Policy Centre (2012) Brain Waves Module 3. In: *Neuroscience, Conflict and Security*, 24. London: Royal Society.

32 Fidler, D. (2007) *op.cit.*, p. 176.

33 UN (1966) *op.cit.*, Article 19

34 Nowak, M. (2005) *U.N. Covenant on Civil and Political Rights: CCPR Commentary*, 2nde, 340. Kehl: N. P. Engel.

35 UN (2004) Report of the Special Rapporteur on the Question of Torture, Theo Van Boven, UN doc. E/CN.4/2005/62, 15th December (2004), paragraph 13; see also: UN Commission on Human Rights (2003) *Study on the situation of trade in and production of equipment which is specifically designed to inflict torture or other cruel, inhuman or degrading treatment, its origin, destination and forms, submitted by Theo Van Boven, Special Rapporteur on torture, pursuant to resolution 2002/38 of the Commission on Human Rights*, UN doc. E/CN.4/2003/69, 13th January 2003.

36 United Nations (1990) *op.cit.*, Principle 3.

12

The Role of International Organisations in Biosecurity and the Prevention of Biological Warfare

Louison Mazeaud[1], *James Revill*[2], *Jaroslav Krasny*[1], *and Vivienne Zhang*[2]

[1] *Weapons of Mass Destruction Programme, UNIDIR, Geneva, Switzerland*
[2] *Weapons of Mass Destruction Programme and Space Security Programmes, UNIDIR, Geneva, Switzerland*

Key Points

1) Numerous international organisations (IOs) have supported biosecurity and the prevention of biological warfare. The United Nations Office for Disarmament Affairs (UNODA), the United Nations Institute for Disarmament Research (UNIDIR), the Food and Agricultural Organisation (FAO), the United Nations Interregional Crime and Justice Research Institute (UNICRI), the World Health Organisation (WHO), the International Criminal Police Organisation (INTERPOL) and the International Committee of the Red Cross (ICRC) are some of the bodies considered in this chapter. This list is, however, not exhaustive.
2) These IOs have played a key role in processing ideas into norms and policies; facilitating discussion around biosecurity and the prevention of biological weapons (BWs); and providing authoritative insight and knowledge related to these issues.
3) IOs continue to support contemporary biosecurity governance through various activities and approaches ranging from facilitating the adoption of legally binding instruments, creating investigative mechanisms and developing practical tools to assist practitioners in prevention.
4) IOs have a unique role to play given their particular status and perspectives. They can shape discussions and facilitate the consideration of specific elements in global agendas. They can also maintain a forum and link between actors even in challenging geopolitical situations. They are, however, also facing numerous challenges, including a lack of resources, mandatory limits and issue-boundaries.
5) At this current juncture, with the rapid advance, convergence and diffusion of biotechnology in a period of considerable geostrategic tension, IOs should not be taken for granted but carefully nurtured through what some see as the biotech century.

Essentials of Biological Security: A Global Perspective, First Edition. Edited by Lijun Shang, Weiwen Zhang, and Malcolm Dando.

Summary

International organisations (IOs) have played an important role in fostering biosecurity and preventing biological warfare through activities ranging from internationalising and institutionalising the norm against biological weapons (BWs), to informing and facilitating multilateral biological disarmament diplomacy, and investigating allegations of the use of BWs. Suitably resourced and mandated by States, some IOs could go much further in preventing biological warfare. However, there are also limits to what IOs can do. To make sense of the role of IOs in biosecurity and the prevention of biological warfare, this chapter looks at the role of IOs in the formation and evolution of the biological disarmament regime, beginning with early efforts to facilitate the development of the Geneva Protocol through to the role of IOs in helping shape the Biological Weapons Convention (BWC) and, subsequently, the wider twenty-first century biosecurity governance regime.

12.1 Introduction

International Organisations (IOs) are defined in this chapter as 'formal interstate institutions that are, or which have the potential to be, planetary in reach'.[1] Examples of IOs working in the field of biosecurity and biological disarmament include the United Nations (UN) through its bodies, such as the United Nations Office for Disarmament Affairs (UNODA), the United Nations Institute for Disarmament Research (UNIDIR), the Food and Agricultural Organisation of the United Nations (FAO), the United Nations Interregional Crime and Justice Research Institute (UNICRI) and the World Health Organisation (WHO), as well as other bodies, such as the International Criminal Police Organization (INTERPOL) and the International Committee of the Red Cross (ICRC).

Such IOs can play an important role in shaping policies. Indeed, contrary to the logic of Weberian bureaucratic thinking, IOs, like their national public counterparts are not simply bureaucratic automatons, blindly implementing the instructions of principles. Rather they can, and have shaped, as well as implemented, policies.

In this chapter, we draw on the work of Lewis and Thakur to illustrate how such IOs have contributed to fostering 'biosecurity' and preventing biological warfare in three different ways, including the following: firstly, IOs can play an important role as 'a funnel for processing ideas into norms and policies and for transmitting information from national sources to the international community'; secondly, they serve as a 'forum for discussion and negotiation of common international positions, policies, conventions and regimes'; and thirdly, IOs can play a role as a 'font of international legitimacy for the authoritative promulgation of international norms, appeals for adherence to global norms and regimes, and coercive measures to enforce compliance with them'.[2]

To make sense of how IOs have fulfilled these functions (or not) in different ways and at different times, this chapter begins with a short historical overview of key stages in the evolution of measures to prevent biological warfare. The chapter subsequently turns to explore the role of IOs in contemporary biosecurity governance before reflecting on the opportunities and limitations of IOs in this field, particularly in a period of geopolitical tensions.

12.2 The Role of IOs in Fostering the Norm Against Biological Weapons

IOs have long played a role in the prevention of BWs. The convening capacity of the League of Nations as a forum for discussion and negotiation of common international positions and policies illustrates the early role of IOs.

Over the course of World War One, chemical weapons were used extensively by both allied and axis powers, resulting in what some estimate to be as many as '1,300,000 casualties, including about 90,000 deaths'.[3] The horrors of gas warfare and the long-term effects of chemical weapons were not lost on politicians or publics around the world. In response to the use of chemical weapons the International Committee of the Red Cross issued an appeal against the use of poisonous gases in February 1918.[4]

Reflecting these wider concerns of the time over gas warfare, the League of Nations picked up on issues related to chemical and biological weapons (CBWs) in the post-war milieu. This was achieved first through the production of a report by the League of Nations' Permanent Advisory Commission for Military, Naval and Air Questions, which was tasked with studying the implications of the use of gas in the First World War and seeking methods to limit or prohibit its use in the future. The subcommittee produced an 'authoritative exposition' on the issue of gas warfare that was one of the first cross-disciplinary international studies on the issue of chemical and biological warfare (see Chapter 2) and included the opinions of 'chemists, physiologists and bacteriologists in various countries'.[5]

The report was rather dismissive of bacteriological warfare, owing to the difficulties in weaponising biology at the time, and the evident risks that the use of infectious agents on the battlefield could backfire. The report thus concluded that bacteriological weapons were not '*at present* capable of paralysing an enemy's defence' (emphasis added).[6] However, the report expressed concern over the potential of gas weapons to the extent that during the conference for the Supervision of the International Trade in Arms and Ammunition and in Implements of War, discussions around the prohibition on *the use* of gas warfare resulted in the 1925 Protocol for the Prohibition of the Use of Asphyxiating, Poisonous or Other Gases, and of Bacteriological Methods of Warfare. The Protocol stated[7]:

> '... the use in war of asphyxiating, poisonous or other gases, and of all analogous liquids materials or devices, has been justly condemned by the general opinion of the civilized world ... To the end that this prohibition shall be universally accepted as a part of International Law, binding alike the conscience and the practice of nations'

The protocol was further extended to cover 'the use of bacteriological methods of warfare', in part on the basis that some delegations to the conference argued that[8]: 'bacteriological weapon[s] can be manufactured more easily, more cheaply and with absolute secrecy'.

12.3 IOs in the Genesis of the Biological and Toxin Weapons Convention

Further work on addressing the challenge of BWs was overshadowed by World War Two, and the subsequent tensions resulting from the onset of the Cold War. When interest in further international work around the prevention of BWs reawakened in earnest in the late 1960s, IOs played an important role, creating a forum for discussion on the prevention of CBWs, a font of authoritative guidance and a funnel for the processing of ideas into norms.

12.3.1 Conference of the Eighteen-Nation Committee on Disarmament

IOs played a role as a forum for discussion through the Conference of the Eighteen-Nation Committee on Disarmament. The Committee facilitated discussions on the prevention of poisonous weapons and bacteriology during the Cold War, and several proposals were submitted related to conventions designed to prohibit CBWs, and later in 1968, a Convention to prohibit the 'use for hostile purposes of microbiological agents causing death or disease by infection in man, other animals or crops'.[9]

The separation of CBWs was not initially welcomed by all States. However, following the United States' unilateral renunciation of BWs on 25 November 1969, momentum built around the idea of tackling BWs first. Over time, States to the Conference of the Eighteen-Nation Committee on Disarmament and later the Conference of the Committee on Disarmament (CCD), both under the auspices of the UN at the time, proceeded to negotiate what later became the BWC.

12.3.2 The UN Secretary-General's CBW Report

IOs also played a role through the provision of authoritative guidance and served as a 'font of international legitimacy'. One early example of such authoritative guidance is the 1969 Report of the Secretary-General: Chemical and Bacteriological (Biological) Weapons (CBW) and the Effects of their Possible Use. This report, which was prepared by a geographically, representative group of 14 consultant experts drawn from defence laboratories and academia, provided an indication of the significance of advances in biology (and BWs) since League of Nations reports and those prior. Specifically, it states that[10]:

> '... scientific and technological advances of the past few decades have increased the potential of [...] biological weapons to such an extent that one can conceive of their use causing casualties on a scale greater than one would associate with conventional warfare [...] Once the door was opened to this kind of warfare, escalation would in all likelihood occur, and no one could say where the process would end'.

Notably, the Report issued a call upon all countries to reach an agreement to halt the development, production and stockpiling of all BWs. This point was not lost in the CCD where it increased the knowledge of delegates on CBW issues and helped move CBW to the forefront of disarmament negotiations of the time.[11]

12.3.3 The WHO Report on Health Aspects of CBW

The first edition of the WHO Report on *Health Aspects of Chemical and Biological Weapons* from 1970 provided a more technically focused assessment of the threat of BWs than the Secretary General's Report. In doing so, it may have created a sense of urgency around the destructive potential of BWs. For example, the WHO report posited that[12]:

> 'if a biological agent such as anthrax were used, an attack on a city by even a single bomber disseminating 50 kg of the dried agent in a suitable aerosol form would affect an area in excess of 20 km^2, with tens to hundreds of thousands of deaths'.

The report was presented by the UN Secretary General who called for a cessation of the development, production and stockpiling of CBW agents as well as their elimination.

Cumulatively, the Conference of the Eighteen-Nation Committee on Disarmament (ENCD), the CCD the UN Secretary-General's CBW report and the WHO Report on Health Aspects of CBW, lent impetus to efforts to the prevention of biological warfare through the creation of a forum for discussion on CBW-related issues, the provision of authoritative reports highlighting the threat posed by CBW, and the bolstering of norms against the development production and stockpiling of such weapons.

12.4 IOs and the Evolution of Biosecurity Governance

In the twenty-first century, several IOs have played an active role in advancing biosecurity and the prevention of biological warfare. To illustrate the depth and breadth of IOs' roles, the following section focuses on the relevant activities of six IOs and related entities: the United Nations Security Council (UNSC) 1540 Committee, the UN Secretary General's Mechanism (UNSGM), the BWC Implementation Support Unit (ISU), the WHO, UNICRI and UNIDIR.

This is not an exhaustive list of IOs bodies working in this area, several others could be included. Nevertheless, it provides an illustrative overview of some of the contemporary functions of IOs working on biosecurity and demonstrates how such entities have continued to support the development of normative, legal, technical and regulatory measures for biosecurity and the prevention of biological warfare.

12.4.1 UNSC Resolution 1540

The events of 11 September and the subsequent 'Amerithrax' attacks increased the salience of BWs in the international security discourse, resulting in renewed interest in the prevention of bioterrorism. Reflecting this change, several new measures were developed to curb the risk of weapons of mass destruction (WMD) proliferation and use in the hands of terrorist groups.

One key example is UNSC Resolution 1540, a legally binding instrument adopted under Chapter VII of the UN Charter with the purpose of curbing the threat of proliferation of WMD posed by non-State actors. It consists of positive obligations requiring States to implement and enforce effective domestic measures to prevent such proliferation. Additionally, UNSC Resolution 1540 requires States to develop appropriate border control and trans-shipment measures on relevant items and services, such as financing. These measures also encompass the creation of necessary jurisdictional grounds and adoption of legislation to prohibit non-State actors from acquiring, possessing, developing, transporting, transferring or using WMDs, including the means of delivery as well as requirements for setting up appropriate controls over related materials (i.e. biological agents and toxins).

The 1540 Committee tends the resolution. Initially, this Committee had a limited mandate restricted to 'examining implementation of the resolution and reporting its findings to the Security Council'.[13] In this sense, the Committee is not mandated to assess States' compliance with treaties. Nonetheless, the Committee is seen as having a powerful nudging effect, as Cupitt notes[14]:

> 'simply prompting national officials to monitor their own legal framework and policies for stopping the spread of WMD-related items has had its own subtle but significant impact on the current international non-proliferation regime ... efforts to meet the very diverse obligations of the resolution generally inaugurate pressure within a state to develop new interagency mechanisms that connect stakeholders from a wide range of authorities, often for the first time, just to prepare a robust report to the 1540 Committee'.

Such nudges, along with other pressures, appear to have resulted in some success. A 2011 Report by the Committee revealed considerable progress in aspects of biosecurity, such as the adoption of legal measures related to production, use and/or storage.[15]

12.4.2 UN Secretary General's Mechanism (Authoritative and Objective Assessment)

The United Nations Secretary-General's Mechanism for investigation of alleged use of CBWs was established in the 1980s to provide a mechanism to systematically investigate allegations of CBWs use. This followed the repeated use of chemical weapons during the Iran–Iraq conflict.[16] The mechanism, which is maintained by UNODA in New York, has been used on a couple of occasions to address cases of chemical weapons use; however, it remains available to the UN Secretary-General in the event of an alleged case of BWs use.

Allegations of association with CBW have long been used deliberately or unwittingly to vilify adversaries. This makes the impartial investigation of CBW allegations extremely important in the process of restoring compliance and deterring non-compliance with the provisions of the BWC. In this context, a competent, geographically representative staff operating in an objective and impartial manner and beholden to no member State remains important. Combined with the development of robust investigative methodologies and calibrated equipment, the UNSGM could play an important role in the future as a 'font' of objective information that can be used to inform judgements related to allegations of BWs use.

12.4.3 Tending the BWC: The Implementation Support Unit (ISU) and the Work of UNODA

The ISU was established in 2007, with its staff located in the UNODA. The ISU has a mandate to provide administrative support for BWC meetings as well as assist with the implementation and universalisation of the BWC amongst other tasks. In this regard, the ISU contributes to the international community's efforts to prevent BWs.

Moreover, the ISU has also played a role in facilitating discussions around biosecurity. For example, the BWC Meetings of Experts in 2003 provided some States with their first exposure to the concept of 'biosecurity' and allowed for useful exchanges on this concept. Subsequent discussions around biosecurity-related issues resulted in BWC States Parties noting their common understanding on biosecurity in 2008, which in the context of the Convention refers to the 'protection, control and accountability measures implemented to prevent the loss, theft, misuse, diversion or intentional release of biological agents and toxins and related resources as well as unauthorized access to, retention or transfer of such material'.[17]

Yet more recently, UNODA's wider biosecurity-related activities have further expanded to include, inter alia, national and (sub-)regional efforts designed to enact BWC implementing legislation; capacity building activities to strengthen biosecurity (including facilitating the development of National Inventories of Dangerous Pathogens[18]); and youth engagement activities designed to raise awareness amongst the next generation.[19]

12.4.4 WHO, Biosecurity and the Governance of Dual-Use Research

As illustrated above, the WHO has played an important role in the provision of authoritative guidance related to BWs and biosecurity. This role has grown over the course of the twenty-first century. For example, in 2020, the WHO published the fourth edition of the *Laboratory Biosafety Manual*. This manual once again recognised the significance of biosafety, whilst also drawing attention to biosecurity stating: 'laboratory biosafety and biosecurity activities are fundamental to protecting the laboratory workforce and the wider community against unintentional exposures or releases of pathogenic biological agents'.[20]

Yet more recently, the work of the WHO has helped advance discussions on biosecurity further in several ways, including firstly, facilitating horizon scanning processes designed to 'identify[ing] emerging opportunities and risks' in the life sciences[21]; and secondly, providing authoritative guidance through the publication of a global guidance framework for the responsible use of the life sciences.[22] The latter, which drew in part from five working groups comprised of geographically representative interdisciplinary sets of experts, serves as a resource for scientists and other stakeholders and includes concrete practical steps and suggestions for stakeholders seeking to understand and mitigate biorisks.

12.4.5 Combatting Biological Crimes: United Nations Interregional Crime and Justice Research Institute (UNICRI)

One final example is the work of UNICRI, which takes a crime prevention and control approach to its work on biosecurity and the prevention of biological threats. Major activities are undertaken by the Chemical, Biological, Radiological and Nuclear (CBRN) Risk

Mitigation and Security Governance Programme, ranging from international collaborations with States and non-State partners, to published studies offering references for States to combat biological crimes. Its initiatives educate and train stakeholders on the evolving risks surrounding BWs and aid law enforcement and judicial assistance in their response.[23]

Of note is that UNICRI manages the International Network on Biotechnology (INB), which gathers experts from academia, research institutions and non-governmental and IOs committed to educating and raising awareness about responsible and secure conduct in the life sciences. The network disseminates biosafety, biosecurity and bioethics education knowledge through its digital navigation platform.[24] It illustrates the risks and opportunities in the field through case studies, scenarios, lab tours, technology briefs and e-learning modules.[25]

To build prosecutorial and judicial capacity for dealing with CB incidents, UNICRI collaborated with the Organisation for the Prohibition of Chemical Weapons (OPCW), the International Association of Prosecutors (IAP) and the BWC ISU to publish *A Prosecutor's Guide to Chemical and Biological Crimes*. This guide, produced within the framework of the European Union CBRN Risk Mitigation Centres of Excellence Initiative (EU CBRN CoE),[26] aims to provide prosecutors and relevant investigative authorities with guidance to support the successful prosecution of incidents involving the deliberate acquisition, stockpiling, production, transfer or use of a chemical or biological agent.[27] UNICRI also recently published the *Handbook to Combat CBRN Disinformation* to equip governmental communication officers as well as CBRN experts with competencies to analyse, understand and respond to CBRN disinformation in the media and on social media platforms.[28]

12.4.6 UNIDIR

Finally, UNIDIR has also played a role in fostering biosecurity and preventing biological warfare. As a voluntarily funded, autonomous institute within the UN focused on disarmament, UNIDIR has published research on a range of issues related to the BWC, providing policy-accessible, politically balanced research outputs that have shaped thinking around multilateral biological disarmament diplomacy.

Examples of UNIDIR's recent work, include a study on gender and the BWC, which facilitated understanding of the gendered impact of BWs and the corresponding implications for the provision of assistance in the event of the use of BWs[29]; work on compliance and verification which shaped thinking around a balanced package of outcomes from the Ninth BWC Review Conference[30]; and activities related to science and technology (S&T), in particular S&T review mechanisms, which has opened policy options for a BWC-specific S&T mechanism.[31]

In this regard, UNIDIR has played a role, particularly as funnel for new ideas and a font of authoritative, balanced insights into building biosecurity and preventing BWs. Current work around the development of a national implementation database for the BWC covering prohibitions, export controls, biosafety and biosecurity and oversight, amongst other elements, will further augment these functions providing a key source of information on all BWC States Parties National Implementation measures and serving as a resource for ideas to advance biosafety and biosecurity.[32]

12.5 The Strengths of IOs in Biosecurity and Prevention of Biological Warfare

IOs have shown leadership in advancing biosecurity and preventing biological warfare in at least three different ways.

Firstly, as a funnel, IOs have demonstrated the ability to raise issues and shape the agenda on elements that might otherwise not receive significant attention. As much is evident in the early work of the WHO and the UN in the genesis of the Biological and Toxin Weapons Convention (BTWC) as well as more recent efforts by UN entities to, for example, shape thinking around biosecurity and engage with life scientists on issues such as biorisks.

Secondly, as a 'forum for discussion and negotiation of common international positions, policies, conventions and regimes' IOs, including UN bodies, such as the ISU, have been critical in facilitating multilateral dialogue around the prevention of biological warfare, while creating spaces for States and civil society to exchange views on good practices and lessons learned in biosecurity. Even in difficult times, such as the collapse of the protocol negotiations in 2001, BWC States Parties have continued to participate in these discussions and engage in multilateral biological disarmament diplomacy. Although concrete, tangible outcomes from such deliberations may not always be clearly visible, sustained discussions have helped build shared understandings around concepts, such as biosecurity, and fostered exchanges on challenges and solutions related to the prevention of bacteriological and biological warfare.

Thirdly, IOs have played and can continue to serve as a 'font of international legitimacy for the authoritative promulgation of international norms'. This has been achieved through, inter alia, the production of authoritative reports by amongst other entities the League of Nations, the UN Secretary-General, UNICRI, UNIDIR and the WHO. The knowledge contained in such reports significantly shaped perceptions of the challenges and responses in this area.

12.6 The Limits of IOs in Biosecurity and Prevention of Biological Warfare

Although IOs can clearly play a role in the prevention of BWs and the building of biosecurity, there remain limitations on the role of IOs for several reasons. Firstly, IOs must be careful to stay within their respective mandates. These mandates are often a carefully balanced compromise that reflects the interest of UN Member States or States Parties to a treaty. As such an abrupt or unsanctioned shift in the activities of an IO can undermine its legitimacy and lead to accusations of "mission creep," something seen by "member States as especially insidious."[33] This can be difficult in the field of biosecurity, where States have indicated differences in perspectives on the concept and its relative importance. Of course, major events can provide a window of opportunity for IOs to adapt and potentially expand their activities.

Secondly, issue-boundaries present a challenge to IOs working in this area. Responsibility for biosecurity falls across several different IOs, yet no single international institution has the mandate and capacity to provide global biosecurity on its own and, until recently, the institutional environment for addressing biorisks was 'fragmented, under-developed and lacking resources'.[34] The prevention of BWs is clearly within the remit of the BWC ISU. However, 'biosecurity' is not an explicit component of the BWC *per se* and not necessarily a priority shared equally by all BWC States Parties. As such, ISU activities related to biosecurity have evolved cautiously. But neither is biosecurity the priority of the WHO nor any other single IOs entity. Indeed, many IOs have long played an important role in biosafety and biosecurity, but none has the monopoly on this issue area. They can nonetheless all provide specific technical expertise on certain aspects of this topic which can collectively enhance biosecurity.

The establishment of the UN Biorisks Working Group has to some extent defragmented work on biosecurity. Indeed, the process has helped 'develop a clear understanding of capacities, mechanisms, and roles and responsibilities within the UN system in order to strengthen the international community's response to biorisks and improve on the prevention of and preparedness for the deliberate use of biological pathogens'.[35] Yet issue-boundaries will likely remain a challenge for IOs operating in the broad, interdisciplinary and evolving area of biosecurity.

A third, interlinked, limitation relates to resources. IOs require new budget allocations to pursue new initiatives; however, to facilitate such budget allocations, member States need to be convinced of the merits of any new initiative.[36] As an example of the overall financial challenges faced by IOs, the budget for the BWC in 2021 was $1,864,700. This amount covers the organisation of meetings and a small staff; however, the ISU budget is a fraction of that of the OPCW or the International Atomic Energy Agency (IAEA). This places limitations on the extent to which the ISU can contribute to biosecurity and the prevention of biological warfare.

12.7 Conclusions

This chapter has demonstrated that IOs have played an important role in fostering biosecurity and preventing BWs through three key approaches. Firstly as a funnel for processing ideas; secondly as a forum for discussion and negotiation, as facilitated by UNODA and later, the BTWC ISU; and thirdly as a font of international legitimacy and the provision of authoritative guidance, something evident in the work of the WHO amongst others.

Despite the value of IOs in this area, many challenges remain in the development and sustainability of a biosecurity culture and prevention of the re-emergence of BWs. At this current juncture, with the rapid advance, convergence and diffusion of biotechnology in a period of considerable geostrategic tension, the sustainability of existing IOs should not be taken for granted but carefully nurtured over the course of what some see as the biotech century.

Author Biography

Ms. Louison Mazeaud is an Associate Researcher in the WMD programme at UNIDIR, where she focuses on the national implementation of the Biological and Toxin Weapons Convention (BTWC). Her areas of expertise include regulatory regimes governing dual-use technologies, multilateral institutional processes and implementation measures. Prior to joining UNIDIR, she worked for various international entities, including the Geneva Centre for Security Policy (GCSP) from 2020 to 2023. She holds a bachelor's degree from King's College London (War Studies Department) and a master's degree from the Graduate Institute Geneva.

Dr. James Revill is the Head of the WMD and Space Security Programmes at UNIDIR. His research interests focus on the evolution of regimes dealing with weapons of mass destruction, and he has published widely on this topic. He was previously a Research Fellow with the Harvard Sussex Programme at SPRU, University of Sussex and completed research fellowships with the Landau Network Volta Centre in Italy and the Bradford Disarmament Research Centre in the United Kingdom. He holds a Ph.D. focused on the evolution of the Biological Weapons Convention from the University of Bradford, United Kingdom. James' areas of expertise include biological weapons, biosecurity, bioterrorism, chemical weapons, chemical terrorism, chemical weapons convention, compliance, verification and improvised explosive devices.

Dr. Jaroslav Krasny is a researcher in the WMD Programme at UNIDIR, focusing on the national implementation of the Biological Weapons Convention. His expertise includes international law, particularly international humanitarian law and its application to weapons of mass destruction. He was previously a Research Associate at the Center for Peace, Hiroshima University (CPHU). His areas of expertise include International Law, International Humanitarian Law, Arms Control and Disarmament. He holds a Ph.D. from Hiroshima University and a second Masters from the Geneva Academy of International Humanitarian Law and Human Rights.

Ms. Vivienne Zhang is a Consultant for the WMD and Space Security Programmes at UNIDIR. Her research interests include WMD verification, risk reduction and nuclear arms control and disarmament diplomacy. Previously, Vivienne was a Junior Research Scholar at the Asia Pacific Foundation of Canada. She has also worked at various Canadian diplomatic missions abroad in Thailand, Laos, Switzerland and Germany. Vivienne holds a master's degree in international security from Sciences Po Paris and a bachelor's degree in Modern European Studies and International Relations from the University of British Columbia.

References

In each chapter of the book, a small number of key references that would be most useful for the reader to follow up are marked with a star*.

1 Weiss, T. G. and Wilkinson, R. (Eds.). (2014) *International Organization and Global Governance*, Routledge, New York, NY, 7.

*2 Lewis, P. and Thakur, R. C. (2004) Arms control, disarmament and the United Nations. *Disarmament Forum*, 1, 17–30.

3 Barnaby, F. (2004) *How to Build a Nuclear Bomb and Other Strategic Weapons of Mass Destruction*, Nation Books, New York, NY, 55–56.

4 ICRC. (1918) *World War I: The ICRC's Appeal against the Use of Poisonous Gases*, ICRC, Geneva.

5 League of Nations. (1924) *Report of the Temporary Mixed Commission for the Reduction of Armaments*. A.16.1924.IX. Page 29.

6 League of Nations. (1924) *Report of the Temporary Mixed Commission for the Reduction of Armaments*. A.16.1924.IX. 29.

7 (1925) *Protocol for the Prohibition of the Use of Asphyxiating, Poisonous or Other Gases, and of Bacteriological Methods of Warfare*, Geneva.

8 Croddy, E. (2002) *Chemical and Biological Warfare: A Comprehensive Survey for the Concerned*, Citizen Copernicus Books, New York, 223.

9 United Kingdom. (2005) *Final Verbatim Record of the Conference of the Eighteen-Nation Committee on Disarmament [Meeting 387]*, University of Michigan Library, Ann Arbor, Michigan.

***10** United Nations. (1969) *Report of the Secretary-General: Chemical and Bacteriological (Biological) Weapons and the Effects of their Possible Use*, United Nations, New York.

11 Goldbalt, J. (1971) CB disarmament negotiations 1920–1970. in *The Problems of Chemical and Biological Warfare* Stockholm International Peace Research Institute (SIPRI), Vol. IV, Humanities Press, New York, 273–274.

***12** World Health Organization. (1970) *Health aspects of chemical and biological weapons: report of a WHO group of consultants*. Goldbalt J. (1971). In *CB Disarmament Negotiations 1920–1970: The Problems of Chemical and Biological Warfare*. Stockholm International Peace Research Institute (SIPRI), Vol. IV. Humanities Press. New York. 273–274.

13 Cupitt, R. T. (2012) *Nearly at the Brink: The Tasks and Capacity of the 1540 Committee*, Arms Control Association.

14 Ibid.

15 For example, The Committee observes that in regard to accounting for materials related to biological weapons, at least 61 States have adopted legal framework measures covering production, use and/or storage, compared to 38 States in 2008. By 1 April 2011, at least 61 States had adopted enforcement measures in these areas, compared to 36 States in 2008. United Nations 1540 Committee. (2011) *Report of the Committee established pursuant to Security Council Resolution 1540 (2004)*, United Nations, New York.

16 United Nations General Assembly. (1987) *General Assembly Resolution A/RES/42/37C*, United Nations, New York.

17 Implementation Support Unit. (22 August 2008) *Biosafety and Biosecurity*. BWC/MSP/2008/MX/INF.1. Meeting of Experts, Geneva; Report of the Meeting of States Parties. (12 December 2008) BWC/MSP/2008/5. Meeting of States Parties, Geneva; Implementation Support Unit. (2022) *Guide to Implementing the Biological Weapons Convention*, United Nations.

18 United Nations Office for Disarmament Affairs. (2022) *Sri Lanka Establishes National Inventory of Dangerous Pathogens in Key Step to Implement the Biological Weapons Convention, with Support from UNODA and the Netherlands*, United Nations, New York.

19 United Nations Office for Disarmament Affairs. (2023) *The Youth for Biosecurity Initiative*, United Nations, New York.

20 World Health Organization. (2020) *Laboratory Biosafety Manual*, 4th ed., World Health Organization, Geneva.

21 World Health Organization. (2021) *Emerging Technologies and Dual-Use Concerns: A Horizon Scan for Global Public Health*, World Health Organization, Geneva.

22 World Health Organization. (2022) *Global Guidance Framework for the Responsible Use of the Life Sciences: Mitigating Biorisks and Governing Dual-Use Research*, World Health Organization, Geneva.

***23** Revill, J. *et al* (eds) (2022) *Stakeholder Perspectives on the Biological Weapons Convention*, UNIDIR, Geneva, 24.

24 UNICRI. (2023) *International Network on Biotechnology (INB): Second Network Partners Meeting*. Accessed 7 June, 2023.

25 Ibid.

26 EU CBRN. (2023) *Centres of Excellence*, European Union. Accessed.

27 UNICRI. (2022) *A Prosecutor's Guide to Chemical and Biological Crimes*, Turin, 11.

28 UNICRI. (2022) *Handbook to Combat CBRN Disinformation*, Turin, iii.

29 Dalaqua, R. H. *et al* (2019) *Missing Links: Understanding Sex- and Gender-Related Impacts of Chemical and Biological Weapons*, UNIDIR, Geneva.

30 Revill, J. *et al* (2022) *Back to the Future for Verification in the Biological Disarmament Regime?* UNIDIR, Geneva.

31 Revill, J. *et al* (2021) *Exploring Science and Technology Review Mechanisms under the Biological Weapons Convention*, UNIDIR, Geneva.

32 UNIDIR. (2024) *Biological Weapons Convention National Implementation Measures Database*. https://bwcimplementation.org/

33 Findlay, T. (2020) *The Role of International Organizations in WMD Compliance and Enforcement: Autonomy, Agency and Influence. WMDCE Series No. 9*, UNIDIR, Geneva, 17.

34 Marelli, F. (2021) *Facing Today's Biorisks: (2021) The Work of the United Nations Biorisk Working Group (UN-BRWG): Towards Action in Strengthening the International Community's Response and in Improving Prevention*, UNICRI, Turin.

35 Ibid.

36 Ghionis, A. (2022) *Change and Continuity in the Organisation for the Prohibition of Chemical Weapons*. PhD, Doctoral Thesis, University of Sussex.

13

Laboratory Biorisk Management as a Key Tool for Scientists to Understand Future Biological Threats and Strengthen the Biological Weapons Convention

Mayra Ameneiros

Centre for Science and Security Studies, King's College London, London, UK

Key Points

1) Laboratory biorisk management has evolved over the years. The static concept of biosafety levels and risk groups for pathogens has shifted to a continuous risk assessment cycle that takes into account the practices conducted and the pathogens involved in those practices.
2) WHO has recently published a new framework, this and other tools are available to scientists to improve their knowledge of biorisk management.
3) Biorisk management involves biosafety and biosecurity, it is crucial to understand the differences between the two and the lack of awareness specifically related to the field of biosecurity, as both are fundamental pillars of a robust biorisk management approach.
4) All of these efforts by the scientific community to achieve a comprehensive approach to biorisk management, need to be shared with other stakeholders and international organisations. Collaboration is an important step and by sharing best practices, we understand better the evolving biological threat landscape.
5) This scientific knowledge must be shared with diplomats and policymakers. If there is a cycle of engagement between these two worlds, we will effectively understand, identify and mitigate future biological threats.
6) Scientists play a crucial role in strengthening how the Biological Weapons Convention addresses advances in science and technology.

Summary

This chapter begins by delineating the evolution of biorisk management practices over the years, drawing insights from earlier WHO guidance and extending to the more contemporary WHO *Global Guidance Framework for the Responsible Use of Life Sciences*. It delves into the distinctions between laboratory biosafety and biosecurity, highlighting

Essentials of Biological Security: A Global Perspective, First Edition. Edited by Lijun Shang, Weiwen Zhang, and Malcolm Dando.

a notable lack of awareness regarding biosecurity measures and putting focus on the shortage of biosecurity professionals within the international scientific community. The chapter also underscores the necessity of implementing customised biorisk management practices on a global scale. Furthermore, it expounds upon the pivotal role of scientists in mitigating biological risks, advocating for adherence to current international guidelines and collaboration with diverse stakeholders throughout research processes. The author introduces case scenarios to encourage a culture of responsibility; awareness and education on dual-use technologies and to provide examples on how to conduct an evidence-based risk assessment. To finalise, the chapter underscores that this collective knowledge and efforts of the scientific community are insufficient if they stay inside the laboratory. It asserts the need of continuous mutual feedback and communication with policymakers and diplomats to understand, identify and mitigate future biological threats, promptly and effectively, and to further strengthen the Biological Weapons Convention.

13.1 History, Context and Current International Guidance

Biorisk management has been known as one of the best approaches for scientists to identify, manage and mitigate biological risks and continuously evaluate the work conducted in the laboratory, to work safely and securely. This chapter will explain how biorisk management practices have changed over the years, and we will base our analysis on the evolution of the World Health Organization (WHO) guidance up to the most recent *Global Guidance Framework for the Responsible Use of the Life Sciences.*[1]

This evolution of biorisk management over the years is due to a combination of factors that have shaped the fields of biosafety and biosecurity, and the way we handle potentially hazardous biological materials. Some of these factors are advances in science and technology (S&T), emerging infectious diseases, globalisation, lessons learned from past events or accidents, the changing biothreat landscape and new biosecurity concerns, national regulations and international standards, codes of conduct and ethics, the concept of culture of responsibility and so on.

WHO launched its first edition of the *Laboratory Biosafety Manual* in 1983.[2] The first edition was intended to provide guidelines and recommendations for safely handling pathogenic microorganisms in laboratories and was a pioneering effort to establish a systematic approach to biosafety practices and management. It laid the foundation for subsequent editions and revisions, which continued to refine and expand on the concept of biosafety, as science, technology and our understanding of infectious diseases advanced. It also encouraged countries to implement these standard biosafety concepts as part of their national strategies.

Ten years later, the second version of the document was released in 1993. The second edition[3] demonstrated a growing awareness of the complexities of biosafety and the need to adapt practices to rapidly evolving fields such as biotechnology. It expanded on risk assessment, refined containment measures and introduced considerations related to genetically modified organisms and as well as biosecurity.

Eleven years later, WHO released the third edition in 2004.[4] This document is the one that has been out, for a longer period, and the one most biosafety professionals were used to working with. When the third edition of the *WHO Laboratory Biosafety Manual* was launched in 2004, it focused on integrating a comprehensive and risk-based approach to biosafety, emphasising the culture of safety and the importance of personal responsibility. In addition, it included a new chapter on enhanced risk assessment and recombinant DNA technologies and tailored containment strategies for various biological agents and activities.

Throughout these three editions, we have seen how WHO started with a basic, containment and maximum containment classification for laboratories up to a more defined classification considering the existence of four Laboratory Biosafety Levels (BSLs). The same happened with risk groups for pathogens and these groups have been increasingly defined and detailed along these editions. Risk groups range from 1 to 4. Risk Group 1 is for pathogens that pose no or low individual and community risk; Risk Group 2 is for pathogens that pose a moderate individual risk, and a low community risk; Risk Group 3 is for pathogens that pose a high individual risk and a low community risk; and Risk Group 4 is for pathogens that pose a high individual and community risk. These risk groups for pathogens were associated with specific biosafety levels: Biosafety Level 1 (basic); Biosafety Level 2 (basic); Biosafety Level 3 (containment); and Biosafety Level 4 (maximum containment). However, there was a general misunderstanding around this concept.

There was an incorrect belief that a Risk Group 2 pathogen should be handled in a Biosafety Level 2 Laboratory, a Risk Group 3 in a Biosafety Level 3 and so on. An example of how these risk groups were related to Biosafety Levels is mentioned in this extract of the third edition:

> '... An agent that is assigned to Risk Group 2 may generally require Biosafety Level 2 facilities, equipment, practices and procedures for safe conduct of work. However, if particular experiments require the generation of high-concentration aerosols, then Biosafety Level 3 may be more appropriate to provide the necessary degree of safety since it ensures superior containment of aerosols in the laboratory workplace...'

This showed that it was not intended to match the agent's risk group number with the same laboratory BSL number. As a general approach, it was adequate, but could change due to a specific practice and after conducting a risk assessment. Previous versions had a more structured approach, and this was the first attempt to be more flexible, however, that flexibility moved around the same or a higher level of containment for a specific risk group pathogen. This demonstrates that such flexibility occurred within a narrow frame.

It took 16 years for the WHO to release a new document, the last version of this laboratory biosafety manual, the fourth edition.[5] This fourth edition changed how we addressed the risks posed by pathogens. WHO has adopted a broader risk-based approach as a foundation, which considers not only the inherent characteristics of the pathogen being handled but also the specific practices carried out in the laboratory. This approach recognises that the level of risk is influenced by a variety of factors, such as procedures, techniques and the overall context of laboratory activities.

This shift allows for a more tailored and nuanced assessment of biosafety and biosecurity measures, ensuring that they are appropriate to the specific situation and the potential hazards involved. This approach is in line with the evolving understanding of biosafety and the need to address risks holistically. It is an evidence-based approach, and it can be applied by several sectors, such as biopharmaceutical companies, research institutions, diagnostic laboratories, food laboratories and basically all laboratories, regardless of the resources they have available.

It can be shaped into the actual conditions under which the pathogens are being manipulated, this makes the process equitable, which did not happen in previous editions of this manual. It eliminates Laboratory BSLs and risk group classifications for pathogens. The fourth edition establishes core, heightened and maximum requirements, and they are defined as:

> '... *Core requirements*: A set of minimum requirements ... to describe a combination of risk control measures that are both the foundation for and an integral part of, laboratory biosafety. These measures reflect international standards and best practices in biosafety that are necessary to work safely with biological agents, even where the associated risks are minimal ...'
>
> '... *Heightened control measures*: A set of risk control measures ... to be applied in a laboratory facility because the outcome of a risk assessment indicates that the biological agents being handled and/or the activities to be performed with them are associated with a risk that cannot be brought below an acceptable risk with the core requirements only ....'
>
> '... *Maximum containment measures*: A set of highly detailed and stringent risk control measures ... that are considered necessary during laboratory work where a risk assessment indicates that the activities to be performed pose very high risks to laboratory personnel, the wider community and/or the environment, and therefore an extremely high level of protection must be provided. These are especially needed for certain types of work with biological agents that may have catastrophic consequences if an exposure or release were to occur...'

For the first time, the WHO added to the Laboratory Biosafety Manual a collection of subject-specific monographs including the following topics: risk assessment, laboratory design and maintenance, biological safety cabinets and other primary containment devices, personal protective equipment, decontamination and waste management, biosafety programme management and outbreak preparedness and resilience. These monographs serve as valuable resources and tools for gaining expertise in these specific topics.

Now that we have discussed how biorisk management has evolved based on the different approaches that the WHO has adopted with its biosafety manuals over the years, it is also important to highlight that recently, in 2022, WHO released a new document called the *Global Guidance Framework for the Responsible Use of the Life Sciences*.

This document highlights the significance of responsible research in preventing and addressing risks in human, animal, plant and environmental domains. Managing these risks in the field of life sciences and converging technologies is a challenging task. While each country's context and requirements may differ, a unified Global Framework is advocated as an inclusive structure to manage a broad range of risks associated with life sciences research, including those arising from accidents and deliberate misuse. It focuses on responsibility and creates awareness of the dual-use potential that emerging technologies and life sciences research could have. These recommendations can be used to strengthen biorisk management systems.

Even though none of these guidance frameworks or manuals are legally binding, they are incredibly helpful as base documents for countries that do not have national regulations in place or laboratories that do not have an institutional biosafety manual or fully developed procedures to properly address biological risks.

The framework contains checklists, scenarios and case studies on biorisk governance and dual-use research, as well as a step-by-step practical implementation process.

The challenge remains now that the document has been released, to see how countries and institutions, will implement the *Global Framework, and take it into action.* Member States and other interested parties should adjust and contextualise the framework to reflect their needs and viewpoints.

13.2 Biosafety and Biosecurity Awareness

Firstly, we need to start by refreshing some key concepts and definitions, such as biosafety and biosecurity. The term biosecurity has been defined by different disciplines, so it has many interpretations. In this chapter, when we refer to biosecurity we are considering laboratory biosecurity. The WHO defined these terms, in the *Laboratory Biosafety Manual 4th Edition:*

> 'Biosafety: Containment principles, technologies and practices that are implemented to prevent unintentional exposure to biological agents or their inadvertent release.'
>
> 'Biosecurity: Principles, technologies and practices that are implemented for the protection, control and accountability of biological materials and/or the equipment, skills and data related to their handling. Biosecurity aims to prevent their unauthorized access, loss, theft, misuse, diversion or release.'

These definitions show how different biosafety and biosecurity are, but they are not acting independently, both are interconnected and they complement each other. Both are fundamental pillars for a robust biorisk management strategy.

Following the previous example of the WHO biorisk management analysis, we could also observe the evolution of biosafety by explaining the transition of these four editions of the Laboratory Biosafety Manual. However, when it comes to biosecurity, WHO has only released one biosecurity document, in September 2006. This document is called *Laboratory Biosecurity Guidance.*[6] It is a very valuable tool, however, it is currently outdated. We anticipate that WHO will release a new edition of this guidance in the near future. In Chapter 17, additional tools for biorisk management and dual-use will be provided.

This reflects that biosafety has always received more attention, which does not mean it is more important than biosecurity. Both are integral components of a comprehensive approach to managing biological risks in laboratory settings. They are designed to work together to minimise the potential for accidents, infections and intentional misuse, and ultimately promote the safety and security of laboratory personnel, the public and the environment.

The lack of awareness specifically related to biosecurity is not solely attributed to the limited availability of documents or guidelines over the years, but is also the result of a shortage of specialized biosecurity professionals. This could happen for different reasons: including the lack of biosecurity employment opportunities and the existence of only a few biosecurity degree programmes worldwide. Biosecurity education usually comes later, in forms of fellowships, master's degrees and certifications. This is why experts in this field are exposed to biosecurity as they advance in their professional careers, resulting in a predominantly non-linear career path.[7,8] Positions in laboratories are usually designed for biosafety officers, who will be responsible for overseen the whole biorisk management system, and even that system needs to include biosecurity practices, it will mainly focus on biosafety issues. The Institutional Biosafety Committee (IBC) is generally designed to address biosafety issues within the laboratory but it could also cover laboratory biosecurity.

In general, all the documents mentioned in this chapter, would allow us to implement laboratory biosecurity practices and address these gaps. However, sometimes is difficult to fully implement these recommendations, mostly in low-middle-income countries where there's also a lack of biosafety officers, IBCs, biosecurity knowledge and resources. That's why the WHO Global Framework appears in this context to provide not only guidance but also tools, values and principles; and overall, several mechanisms that will help an institution to mitigate biorisks. These tools and checklists can be tailored to enhance those areas where gaps have been identified.

Returning to the concept of biorisk management, the Global Framework has one of the most comprehensive definitions, which demonstrates the importance of biosafety and biosecurity as fundamental pillars but it also includes the oversight of dual-use research as another component:

> '... An integrated, overarching approach to address the risks associated with the life sciences research enterprise, from accidents and inadvertent actions to deliberate misuse. Biorisk management relies on three core pillars: biosafety, laboratory biosecurity, and the oversight of dual-use research. Biorisk management involves the quantitative or qualitative forecasting and evaluation of the probability of harm occurring and subsequent consequences (risk assessment), together with the identification and implementation of technologies, measures, or practices to avoid or minimize their likelihood or impact (risk mitigation) ...'

In Section 4 of the Global Framework, it addresses the 'Tools and mechanisms for the governance of biorisks'. This is a very useful section to review and unfrequently to see in other guidelines. Biorisk governance is a multifaceted endeavour that entails the pursuit of varied goals, active participation from diverse stakeholders and the utilisation of several governance tools and mechanisms. The spectrum of goals encompasses tasks such as mitigating biosafety incidents, promptly detecting biosecurity breaches and preventing the malicious exploitation of biological materials.

An extensive range of stakeholders are presented in the framework. This is an interesting point that usually was not incorporated in any other documents, ranging from individual scientists and research institutions to national governments, funding bodies and other relevant entities. Their involvement is essential for biorisk governance.

A diverse set of governance tools is deployed, including legislative frameworks, regulatory measures, comprehensive guidelines, robust oversight mechanisms, expert advisory bodies and thorough training initiatives.

The core foundation of biorisk governance rests on the three previously mentioned central pillars, which involve maintaining high biosafety standards, implementing robust laboratory biosecurity measures, and effectively regulating dual-use research activities.

Another interesting part that could be useful when it comes to biorisk management as a whole is Section 5, which provides tools to implement the Global Framework focusing on national implementation. It delineates the practical steps that Member States and stakeholders can take to operationalise the Global Framework. It is relevant to countries aiming to establish biorisk governance frameworks and those seeking to enhance existing ones, it underscores the need for tailored approaches to implement biorisk management systems, providing a structured six-step approach.

The six steps are Step 1, identify and assess risks and benefits; Step 2, describe values principles and goals; Step 3, undertake stakeholder analysis; Step 4, identify tools and mechanisms; Step 5, implement; and Step 6, review and modify. It is similar to the PDCA cycle (plan–do–check–act); a continuous loop of planning, doing, checking and acting; used previously by the WHO and other international organisations around the world, but this framework, involves into the cycle all relevant stakeholders and their roles and responsibilities in framing, identifying and managing biorisks making this an integral governance approach for biorisk management.

The importance of regular assessment and adaptability is emphasised along the cycle. It offers diverse pathways for implementation, encompassing regulations, policies, guidelines, awareness efforts and the involvement of advisory bodies. These measures address both biorisks and dual-use research from prohibition to informed governance.

Focusing on these two sections, and using the tools and mechanisms provided, could be helpful for an institution to develop or strengthen biorisk management policies and procedures.

13.3 The Role of Scientists: Tailored Biorisk Management Practices

Scientists play a critical role in ensuring the responsible and safe use of life sciences. Their expertise places them in a unique position to assess the potential risks associated with their research activities and develop tailored strategies for effective risk management. What do we understand by risk? Risk is defined as the likelihood of an undesirable event happening, in which the event involves a specific hazard or threat and will result in consequences. A hazard is something that has the potential to cause harm, and a threat could be a person or group with the intent and/or ability to cause harm. Both are not risks without a specific environment or situation.

$$\text{Risk} = (\text{consequences}) * (\text{likelihood})$$

Based on their risk assessments, scientists must create and implement appropriate measures to mitigate those identified risks. These measures may include modifying experimental procedures, using personal protective equipment or establishing specific containment protocols. All the guidelines and tools we described are useful to accomplish risk mitigation, in all the forms that risk could exist, in relation to biosafety, biosecurity and the possible misuse of research. Regular review of research protocols is needed to address any changes or modifications in laboratory practices over time and to ensure that they remain in line with the responsible and safe use of life sciences.

Training and education are crucial aspects of scientists' responsibilities. This includes understanding emergency protocols, using proper handling techniques and adhering to safety guidelines. However, scientists not only have a responsibility to learn biosafety practices, they must also acquire biosecurity knowledge and be aware of the potential misuse or dual-use of their research.

Reporting and communication are essential aspects that scientists need to be comfortable doing in their institutions. Promptly reporting incidents, near misses or violations of biorisk protocols ensures that the system and practices can be modified to avoid these events happening in the future.

Ethical considerations are another important aspect for scientists. Carefully consider the ethical implications of research, especially in areas with dual-use potential. Balancing the benefits of the research and the potential risks ensures that the final purpose of the work conducted serves the common good. Codes of conduct can support this.

Collaboration is crucial. The scientific community must work closely with biosafety and biosecurity experts, Institutional Biosafety Committees and relevant stakeholders to ensure a comprehensive approach to biorisk management. International collaboration must be promoted when it comes to sharing best practices and knowledge. However, research security practices are fundamental to ensure safe and secure exchanges. As stated by the NSPM-33 Implementation Guidance, from the National Science and Technology Council, United States Government,[9] research security is defined as:

> '... Safeguarding the research enterprise against the misappropriation of research and development to the detriment of national or economic security, related violations of research integrity, and foreign government interference...'

Scientists have a critical role in adapting biorisk management practices to the specific contexts of their research. Remembering that scientists' responsibilities do not end there, they have to ensure that they conduct research taking into account not only ethical aspects, but also security aspects, how information is exchanged, what information is shared, to what extent they know all collaborators, and always avoiding by all means any misuse of their research. This not only safeguards the integrity of their work but also contributes to the overall safety, security and well-being of the community at large.

13.4 Case Scenarios: Practical Examples

In this section, we will discuss some case scenarios to cover examples of the topics we have mentioned throughout this chapter. One point was how to evaluate risks and how to conduct a risk assessment basing our approach on the *WHO Laboratory Biosafety Manual 4th edition*.

A good example of how we transitioned to a risk-evidence approach can be explained with SARS-CoV-2: this is one example that is easy to visualise, as we all know most of the techniques that have been used to detect the virus. Using for this example, the WHO *Laboratory biosafety guidance related to coronavirus disease (COVID-19): Interim guidance.*[10] It describes:

> '... Point of care (POC), near-POC assays and antigen detecting rapid diagnostic tests (Ag-RDTs) can be performed on a bench without employing a biosafety cabinet (BSC) when the local risk assessment dictates and proper precautions are in place ...'
>
> '... Propagative work (for example, virus culture or neutralization assays) should be conducted in a containment laboratory with inward directional airflow (heightened control measures/BSL-3) ...'

If we assess the risks, different SARS-CoV-2 laboratory techniques may involve different containment levels. If we conduct an antigen test; Ag-RDT, we can work on a basic biosafety level (what we used to define as BSL-1). However, if with the same virus, SARS-CoV-2, we perform a polymerase chain reaction (PCR) test in the laboratory, we should be working at the equivalent of a BSL-2, if we do a SARS-CoV-2 virus isolation, a BSL-3 would be required, and if we are at a point that the virus needs to be manipulated, for example, performing a gain of function (GoF) technique, enhancing some characteristics of the virus, like its transmissibility, a BSL-4 lab or maximum control measures would be required.

This example showed that a risk-based approach is possible, and containment measures do not depend on a specific pathogen.

Another practical example, of one of the topics we addressed in this chapter, can be explained with the convergence of artificial intelligence and biology. This is a good example of how some technologies are converging, and we still do not know what undesirable outcomes they could have.

Scientists, in order to conduct research in a safe and secure way, need to consider the potential risks and dual-use implications of their work. In this context, the significance of fostering a culture centered around responsibility becomes particularly pronounced and indispensable. Reviewing a publication titled 'Artificial Intelligence in Biological Sciences'[11] it described:

> '... As the field of AI matures with more trained algorithms, the potential of its application in epidemiology, the study of host–pathogen interactions and drug designing widens. AI is now being applied in several fields of drug discovery, customized medicine, gene editing, radiography, image processing and medication management ...'

The benefits of the fusion of these technologies are easy to identify; however, we cannot underestimate the risks. This is why addressing the risks and potential dual-use implications is so important.

In March 2022, a group of academics released an article in Nature Machine Intelligence[12] demonstrating that an AI-powered molecule generator used by a drug development company could have a harmful dual-use. In a matter of hours, the model could create thousands of new biochemical weapons that were as dangerous as, if not more toxic than, the nerve agent VX, known as one of the most lethal agents in the history of warfare.

This is a reminder of how important the role of scientists is, to address potential dual-use concerns in these cases of cutting-edge technologies. We need to continue our efforts, to promote the culture of responsibility, as well as awareness and education related to dual-use research.

13.5 An Ongoing Cycle to Strengthen the Biological Weapons Convention

All of these efforts by the scientific community to achieve a comprehensive approach to biorisk management will help them understand the evolving biological threat landscape. This scientific knowledge cannot be left inside the laboratory, it must be shared with diplomats and policymakers. Scientists need to set long-term mechanisms to engage with the policy and diplomacy world. They need to support the development of national policies and regulations to address biorisks and mitigate the risks posed by emerging technologies. If a cycle of engagement between these two worlds could be implemented, we will be able to understand, identify and mitigate future biological threats promptly and effectively.

The Biological Weapons Convention (BWC) is a legally binding treaty that outlaws' biological weapons. If a S&T mechanism is in place, the scientific community could provide insights about what they consider the next biothreats or risks to come, and influence how the Convention addresses scientific and technological developments and its implications. This will reinforce and strengthen the BWC.

Author Biography

Ms. Mayra Ameneiros is a Research Associate at King's College London's Centre for Science and Security Studies. She is a member of the Technical Advisory Group on Health-Security Interface at the World Health Organization. Ameneiros has over 15 years of experience working at national, regional and international levels, where she has acquired expertise in the fields of biosecurity, biorisk management, biological and chemical threats and global health security. She is Deputy Coordinator and Board of Directors at NextGen, where she serves as liaison to the Global Health Security Agenda Consortium's Steering Committee and the Biological Weapons Convention NGO working group. Ameneiros is a former ELBI Fellow at Johns Hopkins University and an ACONA Fellow at Harvard University. She holds a BSc and an MSc in biochemistry, and she is a Certified Professional in Biorisk Management and Biosecurity from the International Federation of Biosafety Associations.

References

*1 World Health Organization. (2022) *Global Guidance Framework for the Responsible Use of the Life Sciences: Mitigating Biorisks and Governing Dual-Use Research*, World Health Organization, Geneva.

2 World Health Organization (Ed.). (1983) *Laboratory Biosafety Manual*, 1st ed., World Health Organization, Geneva.

3 World Health Organization (Ed.). (1993) *Laboratory Biosafety Manual*, 2nd ed., World Health Organization, Geneva.

4 World Health Organization (2004) *Laboratory Biosafety Manual,* 3rd ed. World Health Organization, Geneva.

*5 World Health Organization. (2020) *Laboratory Biosafety Manual, Fourth Edition, and Associated Monographs*, World Health Organization, Geneva.

6 World Health Organization. (2006) *Biorisk Management: Laboratory Biosecurity Guidance*, World Health Organization, Geneva.

*7 Tessa, A. *et al* (2022) *The next wave of biosecurity experts: young scientists need a better path into global diplomacy. AAAS Science and Diplomacy Magazine.* https://doi.org/10.1126/scidip.ade6807.

8 Lee, Y. J., Chen, X. *et al* (2023) The need for biosecurity education in biotechnology curricula. *Biodesign Research*, 5, 0008.

9 The White House. (2022) *Guidance for Implementing National Security Presidential Memorandum 33 (NSPM-33) on Nation Security Strategy for United States Government-Supported Research and Development*, The white house, Washington DC.

*10 World Health Organization. (2021) *Laboratory biosafety guidance related to coronavirus disease (COVID-19): Interim guidance, 28 January 2021*, World Health Organization Overview. WHO/WPE/GIH/2021.1.

11 Bhardwaj, A. *et al* (2022) Artificial intelligence in biological sciences. *Life (Basel)*, 12 (9), 1430.

12 Urbina, F. *et al* (2022) Dual use of artificial-intelligence-powered drug discovery. *Nature Machine Intelligence*, 4, 189–191.

14

Examples of Biorisk Management National Regulatory Frameworks

Dana Perkins[1] *and Lela Bakanidze*[2]

[1] *Former member of the Group of Experts Supporting the United Nations Security Council 1540 Committee on Weapons of Mass Destruction Non-Proliferation, New York, NY, USA*
[2] *EU CBRN Centers of Excellence Regional Secretariat for Central Asia, Tashkent, Uzbekistan*

Key Points

1) Science saves lives. Life sciences research and biotechnology developments aim to protect against infectious diseases that can harm people, animals, plants, the environment and the economy. Conducting such research safely and securely is as critical as the research itself.
2) The social and economic impact of the COVID-19 pandemic underscored the threat to national security posed by infectious diseases and the need for a proactive whole-of-government and whole-of-society risk management approach rooted in national-level security culture.
3) National biorisk management frameworks encompass laws, regulations, policies and guidelines based on comprehensive risk assessments, integrated into the fabric of national strategies and continuously monitored, reviewed and updated when necessary to anticipate and plan for a broad range of threats and crises.

Summary

Governments around the world face an increasingly complex landscape of threats from State and non-State actors, including those from natural, deliberate or accidental biological threats, outside or inside their countries' borders. Despite the overlapping requirements on biorisk management of international agreements such as the International Health Regulations (IHR), the Biological Weapons Convention (BWC), the United Nations Security Council Resolution (UNSCR) 1540 and the recently renewed Global Health Security Agenda (GHSA), countries' response to COVID-19 has been less than optimal which does not bode well for the mitigation of future biological threats.

Essentials of Biological Security: A Global Perspective, First Edition. Edited by Lijun Shang, Weiwen Zhang, and Malcolm Dando.

Here, we describe the national biorisk management frameworks in the United States and Georgia (with a focus on prevention, i.e. laboratory biosafety and biosecurity, import/export/transportation, genetic engineering and dual-use research oversight and the culture of responsible science) and relevant updates since the start of the pandemic.

14.1 Introduction

Life sciences research and biotechnology developments were indubitably responsible for bringing to an end the COVID-19 pandemic by delivering highly effective vaccines and therapeutics. The social and economic impact of the pandemic underscored the threat to national security posed by infectious diseases and the need for national biorisk management frameworks that enable a unified response to emerging events. Mitigating the risk that events caused by accidents, inadvertent or deliberate misuse of the life sciences may adversely affect the health of humans, animals, plants and the environment requires a web of laws, regulations, policies and guidelines based on comprehensive risk assessments, integrated into the fabric of national strategies and continuously monitored, reviewed and updated when necessary to anticipate and plan for a broad range of threats and crises. World Health Organisation (WHO) describes biorisk management[1] as:

> ... an integrated, overarching approach to address the risks associated with the life sciences research enterprise, from accidents and inadvertent actions to deliberate misuse' which 'relies on three core pillars: biosafety, laboratory biosecurity and the oversight of dual-use research.

By focusing on 'laboratory biosecurity' and not on a broader biosecurity system to ensure biological materials, equipment and technology remain safe and secure, WHO provides a 'short-sighted' approach to biorisk management which requires a proactive whole-of-government and whole-of-society engagement rooted in national-level security culture. As such, national biorisk management frameworks must include, at a minimum, the consideration of obligations imposed by the United Nations Security Council Resolution 1540 (UNSCR 1540) and the Biological Weapons Convention (BWC). Of note, these linkages are recognised by the Global Health Security Agenda (GHSA) Action Package Prevent-3 (APP3) Biosafety and Biosecurity which aims to support various international instruments and agreements, including the IHR, BWC and UNSCR 1540. Mitigation of 'biological proliferation and deliberate use threats' is, however, included in the IHR Joint External Evaluation (JEE) Tool[2], as the target for biosafety and biosecurity is described as:

> a whole-of-government multisectoral national biosafety and biosecurity system with high-consequence biological agents identified, held, secured and monitored in a minimal number of facilities according to best practices, biological risk management training and educational outreach conducted to promote a shared culture of responsibility, reduce dual-use risks, mitigate biological proliferation and deliberate use threats and ensure safe transfer of biological agents; and country-specific biosafety and biosecurity legislation, laboratory licensing and pathogen control measures in place as appropriate.

A national biorisk management cannot be described without a reference to the implementation of UNSCR 1540 which obligates all states to:

1) refrain from providing any form of support to non-State actors that attempt to develop, acquire, manufacture, possess, transport, transfer or use nuclear, chemical or biological weapons or their means of delivery;
2) adopt and enforce appropriate, effective laws prohibiting activities involving the proliferation of such weapons and their means of delivery to non-State actors, in particular for terrorist purposes, as well as any attempts to engage in such activities or assist or finance them; and
3) implement and enforce appropriate controls over related materials in order to: account for and secure items in production, use, storage or transport; physically protect them; detect, deter, prevent and combat the illicit trafficking and brokering through effective border controls and law-enforcement efforts; control exports, transits, transshipments and re-exports, along with the provision of funds and services related to such exports and transshipments that would contribute to proliferation; and penalise violations.

The UN Security Council defines biological weapons-related materials as any materials, equipment and technology covered by relevant multilateral treaties and arrangements or included on national control lists, which could be used for the design, development, production or use of biological weapons and their means of delivery.

While there are no prescribed measures for implementing the BWC, State Parties are required, in accordance with Article IV of the Convention, to

> take any necessary measures to prohibit and prevent the development, production, stockpiling, acquisition or retention of the agents, toxins, weapons, equipment and means of delivery specified in Article I of the Convention, within the territory of such State, under its jurisdiction or under its control anywhere.

The BWC Implementation Support Unit compiles a report for the Review Conferences based on submissions from States, which usually include information about the national implementation of legally binding obligations reflected in the text of the Convention but also the political commitments resulting from past Review Conferences agreements.

14.2 Laboratory Biosafety and Biosecurity in the US

In the United States, laboratory safety is governed by numerous local, state and federal regulations. Laboratorians may encounter a variety of biological hazards stemming from exposure to blood and body fluids, culture specimens, body tissue and cadavers and laboratory animals, as well as other workers, but also chemical and radiological hazards. For the purpose of this chapter, the focus is primarily on biological hazards and the federal oversight framework.

The General Duty Clause of the Occupational Safety and Health Act of 1970 (OSH Act), enforced by the Occupational Health and Safety Administration (OSHA) of the US Department of Labor, requires that employers

> shall furnish to each of his employees' employment and a place of employment which are free from recognized hazards that are causing or likely to cause death or serious physical harm to his employees.

Protection of workers from all hazards (including biological hazards) or hazardous operations is enforceable under this General Duty Clause and may apply if specific guidance is issued by non-regulatory organisations such as the Centers for Disease Control and Prevention (CDC), the National Institutes of Health (NIH) and others. As such, employers would likely need to comply with provisions from a combination of OSHA standards and, for instance, CDC guidance in order to implement a comprehensive worker protection program. The CDC guidance would be enforced via the General Duty Clause. Of note, while various states in the US may impose their own OSHA-approved OSH standards, they are required to be at least 'as effective as' the federal standards. Notably, the General Duty Clause is used only where there is no standard that applies to the particular hazard.

Other OSHA standards that apply to laboratories include:

- The Occupational Exposure to Hazardous Chemicals in Laboratories standard (29 CFR 1910.1450);
- The Hazard Communication Standard (29 CFR 1910.1200);
- The Bloodborne Pathogens Standard (29 CFR 1910.1030), including changes mandated by the Needlestick Safety and Prevention Act of 2001;
- The Personal Protective Equipment (PPE) Standard (29 CFR 1910.132);
- The Eye and Face Protection Standard (29 CFR 1910.133);
- The Respiratory Protection Standard (29 CFR 1910.134);
- The Hand Protection Standard (29 CFR 1910.138);
- The Control of Hazardous Energy standard (29 CFR 1910.147).

The Select Agents Programme was created through the Antiterrorism and Effective Death Penalty Act, 1996. The US Department of Health and Human Services (HHS) was directed at the time to establish a list of biological agents and toxins that have the potential to threaten public health and safety (select agents), procedures governing the transfer of those agents and training requirements for entities transferring select agents. The 2001 anthrax attacks led to the strengthening of the programme through the passage of the Uniting and Strengthening America by Providing Appropriate Tools Required to Intercept and Obstruct Terrorism Act of 2001 (USA PATRIOT Act, 2001) which restricted who may have access to select agents, and the Title II of the Public Health Security and Bioterrorism Preparedness and Response Act of 2002 which:

- Provided legal authority for the current Federal Select Agent Program (FSAP);
- Required security measures in addition to biosafety measures;
- Strengthened the regulatory authorities of HHS; and
- Granted comparable regulatory authorities to the US Department of Agriculture (USDA).

Title II of the Act (enhanced control of dangerous biological select agents and toxins [BSAT]) required the establishment of a list of BSAT taking into consideration the effect of exposure, the degree of contagiousness and method of transmission, availability of effective pharmacotherapies and immunisations and any other criteria determined to be appropriate by the Secretary. The law also required that the regulatory control be established in consultation with other federal departments and agencies, as well as scientific experts representing appropriate professional groups, including groups with paediatric expertise. Moreover, the list must be reviewed/republished biennially, or revised as often as needed. Currently, the FSAP which regulates the possession, use and transfer of BSAT with the potential to pose a severe threat to public, animal or plant health or to animal or plant products, is managed jointly by the CDC, HHS and the Animal and Plant Health Inspection Service (APHIS), USDA. This is a list-based regulatory programme (currently, including 68 agents). Some are HHS-only agents with the potential to affect public health and safety (with HHS as the sole authority and responsibility to regulate), others are USDA-only agents with the potential to affect animal and plant health; animal and plant products (with USDA as the sole authority and responsibility to regulate) and another group consists of 'overlap' agents which are subject to regulation by both agencies because they have potential to affect both humans and animals (requires interagency coordination). A subset of select agents and toxins has been designated as Tier 1, as they present the greatest risk of deliberate misuse with the most significant potential for mass casualties or devastating effects on the economy, critical infrastructure or public confidence.

A number of toxins are not regulated if the amount under the control of a principal investigator, treating physician veterinarian or commercial manufacturer or distributor does not exceed, at any time, certain limits. Regulations also establish a procedure by which an attenuated strain of a select biological agent or select toxin is modified to be less potent or toxic – and therefore does not pose a severe threat to public health and safety, animal/plant health or animal/plant products – may be excluded from the requirements of the select agent regulations. There are also exemptions specified for clinical or diagnostic laboratories (for example, for agents used only for diagnosis, verification or proficiency testing), products approved under a Federal Act, investigational products or for certain public health or agricultural emergencies.

The key FSAP regulatory functions and activities of HHS/CDC and USDA/APHIS involve the promulgation of the select agent regulations; providing oversight of possession, use and transfer, conducting inspections and approving registrations; approving individual access to select agents and toxins; receiving reports of theft, loss or release; taking appropriate enforcement actions; and serving as a resource on compliance with the regulations.

The select agent regulations state that entities registered to possess, use and transfer BSAT must provide site-specific information and training on biocontainment, biosafety, security (including security awareness) and incident response to FSAP-approved individuals before they enter areas where BSAT are handled or stored or within 12 months of the date the individual was approved and to non-FSAP-approved individuals (i.e. escorted visitors) before they enter areas where BSAT are handled or stored.

For FSAP-approved individuals, training must address the needs of the individual and the risks they may encounter. Such training must be based on their access or potential for access to BSAT and the scope of the work and should be designed to mitigate risks to people, BSAT security, the environment and the public, while ensuring that they can do their jobs without causing harm to themselves, coworkers, the public or the environment. In addition, these individuals should also have annual refresher training at a minimum of once per calendar year and whenever significant changes are made to the entity's biocontainment, biosafety, incident response and/or security plans, when the building had renovations or alterations, new agents and/or protocols were introduced, the security system was modified or there were regulatory changes. Training on incident response addresses how to react in emergencies and account for the hazards associated with the BSAT but also how to react to fires or natural disasters, emergency egress, air flow problems and first aid or downed coworkers. There are additional requirements for FSAP-approved individuals with access to Tier 1 agents. They should be trained on the organisational policies and procedures for reporting, on evaluation and corrective actions concerning the assessment of personnel suitability (i.e. the ongoing suitability assessments, peer – and self-reporting); training on procedures for the failure of laboratory intrusion detection systems, reporting suspicious activities and the occupational health program). In addition, training on insider threat awareness should be provided on an annual basis for all FSAP-approved personnel at Tier 1 entities. Escorted visitors (non-FSAP-approved individuals) should also be provided training prior to entry into an area where BSAT are used and/or stored and such training should address risks and hazards of the area and the BSAT and include biocontainment, biosafety, incident response and security, as well as insider threat and security awareness as appropriate.

While CDC and APHIS provide guidance to assist entities in the development of a site-specific ongoing suitability assessment programme for individuals with access to Tier 1 select agents and toxins (BSAT) to meet the requirements of the select agent regulations, the entities should develop the suitability assessment plan according to the specific conditions of their registered space. The ongoing assessment procedures described in the entity's security plan must include procedures for the ongoing suitability monitoring of individuals with access to Tier 1 BSAT. The ongoing assessment procedure includes but is not limited to annual technical, biosafety and security performance evaluations of personnel having access to Tier 1 BSAT; periodic review of Tier 1 BSAT access requirements, as determined by users' duties and responsibilities; annual evaluations as part of an occupational health programme or independent evaluation; and periodic review of criminal records and visa status.

The select agent regulations identify several categories of 'restricted persons' and if an individual falls into such a category during the ongoing suitability assessment, the individual's access to BSAT will be removed and FSAP will be notified. 'Restricted persons' are, for example, fugitives from justice, under indictment for a crime punishable by imprisonment for a term exceeding one year, an alien illegally or unlawfully in the United States, has been discharged from the US Armed Services under dishonourable conditions, is an unlawful user of any controlled substance, has been adjudicated as a mental defective or has been committed to any mental institution, is a national of a country that has repeatedly provided support for acts of international terrorism, or the individual acts for or on behalf of, or operates subject to the direction or control of, a government or official from such a country.

Despite the fact that the select agent regulations are list-based and not risk-based and they only apply to entities and people working with these agents, the programme has been successful in mitigating releases resulting in illness, death or transmission amongst workers or to the outside of a laboratory into the surrounding environment or community, as reported in 2021 amongst the 233 registered entities.

Since 1984, CDC and NIH also publish guidance on *Biosafety in Microbiological and Biomedical Laboratories* (now in its 6th edition) which can be used by FSP-registered entities in developing a biosafety plan but also by non-FSAP laboratories to implement best practices for the safe and secure conduct of work. Adherence to the Biosafety in Microbiological and Biomedical Laboratories (BMBL) is a condition of certain grant awards for recipients of funding from certain federal agencies.

The *NIH Guidelines for Research Involving Recombinant or Synthetic Nucleic Acid Molecules* (NIH Guidelines), last revised in 2019, provide further guidance on risk assessment and biocontainment provisions relating to genetic elements, recombinant nucleic acids and recombinant biological agents and toxins. Compliance with NIH Guidelines is required for federal funding of research. The guidelines describe and designate the responsibilities of institutions, investigators and its Institutional Biosafety Committees may apply to an entire research institution, even if a particular research project/experiment was not funded by NIH.

14.3 Import–Export and Transportation of Infectious Substances in the US

Biological materials that are dangerous goods or hazardous materials, such as infectious substances or select agents and toxins, must be packaged, identified and transported according to the regulations of the Department of Transportation's Pipeline and Hazardous Materials Safety Administration and the Federal Aviation Administration. Import of biological products produced in other countries into the United States for research and evaluation, transit shipment or general sale and distribution, is regulated by the Customs and Border Protection (CBP), the Department of Agriculture, HHS/CDC, Environmental Protection Agency, Fish and Wildlife Services and the Food and Drug Administration, which may require documents or permits for biological materials upon arrival at a CBP port of entry, depending on the nature and condition of such materials. For example, the CDC regulates the importation of materials and infectious biological agents capable (or reasonably expected) of causing illness in humans and vectors of human disease (such as insects or bats) and more than 2500 permits have been issued annually, primarily to laboratory facilities at government agencies and universities or to private and commercial laboratories conducting research studies or diagnostic activities.

Export controls are federal government regulations that restrict the transfer of certain materials, technology or software abroad or to non-US Persons in the United States. There are two sets of export controls that govern biological research and shipments: the Export Administration Regulations (EAR) that govern dual-use items, technologies and software with commercial and military applications and the International Traffic in Arms Regulations (ITAR) that govern materials, technologies and software specially designed for military applications. The list of controlled materials, technologies and software used in biological research, includes human, zoonotic and plant pathogens, genetic elements, genetically modified organisms, technology and equipment, closely aligned with the Australia Group's commitments.

Enforcement of import–export regulations is a daunting task especially as inspection cannot be done comprehensively. According to a presentation by the FBI for the International Working Group on Strengthening the Culture of Biosafety, Biosecurity and Responsible Conduct in the Life Sciences, biological materials were found smuggled in personal luggage of travellers in vials or glass tubes, glass containers or as 'stains' on paper.

National import–export regulations and multilateral efforts in this arena continue to impede the flow of supplies and technology to State and non-State chemical and biological weapons programmes around the world and serve to reinforce and carry out the BWC and UNSCR 1540 commitments and obligations toward national and global health security.

14.4 Genetic Engineering and Dual-Use Oversight in the US

Scientific and technological advances may provide great benefits to society, but they can also be used for malicious purposes. A national biorisk management framework is therefore required to include policies (periodically reviewed and optimised) to minimise risks and maximise benefits of biotechnology research, in addition to addressing public and stakeholders' concerns. While FSAP and the NIH Guidelines implicitly or explicitly address genetic engineering and dual-use oversight, the US Government (USG) also published policies on: (i) research with enhanced potential pandemic pathogens (ePPPs), including the White House Office of Science and Technology Policy (OSTP) Recommended Policy Guidance for Departmental Development of Review Mechanisms for Potential Pandemic Pathogen Care and Oversight (P3CO), and the HHS Framework for Guiding Funding Decisions about Proposed Research Involving ePPPs; and (ii) Dual-Use Research of Concern (DURC), including the USG Policy for Oversight of Life Sciences DURC and the USG Policy for Institutional Oversight of Life Sciences DURC. Of note, while FSAP covers all research involving select biological agents and toxins regardless of the funding source, including research conducted at private institutions and companies, P3CO and DURC policies apply only to research that is funded by the USG. In February 2022, the USG charged the National Science Advisory Board for Biosecurity (NSABB), a federal advisory committee, with evaluating and providing recommendations on the effectiveness of these policy frameworks. NSABB recommended[1] that the US Government takes 13 policy actions, including expanding oversight to non-federally funded research at institutions and private companies and the development of an 'integrated approach to oversight of research that raises significant biosafety and biosecurity concerns, including ePPP research and DURC' while clearly articulating the federal, institutional and investigator responsibilities in the assessment and identification of proposed and ongoing research.

The multiple regulations, policies and guidance are implemented at the institutional level and require a biorisk management approach to determine which regulations and federal guidance may apply to proposed research, including reviews of the knowledge and expertise of the researcher and laboratory personnel and of the proposed research in order to meet the obligations of the institution under federal regulations and guidance and to determine whether experiments can be performed at an acceptable level of safety and security by utilising risk-mitigation measures. Since institutional approaches vary based on available resources, expertise and biosafety/biosecurity cultural norms, an 'overarching federal biorisk management policy that brings together the recommendations, guidance and policies'[3] has been discussed, but it is unclear if that can be accomplished without Congressional action.

14.5 The Culture of Biosafety, Biosecurity and Responsible Conduct in the US

In September 2014, the White House National Security Council (NSC) tasked the Federal Experts Security Advisory Panel to identify needs and gaps and make recommendations to optimise biosafety and biosecurity. Recommendations were published in the 2014 Report of the Federal Experts Security Advisory Panel followed by a 2015 USG plan to implement the FESAP's recommended actions with the expectation that implementing the recommended actions will strengthen biosafety and biosecurity practices and oversight activities. The first recommendation of FESAP was to 'create and strengthen a culture that emphasizes biosafety, laboratory biosecurity and responsible conduct in the life sciences ... characterized by individual and institutional compliance with biosafety and laboratory biosecurity regulations, guidelines, standards, policies and procedures and enhanced by effective training in biorisk management'.[4]

Since 2016, the US Government's efforts to implement this FESAP recommendation materialised in an International Working Group on Strengthening the Culture of Biosafety, Biosecurity and Responsible Conduct in the Life Sciences (IWG). The IWG is co-chaired by HHS/Administration for Strategic Preparedness and Response (ASPR and USDA/APHIS) and is a forum and a community of practice comprised of representatives of governments, academia, industry and professional and international organisations, using crowdsourcing to develop guiding principles and educational/training resources to support and promote a global culture of biosafety, biosecurity, ethical and responsible conduct in the life sciences, based on the culture model and assessment methodology developed by IAEA for the nuclear safety and security culture. The Group conducts periodic webinars and shares information amongst its members and with the GHSA APP3 (Biosafety and Biosecurity) via its Community Corner monthly newsletter. The International Working Group developed a Guide to Training and Information Resources on the Culture of Biosafety, Biosecurity and Responsible Conduct in the Life Science[5] and a Culture of Biosafety, Biosecurity and Responsible Conduct in the Life Sciences – (Self) Assessment Framework[6] with an accompanying data collection tool. This framework provides a measure of the organisational culture of biosafety, biosecurity and responsible conduct to aid in the process of enhancing such culture at the local level through baseline and periodic assessments. The Framework may be used as is or may be customised in accordance with local needs. The Framework has been presented to the GHSA APP3Working Group and to the Biosecurity Working Group of the G7 Global Partnership Against the Spread of Weapons and Materials of Mass Destruction; included in the resource repository for the BWC implementation; and submitted to the WHO for inclusion, as a resource, in the upcoming revision of the WHO Benchmarks for IHR Capacities.

Sustainment and further enhancement of these efforts will be conducted under the National Biodefense Strategy and Implementation Plan for *Countering Biological Threats, Enhancing Pandemic Preparedness and Achieving Global Health* and the Biosafety and Biosecurity Innovation Initiative directed by the *Executive Order on Advancing Biotechnology and Biomanufacturing Innovation for a Sustainable, Safe and Secure American Bioeconomy.*

14.6 The Biorisk Management National Regulatory Framework of Georgia

The COVID-19 pandemic was an example of an infringement on biosafety/biosecurity posing a serious threat to the world. The pandemic had demonstrated that there are gaps in national biosafety/biosecurity in many countries. For example, the lack of awareness of basic biosafety principles was the cause of significant numbers of COVID-19 cases amongst physicians and nurses.

Georgia is a high-risk region for epidemics and pandemics, due to numerous natural foci of especially, dangerous infections and the particular vulnerability in parts of the region to natural hazards, including droughts, earthquakes, river floods and landslides, as well as security challenges arising from insufficient financial resources and the threat of terrorism. Regional instability, including border issues, can also heighten chemical, biological, radiological and nuclear (CBRN) concerns, as there is potential for a 'spill over' of unrest and an increased risk of terrorism. Since Georgia gained its independence from the Soviet Union, there were many attempts to rebuild the country's epidemiological surveillance system and network of laboratories, which, due to economic collapse, were practically fully destroyed. Moreover, Georgia, as a former Soviet Union republic, shared legislation with the other republics, including regulations on biosafety/biosecurity and for quite a long period after its independence, Georgia continued to abide by the old Soviet regulations regarding the handling of dangerous pathogens. These included:

1) The Decree of the Ministry of Health of the USSR 'Concerning Rules of Registration, Containment, Handling and Transfer of Pathogenic Bacteria, Viruses, Rickettsia, Fungi, Protozoa and others, also Bacterial Toxins and Poisons of Biological Origin' approved by the Ministry of Health of USSR, 18.05.1979.
2) The 'Instruction on Regime of Control of Epidemics while Working with Materials Infected or Suspected to be Infected with Causative Agents of Infectious Diseases of I-II Groups' approved by the Ministry of Health of USSR, 29.06.1978.

Later, biosafety/biosecurity was addressed in new Georgian legislation, particularly:

1) Law of Georgia 'on Health Care' (10.12.1997)
   - Cl.70 states: 'Providing safety for public health environment is the responsibility of the State. The Ministry of Health of Georgia elaborates, approves sanitary-hygiene regulations and norms and controls their observance';
   - Cl.72: 'Observance of sanitary-hygiene, sanitary-control regulations and measurements elaborated for avoiding negative effects on the environment or other factors on

public health that are approved is obligatory for any physical or legal body in spite of its proprietary, organizational or legal form or departmental subordination';
- Cl.77: 'Import, export, containment, transfer and work with infectious diseases and causative agents are allowed only under permission of the Ministry of Health of Georgia.'

2) Law of Georgia 'on Public Health' (No. 5069, 27.06.2007) covering:
   - 'Ensuring Biological Safety'
   - 'Restrictions on Owning, Using, Transfer, Transportation and Dismiss of Especially Dangerous Pathogens'
   - 'Dismiss of Causative Agents of Especially Dangerous Infections'
   - 'Import and Export of Causative Agents of Especially Dangerous Infections' 'Functions of Sufficient Services of the Ministry in the Field of Biological Safety'
   - 'Unified Lab System for Identifying, Surveillance and Respond on Causative Agents of Especially Dangerous Infections'

Legislation on biosafety/biosecurity, like all other legislation, rules and regulations, can be ideal and even easily applicable, but the community must be ready to follow them. Furthermore, mechanisms for implementation and enforcement must be established. One of Georgia's most important documents for dealing with especially dangerous pathogens (EDPs) is the Decree of the Minister of Labor, Health and Social Affairs on 'Adoption of Sanitary Norms for Working with Pathogenic Agents (Pathogenic Microorganisms)' (06.12.2005, registered by the Ministry of Justice [MoJ], published in the official bulletin of the MoJ). Though it preceded the Law of Georgia 'on Public Health', it echoes this legislative document. The Decree was drafted based on all abovementioned Western and Georgian documents and contains general principles that must be respected while working with EDPs. Some of the chapter titles, such as 'Risk Assessment and Biosafety Levels' (Chapter 2), 'Good Microbiological Techniques' (Chapter 4), 'High and Maximal Containment Laboratories—Biosafety levels III and IV' (Chapter 5), point to the cogent areas in this document.

Since 2007, Georgia worked to strengthen its national implementational of International Health Regulations (IHR) (2005). However, as detailed in the report of the JEE conducted in 2019, biosafety and biosecurity in the laboratories in the private and academic sectors is still less than optimal and there is a need for a comprehensive biosafety and biosecurity training program. This is particularly important since numerous diagnostic laboratories were established during COVID-19 with personnel often lacking training in biosafety/biosecurity. Such training in basic biosafety/biosecurity, as well as refreshing training were only conducted occasionally, without planning in advance and without oversight from central or regional controlling bodies under national control. Though the Georgian Biosafety Association (GeBSA) carries out series of trainings in basic biosafety and biosecurity, they are not regular, and cannot cover all laboratory workers needing such training. The JEE also noted that Georgia needs to develop a comprehensive biosecurity programme, and this is a gap particularly when it comes to a personnel suitability program.

The 'Law of Georgia on Export Control of Armaments, Military Equipment and Dual-Use Products (28.04.1998) was one of the first Georgian laws, where the meanings of terms, like 'dual-use products', 'list of products subject to export control' and 'rights on results of intellectual activity' are explained. This Law regulates the import–export and transportation of infectious substances. Cl.4d states that products undergoing export control include 'Causative agents of diseases, their genetically modified forms and fragments of genetic materials that can be used for production of bacteriological (biological) and toxic weapons according to the list of international regimes of nonproliferation'. The authorised body issues a license for import–export according to the recommendation of the Standing Interagency Commission on Military–Technical Issues of the NSC of Georgia. Later, the Decree of the President of Georgia No. 408 of 22 September 2002 on some measures for resolution of the issues of export, import, re-export and transit of dual-use items subject to export control was signed. It covers dual-use items subject to export control, which are not designed for military purposes, but can be used in developing nuclear, chemical, biological or any other weapon of mass destruction or any means of transportation thereof, shall not be freely purchased or sold and therefore, State control over them shall be effected through permits, subject to prevailing international practices (international agreement) and permits for export, re-export and transit of dual-use items shall be issued by the Ministry of Economy, Industry and Trade pursuant to the Law of Georgia on 'Export Control of Armaments, Military Equipment and Dual-use Items' and Decree No. 424 of 4 July 1998 and Decree No. 338 of 19 July 2002 of the President of Georgia, except for strategic goods, for which permits shall be issued by the Ministry based on the recommendation of the Standing Interagency Commission on Military-Technical Issues under the National Security Council.

The transportation of dangerous pathogens is regulated by Order #317/n December 6, 2005, of the Ministry of Labor, Health and Social Protection of Georgia on 'Adoption of Sanitary Norms for Working with Pathogenic Agents (Pathogenic Microorganisms)'. Its Article 28 states that '4. Transfer of PBAs [pathogenic biological agents] to those organizations, which do not been granted the right by the legislation to work with sources of specific risk-groups infectious diseases, is prohibited ...'

Aid from western countries to Georgia to strengthen its health security cannot be overestimated. Starting in mid-90s and later, the redirection of so-called 'former weapon scientists' (scientists, having expertise in different areas of making warfare agents – biological, chemical or radiological/nuclear) was implemented by International Science and Technology Center (ISTC) and Science and Technology Center in Ukraine (STCU), started operating in Moscow and Kiev respectively, but covering the whole post – Soviet space. They gave a chance to former Soviet scientists to continue working in their area of expertise but according to Western standards and norms.

Although it is impossible to say how likely it is that there will be a bioterrorist attack or biocrime in Georgia, the likelihood is growing. Protecting dangerous pathogens and toxins from theft and sabotage at the facilities where they are used and stored provides the first line of defense against both biological weapons proliferation and bioterrorism by making it more difficult for proliferators to acquire dangerous biological materials. The risk is higher when EDP research is conducted without appropriate biosafety/biosecurity training and oversight.

Since 2003, the US Defense Threat Reduction Agency (DTRA), through its Cooperative Threat Reduction (CTR) Biological Threat Reduction Program (BTRP) has improved Georgia's capacity to rapidly detect and report dangerous infections. The main achievement of this Program was the Richard Lugar Center for Public Health Research, a state-of-the-art biosafety level 3 research facility constructed by DTRA and handed over to the

Georgian National Center for Disease Control for operation and ownership in 2013. DTRA partners with a variety of Georgian Ministries and agencies. Key intergovernmental collaborators working toward public and animal health goals include the Ministry of Internally Displaced Persons from the Occupied Territories, Labour, Health and Social Affairs of Georgia, National Center for Disease Control and Public Health; and the Ministry of Environmental Protection and Agriculture's Laboratories and the National Food Agency.

Georgia also introduced the concept of One Health and integrated disease surveillance in recent years. Personnel of labs under the integrated surveillance system are trained in biosafety/biosecurity continuously, though COVID-19 pandemic showed that not all the staff in laboratories, like hospital labs, private facilities, physicians, nurses, etc. are knowledgeable of basic biosafety rules.

There were several attempts to introduce the course on dual-use bioethics issues, codes of conduct for biomedical scientists, etc. in universities, but without significant results. Several projects of the EU CBRN Risk Mitigation Centres of Excellence (CoE) were implemented in Georgia, particularly Project 3 – 'Knowledge development and transfer of best practice on biosafety/biosecurity/biorisk management' and Project 18 – 'International Network of Universities and Institutes for Raising Awareness on Dual-Use Concerns in Bio-Technology', etc., besides, numerous occasional trainings were conducted, amongst them by GeBSA, Biosafety Association for Central Asia and Caucasus (BACAC), but these efforts are not enough. Public health laboratories such as NCDC, the Lugar Centre and nine regional laboratories, are well equipped and personnel well trained. Same can be said about the Ministry of Agriculture's network of 11 laboratories across the country.

While Georgia is ranked number 40 overall of 195 countries in the 2021 Global Health Security Index (compared to the US ranked number one), the country continues to improve its biorisk management and health systems capabilities but still scores zero on oversight of dual-use research and culture of responsible science. Strengthening the biorisk management framework in Georgia requires national leaders' commitment and particular attention to laboratories outside the public sector.

14.7 Conclusion

A biorisk management national framework requires more than a one-size-fits-all approach. Countries adopt different approaches based, inter alia, on their national contexts, the type of life sciences research conducted and the established laboratory infrastructure. Even though Georgia and the United States could not be more different in terms of size, geography and governance, both countries deal with challenges of biosafety/biosecurity oversight of public and academic laboratories.

Laws, rules and regulations are only a part of the measures needed to address the whole spectrum of risk and implement international obligations related to BWC and UNSCR 1540. We agree with the conclusion from a recent article[7] that

> 'rules and procedures have no worth if people do not follow them; a culture of responsibility is essential for ensuring that people follow safety and security procedures and that they act responsibly in new or unfamiliar scenarios'.

Training and motivating life scientists to perform proactive biorisk management without intrusive external oversight may be what we need to close the persistent gaps in biosecurity.[8]

Author Biography

Dr. Dana Perkins, former member of the Group of Experts supporting the United Nations Security Council 1540 Committee on Weapons of Mass Destruction Non-Proliferation. Throughout her career, Dr. Perkins worked for the US Government as an Export Licensing Officer with the US Department of Commerce, as a contractor supporting the Defense Threat Reduction Agency (DTRA)'s Biological Weapons Proliferation Prevention Program, and as a Senior Science Advisor with the Administration for Strategic Preparedness and Response, US Department of Health and Human Services. She is also a Microbiologist with the rank of Colonel in the US Army Reserve.

Dr. Lela Bakanidze, professor, RBP, Georgia. For more than 30 years Professor Lela Bakanidze has worked at National Center for Disease Control and Public Health (NCDC) of Georgia. She was the Head of Department of Biosafety and Threat reduction and Head of NCDC Bioethics Committee. Lela Bakanidze is the Vice-President of the Biosafety Association for Central Asia and Caucasus (BACAC). Since 2016 prof. Bakanidze works in Tashkent, Uzbekistan as an Expert for On-Site Technical Assistance to the EU CBRN CoE Initiative Regional Secretariat for Central Asia. Lela Bakanidze is the author of two monographs and more than 80 scientific papers.

References

In each chapter of the book, a small number of key references that would be most useful for the reader to follow up are marked with a star*.

*1 World Health Organization. (2022) *Global Guidance Framework for the Responsible Use of the Life Sciences: Mitigating Biorisks and Governing Dual-Use Research*, World Health Organisation, Geneva.

2 World Health Organization. (2022) *Joint External Evaluation Tool: International Health Regulations (2005)*, 3rd ed., World Health Organisation, Geneva.

*3 Congressional Research Service. (2022) *Oversight of Gain of Function Research with Pathogens: Issues for Congress*, United States Congress, Washington, D.C.

4 U.S. Government (2015) *Implementation of the Recommendations of the Federal Experts Security Advisory Panel (FESAP) and the Fast Track Action Committee on Select Agent Regulations (FTAC-SAR), Washington, D.C.*

*5 International Working Group on Strengthening the Culture of Biosafety, Biosecurity, and Responsible Conduct in the Life Sciences (2021) *A Guide to Training and Information Resources on the Culture of Biosafety, Biosecurity, and Responsible Conduct in the Life Sciences.*

6 International Working Group on Strengthening the Culture of Biosafety, Biosecurity, and Responsible Conduct in the Life Sciences (2020) *Culture of Biosafety, Biosecurity, and Responsible Conduct in the Life Sciences-Self-Assessment Framework*.

***7** Dana, P. *et al* (2019) The culture of biosafety, biosecurity, and responsible conduct in the life sciences: a comprehensive literature review. *Applied Biosafety Journal*, 24 (1), 34–45.

***8** Daniel, G. *et al* (2023) Motivating proactive biorisk management. *Health Security*, 21 (1), 46–60.

15

Lessons from ePPP Research and the COVID-19 Pandemic

Nariyoshi Shinomiya

National Defense Medical College, Tokorozawa, Japan

Key Points

1) To learn how gain-of-function (GOF) studies that create enhanced potentially pandemic pathogens (ePPPs) have been conducted and understand their risks and problems.
2) To learn about the latest developments in GOF research, including discussions at the National Science Advisory Board for Biosecurity (NSABB), and what measures are currently being taken to address these issues.
3) To understand how recent scientific and technological developments are related to the creation of ePPPs, in relation to the COVID-19 epidemic.
4) To review how scientists should consider research, including GOF research, that uses pathogens with pandemic potential or that may pose a threat to our society, and how can they conduct ethical and safe research and contribute to the benefit of society?
5) To stress the need for education for researchers and the importance of transparency and traceability of issues related to research as a way of governance regarding pathogen research, including GOF research, which has the potential for misuse and abuse.

Summary

Advances in life science technologies have brought about major changes in the study of microorganisms and infectious diseases. In particular, the introduction of reverse genetics, genome analysis technology and synthetic biology, which have shown rapid progress since the beginning of the twenty-first century, has made it possible to design microorganisms at the genome level and artificially synthesise nucleic acids based on this design, thereby establishing a research style that leads to the creation of pathogens with actual infectious properties. Specifically, research to change the host range of pathogens and manipulate their infectivity has developed, in particular on influenza

Essentials of Biological Security: A Global Perspective, First Edition. Edited by Lijun Shang, Weiwen Zhang, and Malcolm Dando.

virus research, and this area of research has come to be known as gain-of-function (GOF) research. The range of pathogens handled has then expanded to include the severe acute respiratory syndrome (SARS) and middle east respiratory syndrome (MERS) viruses. Pathogens that humans have never encountered before could lead to a global spread of infection, or pandemic, once an outbreak occurs. We are in the midst of a COVID-19 pandemic, and it is imperative that we avoid similar pandemics in the future from enhanced potentially pandemic pathogens (ePPPs) derived from GOF research. In this chapter, we learn the lessons that concerns stakeholders around the world have about the risks associated with GOF research that creates ePPPs, what measures have been developed as a result, and deepen our understanding of what discussions are currently taking place. Then the ethics that researchers conducting pathogen research should have and the governance of GOF research that could create ePPPs are discussed.

15.1 Advances in Life Science and Technology and the Emergence of 'So-Called GOF Studies' to Create ePPPs

While advances in life science and technology have the potential to enrich our society and revolutionise medicine through the development of the biotechnology industry, they also have undesirable aspects, including misuse and abuse that could provide new tools for the development of biological weapons or become a source of environmental destruction and new pandemics. This is called a dual-use dilemma and many researchers have already recognised that research that could cause such a dilemma is called Dual-Use Research of Concern (DURC). However, it is still not enough to understand this issue. We must also be able to constrain misuse by translating understanding into concrete considerations in actual research and formulate effective countermeasures. To this end, it is necessary to first learn about the specific concerns and problems that have emerged from the progress of life science and technology and how they have manifested themselves in actual research.

Looking back from the perspective of concerns about the development of biological weapons, when the Biological Weapons Convention (BWC) was agreed upon in 1972 (opened for signature), biotechnology-related technologies, including recombinant DNA technology, were still considered to be nascent technologies and advances in life science technology itself were not considered to pose a major threat immediately. Therefore, the main focus of discussion was the handling of highly pathogenic microorganisms and highly lethal toxins, their mass production and the control of their production as weapons. Shortly thereafter, however, genetic manipulation technology, as typified by recombinant DNA technology, began to develop at a rapid pace, and concerns were raised, first from the standpoint of scientific research, about how to control such dangers that could threaten society. This was manifested in the form of the so-called Berg letter of 1974,[1] which declared a moratorium on two types of experimental research for the time being: the introduction of antimicrobial resistance genes and toxin-producing genes (Type 1) and the introduction of genes such as oncogenes and animal viruses (Type 2). The moratorium was immediately followed by the Asilomar Conference of the following year, which formulated a framework of concepts necessary for the safety of recombinant DNA experiments. This has continued to the present day in the form of biosafety.

In the 1990s, however, recombinant DNA experiments shifted to the manipulation of microbial pathogenicity, and it became apparent that the concept of biosafety was an effective measure to ensure safety in conducting experimental research, but it lacked a perspective on DURC. Specifically, concerns were expressed about research results from Russia, such as the creation of a tularaemia bacterium that produces the bioactive substance β-endorphin[2] and the production of a modified anthrax vaccine incorporating the haemolytic toxin Cereolysin AB[3] and suspicions was raised that these research results might be part of an offensive biological weapons development program conducted in the former Soviet Union. A decisive influence on these developments came in 2001 when an Australian research group reported on the creation of a genetically engineered mousepox virus that showed that an IL-4 gene incorporated to enhance antibody production worked to disable existing vaccines.[4] Even if these studies were not intended to develop bioweapons, the result means that the genes of microorganisms can be modified to enhance their virulence. Naturally, if anyone attempted to use these results for bioweapons development, we would have entered a new technological phase that goes beyond the old development scheme (see Chapter 6).

The rapid development of technologies to artificially create microorganisms has occurred in parallel with these recombinant DNA experiments. One example is the technology to create influenza viruses using reverse genetics, and another is synthetic biology, which is the artificial synthesis of genes from scratch at the DNA level and leads to the creation of modified or even novel living organisms. The former is a technique named 'reverse' genetics because it is based on the idea of investigating what kind of trait changes are generated by differences in designed and created genes, as a reverse flow of genetics that analyses genetic differences based on changes in traits or phenotypes. The latter was driven by synthetic biology, the development of DNA artificial synthesis technology, which made it possible to create long-stranded DNA with the desired sequence at low cost and without error. This technology became universally recognised in 2002 when an infectious poliovirus was created *in vitro* from DNA that had been completely synthesised from genetic information alone.[5]

It is quite natural that the introduction of these genetic manipulation technologies will be used to elucidate the pathogenicity of influenza, which has plagued mankind through numerous pandemics in the past. Influenza viruses belong to the family Orthomyxoviridae, and research on influenza A viruses, the source of pandemics, is particularly important to public health. The gene of this virus is composed of eight segments, and it is known that mutations in one segment of the virus or replacement of a segment with that of another virus, result in a major change in virulence, which then becomes a new type of influenza virus and can cause a pandemic. To prove this experimentally, it is better to artificially create new viruses with genetic manipulation of the segments and examine changes in their virulence by infection of cultured respiratory epithelial cells and disease occurrence in experimental animals. Before 2000, the creation of recombinant influenza viruses required the use of helper viruses to assist in the functions necessary for gene expression. However, the method developed by Neumann et al. in 1999 eliminated the need to use such viruses,[6] allowing the direct creation of target viruses from nucleotide sequence information and the systematic analysis of their functions. Such advances in reverse genetics techniques led to the artificial reconstruction of the Spanish flu virus in 2005.[7] Spanish flu is the well-known influenza that caused a pandemic in 1918, but at that time virology was still in its immature phase, the virus had not been isolated, and no laboratory in the world had a sample of the Spanish flu virus, let alone its genetic information. Using the specimens of past infected victims as a clue, the scientists analysed the genetic information of the virus and created the Spanish flu virus from the reconstructed genetic information alone, using reverse genetics technology. This was the first study in the world to artificially revive a pandemic virus.

Based on these technologies, research on the creation of live viruses with altered infectivity, virulence, host susceptibility (host range), etc. by intentionally manipulating some genes of the virus has become widespread. This type of research is called GOF research because the viruses created have acquired new functions that did not exist in previous viruses. The term 'gain-of-function' was originally used to refer to the addition of new functions in genetic engineering in general, and as an antonym for 'loss-of-function', it is not restricted to the field of pathogen research. However, the influenza virus GOF research conducted in 2011 (discussed in detail in the next section) has raised concerns about its potential use as a misused technology for terrorism and bioweapons development, as well as its potential to accidentally spark a pandemic through infection during experiments or leakage from laboratories. GOF research has become an established term in the field of biosecurity, especially in the context of research conducted against PPPs such as influenza, MERS and SARS. Now that we have experienced COVID-19, the term 'enhanced potential pandemic pathogens (ePPPs)' has come to be used to describe pathogens created in GOF research, meaning that they have been artificially created to enhance pathogenic function, rather than simply being called PPPs. However, in consideration of the fact that the impact of a pandemic like COVID-19 would be enormous if a large number of people were infected, the term 'ePPP' is now used to include pathogens that are highly infectious or contagious even if their lethality is not very high.

15.2 Controversy Surrounding GOF Studies on H5N1 Highly Pathogenic Avian Influenza Virus

The first human case of H5N1 highly pathogenic avian influenza was reported in 1997 at a live bird market in Hong Kong.[8] Since 2003, human cases have been reported sporadically in East and Southeast Asia, as well as in the Middle East and Egypt. The total number of infected persons is now more than 870, and the fatality rate in these cases exceeds 50%. In addition, more than 1500 people have been infected with H7N9 avian influenza since 2013, mainly in China, and the fatality rate has risen to nearly 40%. Most of these cases were caused by close contact with live birds in the marketplace, and no human-to-human transmission has been officially confirmed yet. However, it has been pointed out that if such a situation becomes widespread and certain mutations accumulate in the avian influenza virus, a virus that becomes infectious from human to human may emerge, leading to infection of the surrounding community and possibly even a pandemic.

What kind of mutations must accumulate in a virus before it becomes infectious to humans? To answer this question, influenza virus researchers have analysed the genetic information of viruses that have infected humans and examined the differences between viruses infecting birds. However, examining past cases of human infection alone does not provide information on the mechanism of human-to-human transmission, even though it provides information on viruses that multiply in the human body and consequently worsen disease and become lethal. Therefore, the GOF research methodology was introduced as one of the analytical approaches to this problem.

Ron Fouchier et al. at the Erasmus Medical Center in the Netherlands intentionally introduced a genetic mutation into the gene of the H5N1 highly pathogenic avian influenza virus that is frequently observed in human infection, inoculated the virus into the nasal turbinate of a ferret (a kind of weasels), collected tissue from the infected ferret and inoculated it into another ferret. This was repeated five times to allow the virus to multiply and acclimate in the ferrets' body, thereby increasing infectivity and proliferation in the ferrets. Furthermore, the virus was used to simulate infection by sneezing through nasal secretions, which was also repeated five times to passively infect another ferret, and finally, an airborne virus was successfully obtained.[9] In other words, the virus, which was infectious only to birds, was converted into a virus that can infect and spread among mammals, namely ferrets. The virus that can be airborne among ferrets was found to have characteristic mutations in five genes. On the other hand, Yoshihiro Kawaoka at the University of Wisconsin-Madison in the United States created a reassortant (gene reassortment) virus using the H1N1 novel influenza virus that spread in 2009 as a backbone, replacing the segment carrying the hemagglutinin gene with the H5 of the highly pathogenic avian influenza virus. Similarly, this virus with four mutations in H5 HA from an H5N1 virus origin was shown to cause droplet infection.[10] These two studies were conducted in 2011 and submitted to the *Science* and *Nature* journals, respectively, in the fall of the same year.

The editorial boards of both journals raised biosecurity concerns about these studies, which led to a biosecurity review in addition to the usual peer review of the manuscripts, which was referred to the US National Science Advisory Board for Biosecurity (NSABB) under the umbrella of the National Institutes of Health (NIH). The NSABB recommended that some of the content of the paper be redacted because it related to a virus creation method that could be misused. This became a major issue that shook the very foundations of research in the research community. In other words, two ideas clashed head-on: whether to protect the originality of research from the perspective of research autonomy and freedom, or whether to establish the necessary governance structure, even to the extent of limiting research freedom, from the perspective of safety and biosecurity for society.

In January 2012, influenza researchers declared a 90-day voluntary moratorium (on the part of research related to animal infection experiments), during which they proposed to induce discussion among related stakeholders. On the other hand, they also argued that such avian influenza research, including GOF research, is important and urgent in order to prepare for a pandemic, and must be pursued as soon as possible. Influenza researchers held an emergency meeting at World Health Organisation (WHO) in Geneva in February of the same year, where the importance of this study was reaffirmed. Meanwhile, the NSABB held a meeting in March with Fouchier and Kawaoka in person, at which they reconfirmed the contents of these studies and discussed how to handle them in the publication. After deliberations, the final conclusion was that the Kawaoka paper was unanimously approved for publication, and the Fouchier paper was also approved for publication by majority vote, although it was a split decision. Accordingly, both papers were published in May and June of the same year.

However, this did not end the debate over the nature of research that began with the GOF study of highly pathogenic avian influenza viruses. Despite the publication of the papers, the moratorium on the experiment was to remain in place beyond 90 days, causing frustration among influenza researchers. Meanwhile, researchers and policy implementers, who see the dangers of GOF research, called for continued discussion, arguing that risk management was inadequate. Against this backdrop, the US Department of Health and Human Services (HHS) hosted a workshop at NIH in late 2012 entitled 'Gain-of-Function Research on Highly Pathogenic Avian Influenza H5N1 Viruses: An International Consultative Workshop' to comprehensively discuss the state of GOF research. During the workshop, a risk–benefit approach to GOF research using H5N1 avian influenza viruses was discussed, and four hypothetical case studies were also discussed. The final conclusion of this discussion was presented as 'Criteria for Guiding Funding Decisions by the US Department of Health and Human Services for H5N1 Gain-of-Function Research Applications (Table 15.1) in *Science*'s Policy Forum in March of the following year, with seven requirements.[11] To summarise, when research is funded by public research funds, certain standards are necessary to ensure safety, publicise the research and manage the progress of the research. The report also recommends that no blanket research design is acceptable, that the research design must be justified, and that alternative experimental techniques to avoid risks should be explored to the greatest extent possible.

Table 15.1 Criteria for guiding HHS funding decisions for certain H5N1 gain-of-function research proposals.

Description in the paper	Concept
• Such a virus could be produced through a natural evolutionary process.	Justification of experimental design
• The research addresses a scientific question with high significance to public health.	Contribution to Society
• There are no feasible alternative methods to address the same scientific question in a manner that poses less risk than does the proposed approach.	Justification for creating ePPP
• Biosafety risks to laboratory workers and the public can be sufficiently mitigated and managed.	Biosafety Assurance
• Biosecurity risks can be sufficiently mitigated and managed.	Biosecurity Assurance
• The research information is anticipated to be broadly shared in order to realise its potential benefits to global health.	Openness of science
• The research will be supported through funding mechanisms that facilitate appropriate oversight of the conduct and communication of the research.	Appropriate management of research progress

At the author's discretion, annotations are added to the original table.

The discussion on GOF research seemed to have settled down for the moment, but in the wake of the possible exposure of 75 employees to anthrax at US Center for Disease Control (CDC) in June 2014 and the discovery of several vials of smallpox virus that were not supposed to be stored at NIH in July of the same year, the discussion on GOF research was reassessed from the perspective of biosecurity. In October of the same year, the White House suspended the GOF study and ordered that its safety be re-examined and the results reported by the end of the following year (2015). Eventually, the study took further time due to protracted discussions and was reported by the NSABB in May 2016 in the form of 'Recommendations for the Evaluation and Oversight of Proposed Gain-of-Function Research'.[12] In it, many of the same arguments are repeated as before, but a few matters deserve special mention: 'There are many forms of GOF research, not all of which have the same level of risks, and only a small fraction of these studies require oversight,' 'The current oversight regime should be used well, and while there are many feasible issues, the current framework is not always sufficient,' and 'Consideration should be given to ensure that oversight is carried out regardless of the funding source'.

15.3 COVID-19 and GOF Studies on SARS-like Viruses

The COVID-19 pandemic had a major impact that radically changed the nature of society. In particular, until vaccination was initiated in late 2020, many cities around the world implemented lockdowns with penalties, and private rights were severely restricted under the banner of preventing the spread of infection. While economic activity, particularly in the travel and food service industries, fell sharply, information and communication technologies using internet rapidly gained increased social status, and remote activities increased. Although COVID-19 is an emerging infectious disease in our society, even so, the loss of so many lives is unbearable. However, if one were to hypothetically consider that this was planned as part of a bioweapon development program carried out as an act of bioterrorism, or that a leak from a laboratory had caused the spread of infection, such a situation would be unacceptable. Therefore, for similar research, including GOF research, in which such risks are possible, it is essential to examine the content of the research, strictly control the state of research progress, reduce risks and select alternative research methods and provide useful information on the return of research results to society. In some cases, measures, including the postponement or discontinuation of the research should be considered.

Looking at GOF research other than influenza research, particularly research using SARS and SARS-like viruses, it is difficult to say that sufficient governance has been established. Most of the discussions on GOF research started in 2011 with the Fouchier/Kawaoka papers, and the governance of research has been heavily focused on influenza research. In other words, even with the White House's injunction on GOF research in 2014, only research using influenza viruses had been officially recognised as falling under the category of GOF research up to that point, and research using the SARS virus or SARS-like viruses to manipulate infectivity was not recognised as clearly GOF research and was proceeding virtually without governance. While the study of changing the host of infection from birds to mammals in H5N1 highly pathogenic avian influenza research was recognised as having a great impact, the study of manipulating the spike protein gene of a SARS-like virus to increase its affinity for the human ACE2 receptor was already progressed to a certain level as of 2015. Yet there was no particularly active management of the progress of the study. In this study, a chimeric virus expressing the bat coronavirus spike protein in a mouse-adapted SARS-CoV backbone was created, which was efficiently replicated in human airway epithelial cells.[13] It should also be noted that there are many GOF studies that are ongoing in a manner that is not apparent to the third parties, without being manifested as in this case. In general, one imprecise criterion for determining whether a study is a GOF study or not is the degree of enhanced infectivity or transmissibility. In other words, it is quite unclear which level of changes as to the virus infectivity is considered as GOF. From this perspective, there is an urgent need to clearly define the evaluation axis.

Another aspect of GOF studies of SARS-like viruses that must be considered is the justification and necessity of such studies. Researchers have argued that the creation of ePPPs through GOF studies will provide information on the properties and genetic mutations of the virus, which will be useful for the following three purposes: (*1*) detection of early outbreaks through surveillance, (*2*) elucidation of pathogenic mechanisms and creation of therapeutic agents and (*3*) usefulness as a vaccine strain. Unfortunately, however, in light of the COVID-19 experience, there is no evidence anywhere that the GOF study has shown significant advantages in these three areas. (1) Regarding early detection of outbreaks by surveillance, there is no clear surveillance function in areas where emerging infectious diseases occur. Sequencing and matching to existing genome databases are the most effective means of detecting outbreaks. In addition, the probability that an ePPP generated by the GOF study will match an outbreak virus has not been proven at all. (2) In the elucidation of pathogenic mechanisms and the development of therapeutic agents, clinically useful evaluation of pathogenicity and efficacy testing of therapeutic agents is not possible without the use of actual prevalent virus strains themselves, therefore, clinical approval for use of the developed agents cannot be obtained. (3) As for the usefulness of the ePPP as a vaccine strain, similar to the reason (2) above, if the strain differs from the strain actually prevalent, its usefulness as a vaccine cannot be guaranteed, and moreover, with the advent of mRNA vaccines, there is no evidence that ePPP is superior in terms of speed, efficacy or mass production capability as a drug. Thus, GOF research has not been able to demonstrate clear advantages, albeit with some biosecurity risks. Some researchers have cited the short duration of GOF studies as a reason for this, but the dramatic improvement in genome sequencing capabilities, advances in computerised molecular design analysis and the advent of mRNA vaccine technology have only emphasised the risk aspect of GOF research.

15.4 Ongoing Discussions at the NSABB and Governance by HHS

As indicated in the previous section, it is difficult to say that the governance of GOF research has been functioning properly at this time. In particular, there are major problems in the management of the progress of research after the grant review has passed and the research has started, and new risks that emerge during the course of experiments have not been properly assessed. In addition, the flow of research funds is not well understood in terms of what kind of funds are provided to joint research facilities and what kind of experimental safety controls are in place beyond that point. Two major problems have emerged. One is the lack of transparency on the part of researchers in terms of what kind of research they are conducting and what results and problems are emerging as a result. In other words, how timely are the risks and problems that are foreseen in the research plan made visible? In addition, it is not clear what oversight measures should take in the event of unforeseen safety or biosecurity problems. Another issue is traceability: how can those who manage the progress of research follow the progress of researchers' research, and once a problem occurs, can they establish a system for retrospective verification?

After a three-year blank since 2017, the NSABB resumed meetings in 2020. Since then, the NSABB has considered philosophical and operational issues from its May 2016 report (see above) and deliberated on US government policy regarding GOF research to create ePPPs. Symposiums were held and a report titled 'Proposed Biosecurity Oversight Framework for the Future of Science'[14] was compiled after also referring to public comments. The main points of the report are as follows:

- The current definitions of PPP and ePPP are too narrow. So, amend USG policy to clarify the definition of a PPP.
- Remove current blanket exclusions for research activities associated with surveillance and vaccine development or production.
- The responsibility for assessing the applicability of the experimental effects should primarily rest with the investigator and institution.
- Develop principles and guidelines that can be applied and implemented to ensure the removal of unnecessary risks.
- Take additional steps to increase transparency in the review process.
- Continue to facilitate sharing of experiences and best practices regarding DURC policy implementation.
- Renew commitments to international engagement and efforts to harmonise and strengthen international norms, standards, education and training related to the biosafety and biosecurity oversight of ePPP research.

(At the author's discretion, only those items deemed particularly important were excerpted.)

Following the NSABB's submission of this report, NIH, HHS and its interagency partners, including the White House Office of Science and Technology Policy and the National Security Council, will review its contents and will issue new measures for conducting GOF research. In any case, it is necessary to have an easy-to-understand tool that objectively and clearly states the risks involved in GOF research according to the nature of the experiment, without which the legitimacy of the research cannot be ensured. In addition, as the experience of the COVID-19 pandemic has shown, when a readily infectious emerging pathogen emerges, failure to control the initial few cases of infection can lead to a pandemic that spreads worldwide in the blink of an eye. Therefore, it is necessary to strengthen international cooperation, fully recognising that there are no national borders in infectious disease control.

15.5 Future Governance of GOF Research and Prospects

Based on the above description, we would like to summarise our thinking on the governance of GOF research, as it should be in the future. There are three main issues. The first is researchers' understanding of biosecurity and the ideal research philosophy, the second is the limitations of risk–benefit analysis and the third is the viewpoints required for research governance.

Firstly, it is essential that researchers involved in pathogen research, particularly infectious disease research, gain insight into biosecurity and become familiar with case studies that have caused problems in the past. Lessons learned through case studies have many implications, and learning about them has the advantage of preventing accidents and problems before they occur. Another is the importance of a code of conduct. This is a concept common to all research and development in the chemical, biological, radiological, and nuclear (CBRN) field, and it is essential to ensure the legitimacy of research objectives, safe conduct of research, appropriate use of research results, return of benefits to society and prevention of risks, as well as to induce researchers' voluntary involvement and accountability to society. Useful tools have been published to foster the responsibility of life scientists, such as the InterAcademy Panel (IAP) Statement on Biosecurity (2005)[15] and the Tianjin Biosecurity Guidelines for Codes of Conduct for Scientists (2022).[16] In addition, a number of educational materials have been developed on dual-use issues in the life sciences. Efforts should be made to make effective use of such tools and disseminate them among researchers.

Second, there is a need to review risk–benefit analysis, which has always been an essential part of the GOF research discussion. It is undeniable that risk–benefit analysis is useful in its own right when considering the implementation of GOF studies. However, existing analytical methods have resulted in a large divergence in perspectives and ideas between those who emphasise risk and those who emphasise benefits, making it difficult to draw agreed conclusions. In the numerous reports on GOF studies that have been published to date, most are a list of issues to be considered and the conclusions are often inconclusive. Therefore, in the end, it is not clear what should be done as countermeasures. In addition, the philosophy-driven approach lacks specificity as to what and how appropriate measures should be taken. In approving and implementing a research plan that involves a certain level of risk, the two requirements of 'transparency' and 'traceability' are very important in terms of ensuring safety, maintaining a level of oversight from others, responding quickly to problems and accidents and preventing the spread of damage. Adding this perspective to the previous discussion would be a first step toward better governance.

Finally, there is a question of what constitutes true research governance for GOF research. One view is that the decision from a societal perspective as to whether the risk itself is acceptable takes precedence over the freedom to conduct research. Of course, it is understandable that there are some matters where permission for experimental research should be determined depending on the degree of risk. However, at least in light of the COVID-19 pandemic situation, it is essential that, in the unlikely event that the spread of infection by a novel microorganism generated by research could be considered to cause COVID-19-like damage, such research should not be conducted for any reason whatsoever.

Experimental techniques and study designs are constantly evolving, and it is important to be willing to break out of the box and develop better research methods and frameworks. In addition, the search for alternative ways to avoid risk may lead to new innovations.

Taken together, it is important for us to consider the true purpose and significance of research, pursue what the best means are and learn from many past cases to apply the lessons learned to the future in order to create the research results that society desires. The examination of the GOF research case studies that create ePPPs poses this importance to us.

Author Biography

Dr. Nariyoshi Shinomiya, M.D., Ph.D., President, National Defense Medical College, Japan. Nariyoshi Shinomiya is the President of the National Defense Medical College, Japan (from January 2021). He has been long working in the fields of microbiology and immunology, molecular oncology and hyperbaric and diving medicine. He worked for the report on dual-use issues in science and technology as a member of the ad hoc committee in the Science Council of Japan. Regarding biosecurity issues related to the Biological Weapons Convention, utilising his knowledge and experiences in his expertise, he has been involved in education and awareness raising for research scientists to mitigate possible risks associated with dual-use research of concern.

References

In each chapter of the book, a small number of key references that would be most useful for the reader to follow up are marked with a star*.

1 Berg, P. *et al* (1974) Letter: potential biohazards of recombinant DNA molecules. *Science*, 185 (4148), 303.

2 Borzenkov, V. M. *et al* (1993) The additive synthesis of a regulatory peptide *in vivo*: the administration of a vaccinal *CLU tularensis* strain that produces beta-endorphin. *Biulleten' Eksperimental'noĭ Biologii i Meditsiny*, 116 (8), 151–153.

3 Pomerantsev, A. P. *et al* (1997) Expression of cereolysine AB genes in *Bacillus anthracis* vaccine strain ensures protection against experimental hemolytic anthrax infection. *Vaccine*, 15 (17–18), 1846–1850.

***4** Jackson, R. J. *et al* (2001) Expression of mouse interleukin-4 by a recombinant ectromelia virus suppresses cytolytic lymphocyte responses and overcomes genetic resistance to mousepox. *Journal of Virology*, 75 (3), 1205–1210.

***5** Cello, J. *et al* (2002) Chemical synthesis of poliovirus cDNA: generation of infectious virus in the absence of natural template. *Science*, 297 (5583), 1016–1018.

6 Neumann, G. *et al* (1999) Generation of influenza A viruses entirely from cloned cDNAs. *Proceedings of the National Academy of Sciences of the United States of America*, 96 (16), 9345–9350.

7 Tumpey, T. M. *et al* (2005) Characterization of the reconstructed 1918 Spanish influenza pandemic virus. *Science*, 310 (5745), 77–80.

8 Jong, J. C. *et al* (1997) A pandemic warning? *Nature*, 389 (6651), 554.
9 Herfst, S. *et al* (2012) Airborne transmission of influenza A/H5N1 virus between ferrets. *Science*, 336 (6088), 1534–1541.
10 Imai, M., Watanabe, T. *et al* (2012) Experimental adaptation of an influenza H5 HA confers respiratory droplet transmission to a reassortant H5 HA/H1N1 virus in ferrets. *Nature*, 486, 420–428.
11 Patterson, A. P., Tabak, L. A., Fauci, A. S. *et al* (2013) Research funding. A framework for decisions about research with HPAI H5N1 viruses. *Science*, 339 (6123), 1036–1037.
*12 NSABB. (2016) *Recommendations for the Evaluation and Oversight of Proposed Gain-of-Function Research*, NSABB, United States. https://osp.od.nih.gov/wp-content/uploads/2016/06/NSABB_Final_Report_Recommendations_Evaluation_Oversight_Proposed_Gain_of_Function_Research.pdf.
13 Menachery, V. D. *et al* (2015) A SARS-like cluster of circulating bat coronaviruses shows potential for human emergence. *Nature Medicine*, 21, 1508–1513.
*14 NSABB. (2023) *Proposed Biosecurity Oversight Framework for the Future of Science*, NSABB, United States. https://osp.od.nih.gov/wp-content/uploads/2023/03/NSABB-Final-Report-Proposed-Biosecurity-Oversight-Framework-for-the-Future-of-Science.pdf.
15 The Inter Academy Partnership. (2005) *IAP Statement on Biosecurity*, IAP Statement.
16 The Inter Academy Partnership (2022) *The Tianjin Biosecurity Guidelines for Codes of Conduct for Scientists.*

16

The Hague Ethical Guidelines and the Tianjin Biosecurity Guidelines

Yang Xue

Center for Biosafety Research and Strategy, Tianjin University, Tianjin, China

Key Points

1) Commonalities such as international soft law, global public goods and effective multilateralism between the Hague Ethical Guidelines and the Tianjin Biosecurity Guidelines highlight the significance of ethical guidelines and self-regulatory mechanisms to raise the safety and security awareness of relevant practitioners.
2) The Tianjin Biosecurity Guidelines will help cultivate a sense of professional responsibility and historical mission amongst practitioners, highlight the role of the scientist community in self-disciplining their own behaviour and promote the orderly participation of multiple parties in the biosecurity governance process, in order to achieve the utmost goal of preventing biological risks and enjoying full benefits of biotechnology and safe and secure progress.
3) The combination of rule-based governance based on the formal system of the Biological and Toxin Weapons Convention (BTWC) and the governance of scientists' behaviour through self-regulation and consensus to promote cooperation can serve as a reasonable path for the early construction of ethical guidelines and also maximise the restraining power of the existing implementation mechanism and compliance system of the BTWC.
4) In the longer term, we see the need for institutionalisation through the development of national rules, such as the implementation of an education programme based on the Tianjin Biosecurity Guidelines, in line with the conditions in each country.

Summary

The *Hague Ethical Guidelines* and the *Tianjin Biosecurity Guidelines*, as global public goods, attribute named after a certain city in the field of international security governance and jointly created by all parties to achieve security and common governance, coalesce the consensus on the goal of global chemical and biological security

Essentials of Biological Security: A Global Perspective, First Edition. Edited by Lijun Shang, Weiwen Zhang, and Malcolm Dando.

governance and profoundly express the governance concept of jointly addressing global issues such as biological and chemical security. This chapter describes commonalities such as international soft law, global public goods and effective multilateralism between the *Hague Ethical Guidelines* and the *Tianjin Biosecurity Guidelines* to highlight the significance of ethical guidelines and self-regulatory mechanisms to raise the safety and security awareness of relevant practitioners. The author believes, the combination of rule-based governance based on the formal system of the Biological and Toxin Weapons Convention (BTWC) and the governance of scientists' behaviour through self-regulation and consensus to promote cooperation can serve as a reasonable path for the early construction of ethical guidelines and also maximise the restraining power of the existing implementation mechanism and compliance system of the BTWC. It shows the constitution of the *Tianjin Biosecurity Guidelines* from its ideas, principles, elements and path formation, followed by a brief discussion of future challenges and efforts.

16.1 Relations Between the Hague Ethical Guidelines and the Tianjin Biosecurity Guidelines

16.1.1 Commonality in International Soft Law

International soft law is defined as standards, principles and guidelines that are not legally binding but have a significant impact on how nations behave, while not falling into the traditional categories of public international law, such as legally binding bilateral or multilateral treaties.[1] Unlike traditional international law mechanisms, 'international soft law' tends to cooperate in a progressive manner at the level of international governance, and its implementation does not rely on State coercion but is mainly a mechanism oriented by interests, concepts and norms, which helps to enhance the willingness to cooperate. Such international soft laws are used as a general term and include the full range of such documents, from aspirational statements such as the Hippocratic Oath to codes that are enforceable. For example, an ethical guideline is a formal and systematic statement of rules, responsibilities, norms and expectations for appropriate behaviour. For chemical and biological scientists, such guidelines or codes help to raise awareness of dual-use issues and social responsibility, promote best practices and reinforce norms against the use of biological (chemical) agents for terrorism or warfare.[2]

From a public perspective, such guidelines or codes focus on reflecting the interests of scientists as public subjects without contradicting the will of the State and require reflecting the characteristics of pluralism in the creation of subjects, equality in the status of subjects, openness in the formulation process and consensus in the institutional arrangement. From the normative point of view, such guidelines or codes focus on providing guidance for the behavioural choices of practitioners in the scientific field by outlining the background, proclaiming positions, establishing goals, clarifying principles and formulating measures to positively influence public subjects to behave or not to behave in a certain way and prompt them to make behavioural choices that tend to maintain public goals. From the binding force point of view, different types of guidelines or codes reflect varying degrees of binding effects. Ideal codes (codes of ethics) mainly stipulate the scientific beliefs that practitioners should adhere to for biological research, such as standards of integrity, honesty or objectivity; educational (advisory) codes that add to ideal codes by guiding research through guidelines and suggestions; and executable codes that indicate acceptable research practices, and are commonly connected with the legal system to ensure effective implementation.[3]

16.1.2 Commonality in Global Public Goods

A global public good is a good, resource, service, rule system or policy regime with real transnational externalities and non-exclusivity. In general, two main elements influence the legitimacy of global public goods supply: the ability and purpose of suppliers, and the acceptability of consumers. From a participation perspective, the *Hague Ethical Guidelines* and the *Tianjin Biosecurity Guidelines* are deeply open and inclusive in scope, opposing closed exclusivity and coalescing a normative consensus for global security governance.[4] On the one hand, the *Hague Ethical Guidelines* and the *Tianjin Biosecurity Guidelines* apply to all scientists, scientific research institutions and professional organisations worldwide; on the other hand, they are open to participation by scientists from all countries, regions and international organisations worldwide in different forms, such as video conferences, special forums and online calls, without excluding or targeting any party or inclining to any degree of technological development, insisting on win–win cooperation and opposing zero-sum games. It is not inclined to any level of technological development, insists on win–win cooperation and opposes zero-sum games. This openness to participation is consistent with Samuelson's definition of the non-competitive character of public goods in terms of consumption.

From the perspective of the beneficiaries, the globalist value of security without borders guides the Hague Ethical Guidelines and the Tianjin Biosecurity Guidelines and coalesce a consensus on the value of global security governance. For example, the Hague Ethical Guidelines and the Tianjin Biosecurity Guidelines are rooted in the international consensus on responsible scientific research and are based on the diverse practices of various countries, fully reflecting the determination of the international scientific community to further regulate and promote scientific research activities. At the same time, in order to support international authority to strengthen the basic order of stability and ensure the implementation of true multilateralism, both of them have always firmly maintained development under the Organisation for the Prohibition of Chemical Weapons (OPCW) or the Biological and Toxin Weapons Convention (BTWC), in order to coordinate the standards of conduct and concrete actions of multiple subjects and ensure the orderly conduct of multilateral cooperation. The commonality of their global public goods, jointly established by all parties to achieve security and common governance, is in line with the non-exclusive character of international public goods.

16.1.3 Effective Multilateralism

The *Hague Ethical Guidelines* and the *Tianjin Biosecurity Guidelines*, as global public goods were each named after a city where they originated in the field of international security governance and jointly created by all parties to achieve security and common governance, coalesce the consensus on the goal of global chemical and biological security governance and profoundly express the governance concept of jointly addressing global issues such as biosecurity. Both of them are rooted in the consensus of the international community of responsible scientific research and are country-based. They are also rooted in the international consensus on responsible scientific research, based on the diversity of national practices, and fully reflect the determination of the international scientific community to advance regulation and promote scientific research activities. This will help cultivate a sense of professional responsibility and historical mission amongst practitioners, highlight the role of the scientist community in self-disciplining their own behaviour and promote the orderly participation of multiple parties in the security governance process, in order to achieve the utmost goal of preventing risks and enjoying full benefits of technology and safety progress.

Due to the multiplicity and complexity of disciplines involved in ethical guidelines, governance objectives and goals, it is difficult to form a unified system of rules within the international scientific community in the short term. Therefore, the combination of rule-based governance based on the formal system of the OPCW or BTWC and the governance of scientists' behaviour through self-regulation and consensus to promote cooperation can serve as a reasonable path for the early construction of ethical guidelines and maximise the restraining power of the existing implementation mechanism and compliance system of the United Nations (UN) Conventions, so as to solve the current problem of ineffective implementation of ethical guidelines in sciences. In order to ensure the implementation of true multilateralism, the Hague Ethical Guidelines and the Tianjin Biosecurity Guidelines have been developed with a firm commitment to uphold the central roles of the OPCW and the BTWC in the development and promotion of ethical guidelines for chemical and biological scientists, respectively, in order to coordinate the standards of conduct and concrete actions of multiple actors and ensure the orderly implementation of multilateral cooperation. For example, in 2015, OPCW member states reached a resolution on the formulation of OPCW's Hague Ethical Guidelines for to promote a culture of responsible behaviour in chemical research and to prevent the misuse of chemical technologies.[5] The International Union of Pure and Applied Chemistry (IUPAC), the International Council of Chemical Associations (ICCA) and the International Chemical Trade Association (ICTA) have endorsed the Hague Ethical Guidelines to guide the responsible practice of chemistry under the norms of the Chemical Weapons Convention. The Hague Ethical Guidelines, an international model for regulating the conduct of chemical researchers, has benefited from its formulation and implementation in accordance with the principles of the OPCW so that it can be endorsed as a principle of responsible scientific research principle in chemistry.[6] Moreover, the 10 safety guidelines covering the whole process and chain of biological research and development proposed in the Tianjin Biosecurity Guidelines have been endorsed by the International Academy of Sciences (IAP), the World Health Organisation (WHO) and other international bodies and have been incorporated into the BTWC's proceedings.

16.2 BTWC Advances the Formulation of the Tianjin Biosecurity Guidelines for Responsible Scientific Research

16.2.1 Institutional Basis of BTWC for Responsible Scientific Research

Despite the advantages of international ethical guidelines, such as ease of implementation or compliance and low coordination costs, the voluntary nature of soft law, such as ethical guidelines whose implementation focus on self-management by relevant institutions and personnel, usually makes it difficult to guarantee implementation effectiveness because it lacks a strict implementation mechanism.[7] Therefore, it cannot be accomplished solely by individual actions, and it relies on bodies or platforms with executive authority. Dealing with this challenge may require a systematically organised 'web of prevention' that integrates stakeholder demands into a coherent policy and regulatory framework to prevent the unintentional or deliberate release of biological agents and toxins. The BTWC which entered into force in 1975, is the first international treaty banning an entire category of weapons of mass destruction and has irreplaceable special significance in addressing the risk of misuse of emerging technologies, preventing their weaponisation to the greatest extent possible and avoiding impediments to peaceful uses and international cooperation.[8] Since its inception, the BTWC has gradually developed a 'web of prevention' that is universally recognised by all parties to eliminate the threat of biological weapons, prevent their spread and promote the peaceful use of biotechnology. For example, since August 2007, the BTWC used the Implementation Support Unit (ISU), as a conduit to facilitate the flow of information between science and security communities. This has played an essential role in increasing awareness of the Convention and its provisions in policy, technical and public forums. The formulation of the code of conduct for biological scientists within the framework of the BTWC would additionally reaffirm the strong commitment of the State Parties to the objectives of the Convention and strengthen the efforts of all state parties in promoting biosafety education, awareness and advocacy at the domestic level. As biotechnology continues to grow into a more globalised sector, governance should be internationally coordinated through the negotiation of international agreements under the auspices of the UN or other multilateral international organisations.

In addition, the international soft law represented by the ethical guidelines is suitable for solving the problem of uncertainty of the path of international cooperation through 'appropriate international procedures' in Article 5 of the Convention because of its low cost of concluding sovereignty, strong normative experimentation, diverse subjects of formulation and flexible implementation.[9] However, due to the multiplicity and complexity of disciplines involved in the code of conduct for biological scientists as well as the goals and objectives of governance, it is difficult to form a uniform system of rules within the international scientific community in the short term. Therefore, the combination of rule-based governance based on the formal system of the BTWC and the governance of scientists' behaviour through self-regulation and consensus to promote cooperation can serve as a reasonable and practical path for the early construction of ethical guidelines and also maximise the restraining power of the existing implementation mechanism and compliance system of the BTWC, so as to solve the current problem of insufficient effectiveness of the implementation of ethical guidelines in biological sciences.

16.2.2 BTWC Advances the Development of the Code of Conduct for Responsible Scientific Research

As an enduring binding rule for the international community in the field of biological arms control, the BTWC has been shaping the institutional basis for addressing the dilemmas and problems of collective action by defining the roles of actors in State Parties, regulating their behaviour and shaping their expectations of behaviour. The 'legitimate research that could be misused', 'research that could have harmful consequences' and 'research that could be transformed into offensive military applications and raise concerns' as 'dual-use research' have accelerated the development of responsible scientific research under the BTWC. The need for an internationally coordinated set of measures to manage risks has been at the forefront of the consideration of the State Parties to the BTWC for almost two decades.

In October 2001, the UN and its Anti-Terrorism Policy Working Group recommended that the BTWC should develop a code of conduct for biologists. During the 5th Review Conference of the State Parties to the BTWC in 2002, a consensus was reached on the following intersessional work programme to discuss and promote common understanding and effective action by the State Parties on topics including 'the content, promulgation and application of a code of conduct for scientists'. The topic of the 'Code of Conduct for Scientists on the Safe and Ethical Use of Biological Sciences' was dealt with in the 2004 meetings of the Meeting of States Parties. At MX 2005, the States Parties played an active role through the submission of 35 working papers addressing the code of conduct. In the final document of the 6th Review Conference of the State Parties to the BTWC in 2006, the State Parties reached a consensus on 'recognising the importance of a code of conduct and self-regulatory mechanisms to raise biosecurity awareness amongst relevant practitioners and called upon the State Parties to seek to prevent misuse of bioscience and biotechnology research that may be used for purposes prohibited by the BTWC through the development of a code of conduct'. During the 2017–2020 intersessional process, the Netherlands, the United States and Japan made recommendations on the purpose, obligations and procedures of the code of conduct in the context of their respective domestic practices. In 2008, a special topic was established and dedicated to monitoring, education, awareness raising and the adoption and/or development of a code of conduct to prevent potential misuse of BTWC-prohibited substances in the context of advances in bioscience and biotechnology research during the BTWC States Parties meeting. The Experts Meeting at the BTWC in 2012 established 'Voluntary Codes of Conduct and Additional Measures to Encourage Responsible Conduct by Scientists, Academia and Industry' as one of the significant topics. However, progress in developing a code of conduct for scientists under the Convention has been limited, hampered by a mismatch between the rapid pace of technological shift and the relatively low pace of multilateral negotiations on the Convention, making it difficult to develop a unified consensus amongst all Parties.[10]

During the 2015 Meeting of the Parties to the BTWC, the Chinese government presented a working paper on the development of a model code of conduct for biologists under the BTWC. The Chinese government also proposed that States Parties should reintegrate the topic 'into the Eighth Review Conference of the States Parties and its intersessional process'. Meanwhile, the working document 'Code of Conduct for Biologists (Model)' developed by the Tianjin University Center for Biosafety Research and Strategy (TJU-CBRS) was formally submitted to the 8th Review Conference of the State Parties to the BTWC in 2016, jointly by the Chinese and Pakistan governments as Working Paper 30 of the BTWC. Subsequently, the State Parties reached a critical consensus to 'encourage the promotion of a culture of responsibility amongst relevant national professionals and the voluntary development, adoption and promulgation of codes of conduct'.

Later, these efforts were expedited in the 2017–2021 interdepartmental work programme of the 9th BTWC Review Conference. The actions included:

a) Recommendations made during the 2017 Annual Meeting of the State Parties to the BTWC to strengthen the scientific and technical review process of the BTWC, such as the establishment of an appointed review body and the formulation of a code of conduct for biologists;
b) The Chinese government and the ISU of the BTWC and the CBRS jointly organised an international workshop on '*Building a Global Biosafety Community with a Shared Future: Developing a Code of Conduct for Biologists*' in Tianjin, China in June 2018;
c) As a result of these discussions, in 2018 the governments of China and Pakistan submitted a revised version of the model code of conduct for discussion at the Convention's Expert Group Meeting (MX2).[11]
d) A side meeting on '*Developing a Code of Conduct for Biologists*' was also presented by a group of scientists and policy researchers from China, the United Kingdom and the Netherlands. Recently, experts from the TJU-CBRS also shared their thoughts on the urgency, significance and implementation of codes of conduct at the 2020–2022 Annual Meeting of Experts on Developments in the Field of Science and Technology Related to the BTWC (MX2).

16.3 Constitution of the Tianjin Biosecurity Guidelines: Ideas, Principles, Elements and Path Formation

16.3.1 Ideas

The community of human destiny is the greatest ideological impetus for the global supply of biosecurity public goods, such as the *Tianjin Biosecurity Guidelines*. As a global security initiative, the *Tianjin Biosecurity Guidelines* guide each country to consciously align its own biosecurity with global biosecurity and its own biotechnology development with that of the rest of the world. The threat of biosecurity risks the survival and development of humanity and transcends national sovereignty. Therefore, the corresponding global biosecurity public goods should also be set up around how to strengthen the shared responsibility for security, common prevention of biological risks and common sharing of biological science and technology achievements and security achievements amongst all parties. The public goods thus designed must have both moral and practical values and help build consensus in the international community to address global challenges. On the one hand, the code of conduct for scientists, which has been repeatedly developed by professional associations and organisations at various levels around the world, carries a certain cosmopolitan concept and implies an ideal moral vision for individuals and an innovative mechanism for international governance. On the other hand, as a public good of global security in the process of collective action, its concrete expression implies the process of international consultation and interaction in dealing with risks or providing public goods, as well as the resources and resources for human society to deal with certain common challenges. Global security public goods, on the other hand, are implicitly articulated as a process of international dialogue and engagement to address risks or provide public goods, as well as resources and programmes for human communities to address specific common concerns.

16.3.2 Principles

We need to maintain coordination between development and security. Practice shows that the critical factors inducing the risk of misuse and misapplication of biotechnology are influenced by the policy orientation of seising the high ground of scientific and technological development, while an overemphasis on prevention and resolution of risks tends to over-suppress and constrain the activities of cutting-edge and breakthrough scientific research. To this end, a strategy of classifying and grading technologies or biological toxins under the principle of maintaining coordinated development, and security is a common choice adopted by major biotechnology powers. Classification and grading management strategies have the advantage of highlighting management priorities, clarifying key risk prevention and control objectives and promoting the healthy and sustainable development of biotechnology. It is worth noting that the ability to assess the risks of biotechnology for dual use is a defining condition for its effective governance.[12] This capability currently relies heavily on the subjective judgment of experts, influenced by their personal knowledge, experience and vision. In this regard, biological scientists should adhere to a careful grasp of current controversial research directions and carefully study and judge the possible ethical and safety risks of biotechnology to minimise harm.

The principle of human orientation highlights humanity as the key governance object of the security concept. The key point in biosecurity risk management is that scientists and biotechnology risk prevention also need to be internalised in the relevant disciplines. That is, to effectively reduce the potential risks arising from the knowledge, tools and technologies required to conduct cutting-edge biotechnology research, ensure that the life sciences are used only for the purposes of peace and development and achieve sustainable development, is the historical responsibility and great mission that all relevant subjects engaged in biological research activities should assume. However, the uncertainty of biosecurity risk is considerably exacerbated by the fact that, throughout the development and application of cutting-edge biotechnology, multiple actors frequently visualise cognitive biases based on their own interests. In this regard, the Tianjin Biosecurity Guidelines not only focus on strengthening the biosafety education and safety awareness of relevant subjects in the process of biosafety Research and Development but also emphasise the sense of professional responsibility and historical mission of practitioners at the institutional level, highlighting the subjectivity and spontaneity of the scientists' group to restrain their own behaviour. It also focuses on strengthening human rationality to curb the misuse of technical capabilities and illicit acquisition of economic benefits and promoting the orderly participation of multiple stakeholders in the biosecurity governance process.

The precautionary principle requires the preventive regulation of risky behaviour 'before the danger is imminent'. Compared with 'crisis prevention', the former emphasises 'decision making before it happens' and 'prevention before it happens', and no longer relies exclusively on 'cost-benefit' analysis and utilitarian value orientation. Institutional tools should reduce the risk potential of related disciplines in the initial stages of security risk arising from disruptive scientific discoveries and technological inventions. The precautionary principle holds that uncertainty in science and technology (S&T) should not be used as an excuse to delay the adoption of measures to prevent harm or threat, arguing that whoever pursues the development of biotechnology is obliged to bear the burden of proof that there is no harm. One of the premises is that if there is a societal obligation to protect the public or the environment, that responsibility should be reduced only when research shows that harm is unlikely.[13]

16.3.3 Path Formation

Tianjin Biosecurity Guidelines integrates multiple considerations such as international rules, experience in S&T governance and the needs of multiple stakeholders, and initially forms a multi-level supply model based on the multilateral framework of the BTWC, supplemented by international organisations of scientists' associations and supported by bilateral and multilateral cooperation. To avoid overplaying the dominant role of sovereign States, the *Tianjin Biosecurity Guidelines* do not adopt the model of self-regulatory norms drawn up by State institutions and non-State public organisations with additional powers. On the one hand, the deeply consultative nature of the extensive participation of experts in life sciences, social sciences and industry is highlighted in the process of model drafting, discussion and consultation, and text polishing, which will reflect the consensus of self-regulation with scientists as the right subjects of self-creation and self-implementation in all aspects of creation subjects, formulation procedures, rule implementation and rule compliance; on the other hand, every effort is made to maintain the subjectivity of the State, the drafting institutions tried to maintain the relevance of State subjectivity and non-State actors represented by scientists, and actively promote the role of international biosafety and biosecurity think tanks embedded in the decision-making mechanism in the process of communication, mediation and endorsement.

The CBRS at Tianjin University successfully registered as a BTWC NGO in 2017 and was responsible for organising Chinese biologists, sociologists and jurists to jointly draft the 'Code of Conduct for Biologists (Model)' at the embryonic stage. China, together with Pakistan, formally submitted a working paper titled 'A Model Code of Conduct for Biological Scientists' developed by scholars at TJU-CBRS to the BTWC's 8th Review Conference in 2016 and a follow-up document was submitted in 2018. Then, in close liaison with the Chinese Ministry of Foreign Affairs and the US Department of States, the TJU-CBRS of China established a 'track II' dialogue with the Johns Hopkins Center for Health Security in the U.S. It is through the 'track II' interaction between the two major biotechnology powers of China and the U.S. that the foundation was laid for the subsequent high-quality consensus and action in the Tianjin Biosecurity Guidelines. Since then, and especially since January 2021, TJU-CBRS and Johns Hopkins Center for Health Security have been working with the Interacademy Partnership (IAP), as well as scientists from more than 20 countries on four continents, in order to reached a balance between the local interests of biotechnology systems and the public interests of global biosafety, and transformed the code of conduct for biologists into a common interest acceptable to multiple subjects through repeated interactions and practices. The IAP formally endorsed the Tianjin Biosecurity Guidelines in July 2022, which are expected to be followed by individual scientists and institutions in the IAP network.

16.4 Future Discussion

Firstly, there is a growing awareness amongst national parties that there should be a code of conduct within the framework of the BTWC for professionals in life sciences research from government, academia and industry to enhance biosecurity education, awareness and advocacy. The current trend in bioscience and technology risk governance is to establish national-level biosecurity measures. Governance tools include soft laws and informal measures such as professional standards, codes of ethics and education and awareness measures. Practitioners should adhere to scientific beliefs, and the characteristics of the *Tianjin Biosecurity Guidelines* as an ideal code are mainly reflected in the declarative provisions. Therefore, the task of implementing institutionalisation is to develop rules for implementing a code of conduct for biological scientists in accordance with the *Tianjin Biosecurity Guidelines*, in line with national conditions.

Secondly, challenges remain for the promotion, adoption and implementation of these codes. To overcome this challenge, relevant stakeholders should assume specific responsibilities for supervision, evaluation and education, which need to be further defined. Regulators should promote legislation at the national level to avoid temporary, ineffective and uncoordinated regulation and use their special organisational capacity to actively engage in dialogue and communication with stakeholders. In these dialogues, regulators, scientists and other stakeholders should openly discuss the potential risks of scientific research and propose measures that can mitigate those risks, including responsible research by scientists and sound regulatory policies by regulators. National scientific institutions and scientific societies should actively work with their national regulatory agencies to develop codes of conduct for biologists and provide independent expert assistance to government regulatory decisions, while publicly condemning any academic misconduct by domestic biologists on behalf of domestic academia. Research institutions should also develop educational and training programmes by providing a guiding framework to define the role and responsibilities of professional researchers in risk prevention and control and to assess individual performance.

Third, the current international organisations and institutions for biosecurity governance are characterised by multiple and fragmented governance, and the multidimensional international public law organisations, consisting of the BWTC, the investigation mechanism of the UN Secretary-General on the use of biological and chemical weapons, UN Security Council resolution 1540 (2004) and the WHO have not yet formed a multidimensional international public law organisation cohesive mechanism. The loss of governance authority due to the fragmentation of the global biosecurity governance mechanism has led to the loosening and confusion of 'self-regulation' in this field, and it is difficult to cooperate with each other. Ideally, the Tianjin Biosecurity Guidelines should be formally endorsed by the Review Conference of the Parties to the Convention as a consensus resolution for the States Parties to develop their national codes based on their own needs and circumstances. However, as the Convention is essentially a disarmament treaty, it can hardly serve as an effective instrument for such cooperation. In 2022, China and Pakistan, with co-sponsorship from Brazil, submitted the latest version of the Tianjin Biosecurity Guidelines to the 9th Preparatory Meeting of the BTWC.[14] If the States Parties adopt the Tianjin Biosecurity Guidelines at the 10th BTWC Review

Conference in the future, it will be a significant step forward. In this sense, the Tianjin Biosecurity Guidelines adequately highlight the original narrative as a global biosafety public good in the process of formation, construction and improvement and can transcend the traditional shackles of power struggles and geopolitical conflicts amongst sovereign states under the framework of a certain single public international law or organisation, and use the principle of common norms to bring together the fragmented while diverse participants can share the same goals and comply with common norms.

Author Biography

Dr. Yang Xue is a Senior Fellow at the Center for Biosafety Research and Strategy, Tianjin University, China. Yang has already been honoured with China's top social science award for young people. He was involved in the drafting of the PRC Biosecurity Law and other related legislation. Additionally, Yang contributed to the writing of the 'Proposal for the Development of a Model Code of Conduct for Biological Scientists' (BWC/CONF.VIII/WP.30), represented China at the 8th/9th BTWC Review Conference and the Meeting of Experts in 2017–2022. He played a key role as the main Chinese drafter of 'The Tianjin Biosecurity Guidelines for Codes of Conduct for Scientists'.

References

1 Kern, A. *et al* (2005) *Global Governance of Financial Systems: The International Regulation of Systemic Risk*, Oxford University Press, Oxford.

2 Fifth Review Conference of the States Parties to the BWC. (2002) *Final Document.* BWC/CONF.V/17, United Nations, Geneva.

***3** Rappert, B. (2004) *Towards a Life Sciences Code: Countering the Threats from Biological Weapons*, Bradford Briefing Papers (2nd Series), No. 13, University of Bradford, Bradford, UK.

***4** Tianjin University, Johns Hopkins and IAP (2021) *The Tianjin Biosecurity Guidelines for Codes of Conduct for Scientists.*

5 OPCW. (2015) *The Hague Ethical Guidelines*, OPCW, The Hague.

6 Hartmut, F. *et al* (2018) Chemical weapons: what is the purpose? The Hague Ethical Guidelines. *Toxicological & Environmental Chemistry*, 100 (1), 1–5.

7 Schuurbiers, D. *et al* (2009) Implementing the Netherlands code of conduct for scientific practice—a case study. *Science Engineering Ethics*, 15 (2), 213–231.

***8** Novossiolova, T. *et al* (2019) *Strengthening the Biological and Toxin Weapons Convention: The Vital Importance of a Web of Prevention for Effective Biosafety and Biosecurity in the 21st Century*, University of Bradford, Bradford Briefing Paper.

9 Goldblat, J. (1997) The biological weapons convention: an overview. *International Review of the Red Cross*, 318, 251–266.

10 Chyba, C. F. (2006) Biotechnology and the challenge to arms control. *Arms Control Today.*, 36 (8), 11–17.

***11** China and Pakistan (2018) *Proposal for the Development of a Model Code of Conduct for Biological Scientists under the Biological Weapons Convention*. BWC/MSP/MX.2/WP.9, 9 August 2018, Geneva.

12 Tucker, J. B. (2012) *Innovation, Dual Use, and Security: Managing the Risks of Emerging Biological and Chemical Technologies*, MIT, Cambridge, MA.

13 Forge, J. (2010) A note on the definition of 'dual use'. *Science Engineering Ethics*, 16, 111–118.

***14** China and Pakistan, Co-sponsored by Brazil. (2022) *The Tianjin Biosecurity Guidelines for Codes of Conduct for Scientists*, BWC/CONF.IX/PC/WP.10.

17

Engaging Scientists in Biorisk Management

Yuhan Bao[1,2] *and Alonso Flores*[1]

[1] *iGEM Foundation, Cambridge, MA, USA*
[2] *Tsinghua University, Beijing, China*

Key Points

1) Engaging scientists into biorisk management and the assessment of management systems are essential for its effective implementation and success, especially on the efforts to deal with dual-use risks.
2) The key issues of engaging scientists into biorisk management are mainly about the awareness raising, self-assessment, monitoring and report of research risks or hazards.
3) Engaging scientists can not only contribute to the daily governance of biorisks but also contribute to the improvement of whole risk management system.
4) User-friendly frameworks and concrete instructions for assessing risks and reporting are important tools to engage scientists and improve their awareness.

Summary

Scientists' engagement in biorisk management can not only contribute to the daily governance but also contribute to the improvement of the whole risk management system. This chapter analyses three tools and cases of biorisk management, including the IWG Assessment Framework, Dual-Use Quickscan of Netherlands and the Responsibility Program of iGEM, exhibiting the key elements for engaging scientists in governing dual-use research risks and also demonstrating how people can use specific tools and frameworks to help engage sciences in biorisk management. The awareness of biorisks, self-assessment, report and improvement are the key expected performance of scientists' engagement. To promote scientists' engagement in biorisk management, it is evident that easy tools that guide scientists into a self-assessment, the availability of 'background/reference material' or clear guidance and rules, the effective encouragement and trust within the management system, and also the supplementing arrangement will allow scientists to actively and effectively contribute to the governance of complex risks.

Essentials of Biological Security: A Global Perspective, First Edition. Edited by Lijun Shang, Weiwen Zhang, and Malcolm Dando.

17.1 Introduction: Scientists Engagement and Biorisk Management

Life science researches can bring harm to the environment and society by involving biological agents. Such kinds of harm include accidental exposure, deliberate misuse or the harm induced from dual-use information sharing. The effort to assess, prevent, mitigate or reduce these biorisks all constitute the biorisk management.[1]

Biorisk management requires a set of elements ranging from formal regulation, including institutional, national and international policies, management systems and technical standards, to informal settings such as responsible culture, especially at organisation level.[2] Engaging scientists is always important in achieving high-quality biorisk management, no matter whether implementing formal regulation or shaping the culture of responsibility. For the accidental exposure and traditional safety related issues, these are mainly regulated by formal policies and also the technical standard for laboratories and bioagents, a series of standard procedures within laboratories have been applied to the researchers. And within such a context, the scientists' compliance with relevant policies and their awareness of biosafety issues are essential for tackling these traditional laboratories biosafety risks.

However, when faced with dual-use dilemma arising within life sciences, the engagement of scientists will play a more important role, especially when dealing with emerging biotechnologies. The uncertainty and ambiguity of these life sciences make formal regulations and policies that cannot predict the cover of all the possibilities of research, while these researches might bring some more harmful pathogens even unintentionally. Therefore, scientists are expected to be more proactively reflective and responsible for the research process and also its product beyond passive compliance with policies or other ethical rules. The awareness, active action and behaviour of scientists of monitoring, assessing, reporting and reducing the potential risks arising from these researches are in a core position in dealing with the dual-use research.

Such an idea has been widely accepted from governance to academia, many policy documents and scholars have recognised that scientists should be more active in monitoring, assessing and reducing potential risks, as well as seeking possible solutions. But, it is not an easy task to engage scientists in such a more proactive and reflective manner in biorisk management. Both the motivation and ability of scientists in biorisk management are the key gaps.1 This has resulted in many explorative tools and cases of which engage scientists in the biorisk management at different levels, providing the techniques to scientists to participate in biorisk management as well as strengthening their motivation. The World Health Organisation (WHO) Global Guidance Framework for the Responsible Use of the Life Sciences: Mitigating Biorisks and Governing Dual-Use Research and the Culture of Biosafety, Biosecurity and Responsible Conduct in the Life Sciences: (Self) Assessment Framework published by International Working Group (IWG) on strengthening the culture of biosafety, biosecurity and responsible conduct in the life sciences3, the Netherlands Biosecurity Office Quickscan Series and also the responsibility programme in International Genetically Engineered Machines (iGEM) can be seen as some examples. These different cases or tools might have different purposes and application scenarios, but all involve scientists' engagement in biorisk management.

This chapter will analyse some of these tools and cases of biorisk management, exhibiting the key elements for engaging scientists in governing dual-use research risks and also demonstrating how people can use specific tools and frameworks to help engage sciences in biorisk management.

17.2 Engaging Scientists in Biorisk Management at International Level: Case from IWG Assessment Framework

Promoting biosafety and biosecurity amongst life science researchers or engaging scientists should be implemented finally at organisational level such as the research institutes, universities and other organisations, where researchers are embedded in. For achieving this goal, some international organisations or professional associations are also trying to promote the awareness of biosafety and security amongst life sciences researchers by exploring and providing more useful tools.

The IWG on strengthening the culture of biosafety, biosecurity and responsible conduct in the life sciences is such a collaborative platform for exploring the toolkits. The IWG is convened by the US Department of Health and Human Services (HHS) and the US Department of Agriculture (USDA), and it includes representatives of governments, academia, industry, professional and international organisations, including, but not limited to, INTERPOL, WHO, the World Organisation for Animal Health. One of the most important outputs of IWG is the (self) assessment framework for culture of biosafety, biosecurity and responsible conduct in the life sciences (self) assessment framework. This tool draws on the experiences and insight of guidance on nuclear safety and security culture and extends this through to the context of life sciences. This framework is developed based on thorough research about specific elements and metrics which can be used to identify and evaluate the organisational culture of responsibility.[4]

Although this framework is designed to evaluate and promote the organisational culture of responsibility, in reality, it provides a detailed guidance for engaging scientists in biorisk management from its contents and also its usage.

17.2.1 Building the Culture of Responsibility: The Key Elements and Approaches of Engaging Scientist in Biorisk Management

From its content side, although the focus of this self-assessment framework is the on culture of biosafety, biosecurity and responsible conduct, it intrinsically requires the scientist's engagement. And, the key elements of this culture listed in the framework delineate the map for engaging scientists in biorisk management.

According to its formal statement in this framework, the definition of culture of biosafety, biosecurity and responsible conduct focuses on 'an assembly of beliefs, attitudes and patterns of behaviour of individuals and organisations', these beliefs, attitudes and patterns of behaviour can 'support, complement or enhance' the formal rules, procedures and standards to prevent the occurrence of biorisks, including the dual use of biotechnologies. Such an emphasis on culture is, in essence to engage scientists in biorisks, as the beliefs, attitudes and patterns of behaviours are oriented at scientists themselves, and only with the participation of scientists can this culture of responsibility be successfully built and maintained. This is the lesson people taken from the governance of nuclear and chemical risks, where the people handling the materials and conducting research will heavily influence the effectiveness of biorisk management.

Moreover, the key elements in this framework are further mapping the keys and approaches of engaging scientists in each organisation. In the assessment framework, it specifies four elements of the culture of biosafety, biosecurity and responsible conduct in the life sciences: the management systems, behaviour of leadership and personnel, principles for guiding decisions and behaviours, beliefs, opinions and attitudes. All of the four elements require the participation and contribution of scientists at different levels.

1) Management Systems. The management systems need to prioritise the biosafety and security as a top topic within an organisation by imposing policies, processes, procedures and programmes on all individuals in the research organisation. The examples of management systems include clear roles and responsibilities for biorisks management, visible safety and security policy, performance measurement, feedback process, compliance checks and competency-based training. The framework further lists 20 items which cover the different aspects to evaluate the performance of management systems. These aspects of management systems are important to engage scientists in governing dual-use research through the following mechanisms besides the usual regulation of toxic bioagents:
   - Information and explanation: the clear roles and responsibilities, the visible safety and security policy, explanation on procedures and rules of conduct can help to strengthen all scientists' awareness and understanding in biorisk and its governance by providing the information about biosafety and security. Such a mechanism can basically lay down the cognitive foundation of engaging the scientists in biorisk management. This information is not limited to traditional lab safety but will also cover the risks of dual use, intentional release, the mitigation procedures and preparedness for emergencies. A good system for information communication also includes an established internal communication procedure to inform safety and security incidents.
   - Channel of participation and report: well-established management systems can engage scientists by providing a channel for researchers to participate in mitigating risks, including to report unusual behaviour amongst co-workers. The abnormal or unusual behaviours of researchers are the roots of many risks; researchers are in a very important position to discover and report these potential problems. This channel mechanism can engage scientist in a more active way.
   - Systematic risk benefit analysis for dual-use research: For dealing with dual-use research, it will be quite essential to develop a systematic risk–benefit analysis process in place for dual-use research. Such an analysis process will provide the opportunity of engaging scientists in assessing the specific risks of relevant studies and make decision on the following steps, such a process will put the scientist in a proactive position to forecast the potential risks of their research.
2) Behaviour of Leadership and Personnel. Leadership behaviour means the actions of leaders which are designed to foster more effective biorisk management, it emphasises the expectations, decision-making, oversight and communication about biorisk, motivation

and inspiration. Personnel behaviour is the desired outcomes of leadership efforts and operation of above management systems, it underscores the professional conduct, adherence to approved procedures/protocols, team work and vigilance. The leaders of a research organisation include the different level superiors and managers; it should also include the leading scientists such as principal investigator of laboratories. Leadership behaviour can be regarded as the supplement to management systems, this can help engage scientists by strengthening the implementation above mechanism of management system and also the following mechanism:

- Communication: the regular communication about the potential biorisk, reduction measures and expectations of performance between leaders and researchers will effectively enhance the awareness and understanding of scientists. This will also make researchers attach greater importance to these issues.
- Trust and encouragement: another key mechanism of leaders in engaging scientists is inspiration and encouragement. The scientists self-monitoring, assessment and reporting behaviour will most likely happen within the context of sufficient trust and inspiration. The leaders should take more actions to encourage, recognise and reward commendable attitudes and behaviour.
- Support: leaders also need to provide the required means and resources for researchers to implement biosafety and biosecurity measures. This can supplement the above mechanisms.

Personnel behaviour of researchers are the pillars of biorisk management no matter for traditional biosafety issues or for the security issues arising from dual-use research. Within the context of encouraging and engaging scientists in dealing with these complex risks, the behaviours of researchers are expected to be more proactive and reflective besides compliance with relevant policies and rules.

- Vigilance on biosafety and biosecurity: the risks rising from dual-use research requires researchers be more vigilant for the potential misuse and other risks. This should be valued by each organisation. So, vigilance on biosafety and security is the key metric for measuring the outcomes of engaging scientists in the framework.
- Report and self-report without fear: report of suspicions from researchers can greatly help to identify and mitigate risks at early stages, these are quite necessary for dual-use research. These include the report about any concerns or suspicions within the organisation, and unusual behaviour of colleagues which may induce risks, and also the self-report of illness or other conditions that may affect biosafety/biosecurity. All of these reports should be secured with a trusted and supportive environment, which allow researchers to be confident to tell anything they find without fear.
- Involve in the risk assessment and decision-making process of risks reduction: researchers need to take a more active role in shaping the design of research to reduce its potential risk, and they should participate in the risk assessment for all research activities. As scientists are at an advantageous position of knowing what happens in the labs, and they will have a range of tacit knowledge about research detailed process and biorisks. So, the engagement of scientists can contribute a lot to organisation decision in risk mitigation.

3) Principles for Guiding Decisions and Behaviours. The principles for guiding decisions and behaviours are the foundation for the above two elements of culture of responsibility, the management system and the personnel behaviour. These principles range from influence from leadership, commitment and responsibility, professionalism and competence, learning and improvement and maintaining public trust to codes of ethics. Many guiding principles in this framework are directly related to the expected performance of engaging scientists in dual-use researches:
   - Organisational guiding principle in reinforcing the scientists' engagement: organisations need to reinforce ethical norms and professional codes of conduct to make sure researchers are able to consider the potential risks of researches and also encourage researchers to participate in biorisk management. Focusing on improvement by collective brainstorming when an incident or near miss occurs rather than by blaming the staff. All of potential misuses by researchers are considered at all stages and take necessary actions with the contribution of scientists.
   - Personnel guiding principles in dealing the diverse biorisks: by encouragement and training, researchers are expected to be aware of the concept and implications of dual-use research of concern, cognisant of the threat of bioterrorism and biological weapons (BWs), motivated to minimise risks of misuse of science and consider the possible applications of work and the balance between the pursuit of science and ethical responsibilities to society. All of these principles will guide researchers be more proactive and reflective in daily work and biorisk governance.
4) Beliefs, Opinions and Attitudes. Beliefs, opinions and attitudes of individuals on biosafety and security can be embodied in scientists' awareness on consequences and mitigation strategies of risks associated with working in a laboratory with biological materials, compliance with relevant regulations, policies and procedures. Researchers' beliefs, opinions and attitudes for biorisk, especially dual-use research, are another important feature which can exhibit the effectiveness of engaging scientists besides their personnel behaviour.

Within this framework, researchers are expected to show their awareness of the risk of bioterrorism or an attack with a BW, attach great importance to risk assessment for identifying and reducing risks and understanding the necessity of controlling access to sensitive information for the reason of biosecurity.

17.2.2 Improving the Culture of Responsibility: Engaging Scientist in the Assessment of Biorisk Management Systems

The usage side of this assessment framework will also involve the scientists in the assessment process. This framework is designed to enable to conduct the self-evaluation for the research community, which hopes to share the understanding of how basic assumptions influence behaviours and performance, as they relate to biosafety and biosecurity, provide information on behaviours that contribute to or undermine biosafety and biosecurity and gain an understanding of the impact of culture on organisational performance.

Targeted at the above four elements of the culture of responsibility, this framework develops a series of surveys which can be answered by the researchers as well as the managers of the organisation in order to conduct the self-assessment. And, this framework also recommends that each organisation can further combine interviews, focus group discussion to collect more information. Not only the information and feedback provided by scientists will help improve the biorisk management, but these processes will actually involve scientists in the biroisk management deeply, inducing their reflection about safety and security.

And this framework also encourages that leadership and scientists' engagement at all levels should foster self-exploration and learning about biosafety and biosecurity perceptions and patterns of behaviours to benefit from the assessment process.

17.3 Engaging Scientists in Biorisk Management in National Institutional Oversight: Case from the Netherlands

Engaging scientists has been emphasised to be integrated into the institutional oversight of biorisks at different levels, supplementing the traditional policies and regulations. Some countries proposed some tools for engaging scientists such as the Netherlands. The Netherlands' Biosecurity Office is a national centre focused on providing information and knowledge related to biosecurity for the government and organisations working with high-risk biological material. Amongst other activities, the Biosecurity Office aims to raise awareness about biosecurity by giving lectures, workshops as well as knowledge and web applications.

Amongst the really valuable resources that the Biosecurity Office has developed, three web-based applications shine above all other resources as meaningful ways in which the Office has enabled risk identification and management engagement at a national level amongst their scientists. The 'self-scan toolkit', the 'Vulnerability Scan', and the 'Dual-Use Quickscan'.[5]

All of the previous tools engage scientists in biorisk management by offering a set of easy-accessible tools that allow them to perform a self-assessment of different kinds of biosecurity hazards, while also providing educational and reference material that scientists can use in order to better understand the rationale behind risk identification and management.

Both the 'Self-scan toolkit' and the 'Vulnerability scan' are tools meant to aid in the identification of weak or strong biosecurity aspects in organisations. This evaluation is based on the eight pillars of good practices, which represent key areas of biosecurity, such as biosecurity awareness, personnel liability, information security, management, accountability for materials, physical security, emergency response and transport security. The main difference between the two tools is that the Self-scan tool offers a quick analysis, while the Vulnerability scan aids a more in-depth analysis of potential vulnerability.[6]

Amongst these three tools, the Dual-Use Quickscan presents an interesting case since this is one of the first widely available tools developed to promote scientific engagement in the topic of dual use. This tool was published in December 2021, inspired by the necessity that researchers and the broader scientific community have for concrete instructions and guidance on how to assess dual-use risks. The Dual-Use Quickscan is a tool that can help identify dual-use risks by answering a set of 15 yes/no questions, organised into three main categories: characteristics of the biological agent, knowledge and technology about the biological agent and consequences of misuse (see Table 17.1).

Table 17.1 The 15 themes of Dual-Use Quickscan.

No	Themes	No	Themes
1.	High-risk biological agent	9.	Detection methodology and diagnostics
2.	Host range and tropism	10.	Reconstruction
3.	Virulence	11.	Harmful effects
4.	Stability	12.	Knowledge and technology
5.	Transmissibility	13.	Ecological consequences
6.	Absorption and toxicokinetics	14.	Economic consequences
7.	Drug resistance	15.	Consequences for society
8.	Population immunity		

The Dual-Use Quickscan is designed for users in the field of the life sciences, the staff working with microorganisms or performing laboratory research and development process. As its name imply, this tool is not meant to offer an in-depth analysis of dual-use risks, but rather allow scientists to quickly identify if there is potential for dual use in their research.[7]

By answering these 15 questions, the researchers will get a basic awareness about the project and if it contains dual-use issues. Such a process engages scientists in order to raise the awareness on dual-use risks and then lead to more discussion or thorough risk assessment for this project. Thus, producing a practical tool, which allows scientists to identify potential dual-use risks through a set of closed questions that aim to communicate potential findings in an easy and clear way, capable of being used periodically to facilitate assessing dual-use risks in their specific research.

Within the context of biorisk management system of the Netherlands, the output of this Dual-Use Quickscan will be used as a starting tool to perform the risk assessment and a management cycle, such as formal risk assessment and management for research projects. Engaging scientists with such an easy tool at the beginning of biorisk management will make the biorisk management system more effective and vigilant.

Thanks to these tools being able to promote self-assessment of risks, while also providing a set of reference materials to understand the rationale followed by the Biosecurity Office to categorise risks, might promote awareness and scientific engagement in dual-use risks as well as biorisk management at a national scale.

17.4 Engaging Scientists in Biorisk Management in Community: Case from iGEM

The iGEM competition is a yearly international synthetic biology competition that enables multidisciplinary students' teams ranging from high school to over graduate programmes to design, build and test synthetic biology projects, while recording their findings via webpages (referred as Wikis within iGEM), videos and presentation that are evaluated by judges at the end of the competition cycle. Currently, the iGEM competition has an average participation of 350 teams per year, from 46 different countries, resulting in the participation of over 70 000 individuals from around the world across 20 years.[8]

As synthetic biology research also has the possibility of bringing risks to the society and nature. iGEM takes biosafety and biosecurity as the core requirement in the community from the very beginning. In order to effectively govern the risks raised from iGEM competition and all students' research, iGEM established the responsibility programme to implement the risk management policies and rules and also promote the culture of responsibility. The responsibility programme also integrates the scientist's engagement deeply through the safety and biosecurity programme and Human Practices programme.[9]

17.4.1 Engaging Scientists Through iGEM Safety Rules Checklist and Safety Screening System

As iGEM has to identify risks for over 300 projects per year, its safety screening process had to fulfil to main requirements, being scalable and engaging for participants of the competition. Thus, iGEM's responsibility programme has developed two resources, a forms-based safety screening system and a set of safety policies and rules that help guide participants who do not necessarily have enough background information in biosafety and biosecurity.[10]

iGEM's rules and policies are a series of mandatory to-follow clarifications on different aspects of laboratory work, activities outside the laboratory, and the responsible use and communication of information (Safety Committee rules and policies). This allows people without background knowledge of risk assessment to be able to identify potential risks in their own projects and work.

This proves to be especially effective since iGEM's rules and policies are not only a 'Do and Do not's list', but rather a place in which scientists can find further resources and guidance on where to begin identifying potential risks. This is particularly true for the 'White List Policy', which at first glance might look like a list of 'do's and do not's', but once looked into in further depth, would prove to be a compendium of resources and links to different webpages and even iGEM's own policies that not necessarily tell you what you 'cannot do' within the iGEM competition but rather what biological material or activities would need you to do a further and more conscious assessment of risks for your own safety. Having students read through and understand these rules and policies might prove to be an excellent way to engage mainly young scientists in risk assessment and mitigation (White List).

The second way by which iGEM accomplishes this engagement at the community level is through the use of three different forms, the 'Project Safety Form', which are mandatory forms that each team must fill and which serves as iGEM's main source of risk assessment throughout the year, the 'Check-In forms', which are tightly related to the White List Policy, since anything that is not present in the White List must first be checked-in and approved by iGEM, and finally, the 'Animal Use Form', which as it name implies, is necessary for any team wanting to use animals in their experiments.

For the purpose of explaining iGEM's method for scientific engagement in risk assessment, we will focus on the mandatory form, the Project Safety Form. This form is a set of open and closed questions (31 in total) that aim to guide team members participating in the iGEM competition to identify any potential risks in their projects, it is important to note that this form not only helps to identify biological risks but also dual-use, chemical and information handling amongst other risks. This form is segmented into six sections: (1) following iGEM's rules and policies, (2) about your lab, (3) about your project, (4) identifying project risks, (5) anticipating future risks and (6) managing risks.

It is important to know the rationale behind this order since the first point ensures that any team member trying to perform an assessment of their own project has first read iGEM's rules and policies, thus allowing them to have the necessary background to understand what iGEM considers as a risk. Then points 2 and 3 allow both the team and the safety screeners to understand more about both the team's project and their facilities. Followed by points 4 and 5, where team members are required to self-assess the risks of their project, both currently existing risks and potential future risks. And finally, closing with point 6, where teams are asked to identify the safety and security resources the team has access to, like experts, local regulations, national or international frameworks and even to clearly identify if they have been previously trained in safety and security; this allows team members to clearly note if they have the necessary mitigation measures for their previously identified risks, and if they do not, then where they can find help closest to them.

After the team has finished the Project Safety Forms, it gets reviewed by external safety screeners, which help iGEM identify any under- or over-estimation or risks, allowing teams to have a better understanding of whether or not their self-assessment was accurate.

17.4.2 Engaging Scientists Through iGEM Human Practices Programme

Human Practices programme is another programme in iGEM which can engage scientists proactively in doing research responsibly and dealing with complex risks. Human Practices programme is one of the special features of iGEM competition since 2008 and has been incorporated into the completion recruitment, it asks every team to think deeply and creatively about whether their project is responsible and good for the world. Such a programme provides a motivated and voluntary approach to engage scientists in biorisk management.[11]

The Human Practices programme is not any specific activity but it also requires all researchers to exhibit they have considered these principles while doing their research: reflection, responsibility and responsiveness. This requires participants not only to consider the scientific research but also to expand their thinking to explore the contexts and societal impacts of projects, anticipate the risks and possibility of misuse of their research and reflect the project design while incorporating diverse stakeholders' views inclusively.

When the technique/knowledge developed by the project is likely, no matter how remotely, to be misused or create ethical, social and legal problems, the Human Practices programme requires the team to demonstrate the willingness, capabilities and skills to either alter their project or develop sufficient countermeasures to mitigate such risk. The team is able to engage with relevant stakeholders to find alternatives that would reduce risk whilst increasing benefits, allowing project to progress. The active detection and report by researchers themselves will not only improve the research projects but also will help to update the biorisk management system within iGEM.

An exemplary case is the development of safety polices and rules on gene drives in iGEM around 2016, this is also a case of engaging researchers in biorisk management.[12] RNA-guided gene drives research is regarded to bring some potential impacts to ecology and society, but this kind of research did not fall into the traditional category of high biorisks in the 2016 and was neglected by many national oversights at that time. So, when the first team in iGEM did a project of gene drive, it had not been discovered by the screening system at the first instance. However, the team researchers were aware of the potential environmental risks for this kind of research and they took the safety and security issues seriously while designing their research and developing some mitigation measures, also they gave attention to the ethical and social impacts for the use of this kind technology. Such actions were expected by iGEM, especially from the Human Practices programme.

With the teams broader considering of their technology impacts, the iGEM Safety and Security Committee started to develop new policies and rules to respond this kind of research. The student researchers of this team were invited to join the review process and discussion about the future rules together with other experts. Such a discussion with engagement led to the first policy release on gene drives of iGEM within a half year. And very quickly such a policy was successfully applied to the screening process of similar research in the second-year completion. The motivated anticipation and reflection amongst the community contributed a lot to the quick identity and update for biorisk management.

In iGEM's case, having a well-established framework of what is considered a risk or not within the competition as well as having a tool that allows and guides team members through a self-assessment process has proved to be an effective and scalable process for the identification of biological, chemical, dual use, information and other hazards regardless of differences in training, culture and previous knowledge on biosafety and biosecurity.

17.5 Conclusion: How to Engage Scientists in Management of Biorisk and Other Emerging Fields

From the governance of nuclear risks to nanotechnologies, from governance of AI risk to biorisk, the engagement of researchers is regarded to play an important role, especially in these emerging fields. All of these risks arising from emerging technologies all have a great degree of uncertainty and even ignorance[13] which means that the formal regulations and rules cannot catch all possibilities of science and technology development, and then highlight the awareness and responsible behaviour of scientists.

The three cases discussed in this chapter represent common but differentiated approaches to engage researchers in biorisk management. The awareness of biorisks, self-assessment, report and improvement are the key expected performance of scientists' engagement. Scientist engagement can not only contribute to the daily governance of researchers but also can contribute to the improvement of the whole risk management system. To promote scientists' engagement in biorisk management, it is evident that easy tools that guide scientists into a self-assessment, the availability of 'background/reference material' or clear guidance and rules, the effective encouragement and trust within the management system, and also the supplementing arrangement will allow scientists to actively and effectively contribute to the governance of complex risks.

Author Biography

Yuhan Bao is the Liaison Officer and the Human Practices Programme Coordinator at the International Genetically Engineered Machines (iGEM) Foundation. His role in iGEM is aiming at motivating young researchers to do synthetic biology responsibly, including incorporating diverse perspectives, anticipating risks of projects and then improving the research to meet the expectations of society. He is also the vice director of the Youth Working Group of the Synthetic Biology Branch under the Chinese Society of Biotechnology. He is a Ph.D. researcher in public policy and management at Tsinghua University. His personal research focuses on the governance of emerging technologies responsible research and innovation, especially in the area of synthetic biology.

Alonso Flores is the Safety and Security Programme Officer at the International Genetically Engineered Machines (iGEM) Competition. His role involves the assessment and mitigation of biosafety and biosecurity hazards for teams participating in this competition, developing and updating tools to evaluate hazards, helping develop frameworks for risk assessment as well as developing educational material to raise awareness and engagement in biosafety, biosecurity and responsible science. His personal research focuses on the implementation of dual-use assessment tools in Latin American countries, in addition to working towards strengthening the development of risk evaluation tools in Mexico.

References

***1** Greene, D. *et al* (2023) Motivating proactive biorisk management. *Health Security*, 21 (1), 46–60.

2 Novossiolova, T. A. *et al* (2021) The vital importance of a web of prevention for effective biosafety and biosecurity in the twenty-first century. *One Health Outlook*, 3, 17.

***3** International Working Group on Strengthening the Culture of Biosafety, Biosecurity, and Responsible Conduct in the Life Sciences (2020) *Culture of biosafety, biosecurity, and responsible conduct in the life sciences: (Self) Assessment Framework.*

***4** Perkins, D. *et al* (2019) The culture of biosafety, biosecurity, and responsible conduct in the life sciences: a comprehensive literature review. *Applied Biosafety*, 24 (1), 34–45.

***5** Biosecurity, B. (2021) *Biosecurity – Dual-Use Quickscan*, Biosecurity Office, Bilthoven, The Netherlands.

6 Meulenbelt, S. E. *et al* (2019) The vulnerability scan, a web tool to increase institutional biosecurity resilience. *Frontiers in Public Health*, 7, 47.

7 Vennis, I. M. *et al* (2021) Dual-use Quickscan: a web-based tool to assess the dual-use potential of life science research. *Frontiers in Bioengineering and Biotechnology*, 9, 1–9.

8 iGEM. (2023) *iGEM: The Heart of Synthetic Biology*, iGEM.

***9** Millett, P. and Tessa, A. (2021) Implementing adaptive risk management for synthetic biology: lessons from IGEM's safety and security programme. *Engineering Biology*, 5 (3), 64–71.

10 Millett, P. *et al* (2019) Developing a comprehensive, adaptive, and international biosafety and biosecurity program for advanced biotechnology: the IGEM experience. *Applied Biosafety*, 24 (2), 64–71.

11 iGEM. (2023) *iGEM Human Practices Program*, IGEM Human Practices Program.

***12** Millett, P. *et al* (2022) IGEM and gene drives: a case study for governance. *Health Security*, 20 (1), 26–34.

13 Rotolo, D. *et al* (2015) What is an emerging technology? *Research Policy*, 44 (10), 1827–1843.

18

The Role of Ethics in Dealing with Dual Use

Leifan Wang

Center for Biosafety Research and Strategy, Tianjin University Law School, Tianjin, China

Key Points

1) To understand the changing concept of dual use and concerns about its potential threat to public health and safety.
2) To understand the pros and cons of ethics in mitigating biorisks under a triangular framework of governance on dual use.
3) To learn how to employ existing and complementary ethical guidelines to safeguard against the misuse of dual use at all levels.
4) To think through the possible steps to promote socially responsible dual use for all through ethical education and training.

Summary

The potential for misuse to cause harm is inherent to the dual-use nature of the life sciences. To mitigate such risks to public health and the environment, ethics plays an irreplaceable role in increasing vigilant stewardship of both researchers and institutions to science and biotechnology and promoting responsible practices in dealing with dual use, but it has limitations and should be combined with other governing instruments, such as regulations and policies, complementary and synergistic, to minimise the potential misuse of dual use while not hampering the benefits and advancements of science and technology (S&T). In addition, ethics as a subject of biosecurity education is vital to raise awareness and cultivate a responsible culture in science in general and the life sciences in particular.

Essentials of Biological Security: A Global Perspective, First Edition. Edited by Lijun Shang, Weiwen Zhang, and Malcolm Dando.

18.1 The Dual-Use Concept and Concerns

The concept of dual use encapsulates the potential of some well-intentioned and beneficial scientific research that could be misused to do harm to public health and environment. In different political, legal and ethical discourses, the term 'dual use' could have different definitions.[1] A narrow understanding of dual use refers to the same knowledge, information, methods, products or technologies that could be used in both civilian and military domains. The main concern about this dual use is how to restrict its spreading to unwanted hands and control undesirable consequences, particularly in the context of non-proliferation. In recent years, a broader understanding of dual use focuses on the research that can be reasonably anticipated to generate knowledge, information, methods, products or technologies that could be misapplied, intentionally or accidentally for malicious purposes, and the main concern about this dual use is how to mitigate the risk of misuse that could either directly threaten the peace, health, safety, security and well-being of citizens or lack of responsible regard to such potential consequences. The application of distinct focuses of dual use reflects the shift of concerns and approaches to the dual-use issues in different settings. This shift is largely driven by the evolution of the notion of 'security' over the last several decades.

We focus here on the dual-use research conducted for peaceful and beneficial purposes that has the potential to produce knowledge, information, methods, products or technologies that could also be misused to do harm to public health and the environment. Our concern is how to maximise the potential benefits of the dual-use research while minimising its risk to cause harm, through appropriate oversight mechanisms, such as ethics, regulations and policies. Based on current understanding, this kind of 'dual-use' research has become a challenge worldwide to public safety and security. The World Health Organisation (WHO) uses a specific term 'dual-use research of concern' to describe some types of dual-use research that could easily be misapplied to do harm with no, or only minor, modification, particularly in the life sciences.[2]

In dealing with this dual use of biological research, we confront two competing ethical values: individual's freedom of scientific research versus the need to avoid causing harm to public health and the environment. The former brings great progress and benefits to mankind, and the latter stresses that such progress and development should occur within a framework of oversight that safeguards against any unacceptable harm to public health and the environment. The tension between the two competing values is sometimes characterised as the dual-use dilemma. It poses an ethical dilemma not only for researchers but also for those who are in the positions to regulate or supervise the researcher's work, as 'it is about promoting good in the context of the potential for also causing harm' and the 'dilemma arises for the researcher because of the potential actions of others'.[3] Thus, dealing with the dual-use issues demands the engagement of all stakeholders in the whole lifecycle of research, supervision and administration of dual-use and related technological developments, where bioethics can play an important role as a governing instrument in guiding stakeholders' decisions and practices to address relevant concerns about the dual use.

18.2 Ethics as an Instrument on Dual-Use Governance

Risk mitigation is a shared responsibility amongst all stakeholders in the life sciences. Under a dynamic, triangulated governance framework of dual use, the knowledge, information, methods, products or technologies with the potential of dual use is always changing and evolving, and the two main governing instruments should respond and co-evolve. One is governmental oversight through regulations and policies, and another is self-regulation through ethical requirements from professional associations, research ethics committees or boards at all levels. Ethics may play an irreplaceable role in promoting responsible practices on dual use, particularly in increasing vigilant stewardship of both researchers and institutions to identify, assess and mitigate risk related to dual use.

18.2.1 Ethics Promote Responsible Practices on Dual Use

The United Nations Educational, Scientific and Cultural Organization (UNESCO) framed bioethics as a facet of responsible research and innovation and highlighted the role of ethics in science applications to improve humanity in its Declaration on Science and the Use of Scientific Knowledge.[4]

> 'The practice of scientific research and the use of knowledge from that research should always aim at the welfare of humankind, including the reduction of poverty, be respectful of the dignity and rights of human beings, and of the global environment, and take fully into account our responsibility towards present and future generations'.

To this end, researchers engaged in scientific activities are expected to take a special responsibility that should go beyond the traditional views of misconduct related to plagiarism or consent for their experimentation, and extend to being deeply alert on how their work and relevant products might be used in society. To fulfil this responsibility, ethics may provide a useful normative framework for them to take into account relevant ethical values in the identification and assessment of potential risks, and the allocation of the responsibilities of other stakeholders to mitigate them. The open discussion of ethical issues in this process contributes to consensus-building amongst all stakeholders, particularly regarding the following important issues on dual use:

- how to weigh the potential benefits of dual-use research against the potential risk of misuse?
- how to weigh the interests of researchers against the common good for public health and security?
- how to best manage the biorisks associated with dual use, without unduly hampering the advancement of biosciences?
- what are the responsibilities of individual researchers and of the scientific community as a whole to protect dual-use knowledge, products or technologies against misuse; and
- What mechanisms and tools are effective to support all stakeholders to fulfil their responsibilities, respectively, and collectively, in dealing with dual use?

Table 18.1 Main principles on bioethics.

Articles	Main Principles on Bioethics
Article 3	Human dignity and human rights and fundamental freedoms are to be fully respected.
Article 4	Benefits should be maximized and any possible harm to such individuals should be minimized.
Article 5	Individual autonomy to make decisions should be respected.
Article 6	Free and informed consent of the person concerned should be warranted.
Article 7	Special protection should be given to persons lack of the capacity to consent.
Article 8	Individuals and groups of special vulnerability should be protected.
Article 9	The privacy of the persons concerned should be respected.
Article 10	Equality of all human beings in dignity and rights should be respected.
Article 11	No individual or group should be discriminated on any unethical grounds.
Article 12	Cultural diversity and pluralism should be given due regards.

In an attempt to answer these questions, considerations should also be taken into account regarding what ethical values can be ascribed to notions such as security, liberty and research benefit, as well as how those are prioritised, traded off and balanced. The principles enshrined in the UNESCO Universal Declaration on Bioethics and Human Rights of 2005 provide guidance to all individuals, groups, communities, institutions and corporations, public and private, with respect to their decisions and practices in scientific research. These principles are closely related to the ethical issues of dual use and more specifically as shown in Table 18.1.

These principles remain inspirational and compelling in guiding ethically responsible practices in life sciences nearly 20 years later. They should be taken as fundamental ethical values in dealing with dual use. They are to be understood as complementary and interrelated, as appropriate and relevant in the circumstances. Ethics facilitates responsible research and innovation in a manner that places these values at the centre of decisions and practices in research and innovation. It moves the responsibilities of researchers and other stakeholders beyond the dominance of legal compliance and addresses a broader range of societal concerns. As steward of science and innovation, they have a shared moral responsibility to ensure that the knowledge, information, methods, products and technologies that are generated from their research are used to enhance human welfare and not to do harm.

These ethical values should be taken into account in decision-making and practice at every stage of the research by stakeholders at all levels (i.e. individual, institutional, national and international), from initial conceptualisation and development of a proposal to provision of funding, to conduct of the research, analysis of results, storage and potential use of material results and dissemination of findings. Individual researchers are viewed as the centre of the ethical discussion in dealing with dual use. They propose and implement ideas for research and are the first line of identifying, assessing and mitigating risks in their research. They also carry a special responsibility to prevent the knowledge, information, methods, products and technologies that they develop and disseminate from being misused for harmful purposes. Research institutions where the research take place are the second line of risk mitigation for dual use. National governments are ultimately responsible for defining and implementing the standards of risk mitigation, regulating behaviours of each stakeholder and addressing the impact of the life sciences in their jurisdiction. Other stakeholders, including funding bodies, national academies, professional societies, editors and publishers, educators, international organisation, private sector, civil society and the media also have their responsibility to contribute to risk mitigation from dual use.

The ethical issues raised by dual-use research are often highly controversial and contextualised. Some countries and institutions have developed oversight mechanisms to manage dual-use misuse. Many have not done so or done insufficiently, owing to limited awareness, lack of capacity or competing demands on resources. Dealing with ethical issues on dual use involves evaluation of competing values, principles and interests, and thus relevant stakeholders that have different priorities in goals relating to biological research could reach different outcome of a bioethical analysis with respect to the dual-use dilemma in a specific situation. They should assume the responsibility to avoid doing harm, seek cooperation and exhibit respect to all relevant ethical principles in good faith.

18.2.2 Limitation of Ethics in Dealing with Dual Use

Ethics as a governing instrument on dual use is generally more agile, flexible, and adaptive to specific contexts than regulations, but it has limitations in dealing with dual-use concerns. Often researchers are not well equipped with expertise and resources in assessing all consequences of their research. They cannot entirely guarantee that their work will never be applied intentionally or accidentally to do harm. They can only take reasonable steps to minimise the chances that such harm will occur within their awareness and professional capacity and ability. Meanwhile, they have a vested interest in engaging with risky research to pursue opportunities of career advancement. It is, therefore, not surprising that there are limitations to what researchers can do on their own to prevent against misuse of their research. In those circumstances, governmental oversight of dual-use research through legally binding regulations and policies is critical to preserve the benefits of the life sciences research while minimising the risk of being misused. In addition, in some areas of dual use, regulations and policies are necessary to prohibit certain types of undesirable activities, such as export control of certain sensitive and dangerous items, technologies and information; and protection of certain plants, agriculture and the environment, where legally non-binding ethics are deemed to be insufficient to shape behaviours of those who have a vested interest in them.

The degree of governmental oversight on dual use largely depends on how the relevant stakeholders respond to opportunities for self-regulation presented by dual-use dilemmas. Normally, national governments would endorse a principle of regulatory parsimony, imposing only as much oversight as is truly necessary to pursue public good and avoid unduly impeding the freedom of scientific research and the beneficial applications. If, however, a scientific community fails to take appropriate steps towards self-regulation through developing appropriate professional codes, journal policies or institutional oversight mechanisms or its members seriously violate such requirements, a higher degree of governmental oversight may result. For instance, after the CRISPER'd babies experiment occurred in November 2018, China legalised certain ethical principles on the research in its newly enacted Biosecurity Law as a response to a researcher's breach of relevant ethical requirements and policies not to use gene editing for reproduction purposes that caused a public crisis. Article 33 of the Chinese Biosecurity Law stipulates that those engaged in research, development and application of biotechnology should 'conform to ethical principles'. The open-ended phrase 'ethical principles' used in the provision leaves much discretion for the judiciary to decide on contextualised interpretations in response to the rapid changes of the life sciences and converging technologies in China.[5] This also creates a situation where a procedure of legal interpretations replaces ethical reflections on dual use when a case under this provision is brought to courts in China.

18.3 Existing and Complementary Ethical Guidelines on Dual Use

Dealing with dual use in the life sciences is a continuous challenge to policymakers in all States. One problem is how to translate the responsibilities of researchers and other stakeholders into effective means of governance on dual use? Besides existing regulations and policies, there are some ethical guidelines designed to address risks associated with dual use by a multiplicity of stakeholders.

Already in 1974, UNESCO adopted the *Recommendation on the Status of Scientific Researchers* which was revised in 2017. The revised version is an important standard-setting instrument that codifies the goals and values for science to operate and to flourish. It enumerates all scientific researchers' rights and responsibilities in delivering high-quality science in a responsible manner. The implementation of the Recommendation in Member States is monitored and evaluated by the UNESCO governing bodies periodically. Some national academies of sciences and professional associations released their professional codes of conduct or ethics, such as American Society for Microbiology, the Chinese Academy of Sciences and the Royal Society in the United Kingdom, that provide specific guidelines with respect to what is considered appropriate behaviour in context. At the international level, the Hague Ethical Guidelines and Tianjin Biosecurity Guidelines for Codes of Conduct for Scientists laid out a set of ethical principles and standards of responsible conduct to guard against the misuse of chemistry and biosciences, respectively, so as to facilitate the implementation of the vital dimension of ethics in relation to the Chemical Weapons Convention (CWC) and the Biological and Toxin Weapons Convention (BTWC).

Although these ethical guidelines are voluntary and aspirational in nature, the ethical guidelines help to clarify minimum social expectations for the responsible conduct of researchers and other relevant stakeholders that can be viewed as standards of baseline performance to fulfil their responsibility in dealing with dual use and the broader issue of responsible scientific conduct. When researchers and other stakeholders navigate in gray ethical areas related to their work, these guidelines may be used in a complementary way to set direction in the ocean of uncertainties and controversies. These ethical guidelines are mainly built upon the precautionary principle, a conceptual tool applied to guide decision-making in risk management under uncertainty, while leave considerable discretion to researchers and institutions in response to changes and uncertainties in different settings. They are not merely symbolic statements of values but useful guidance for members of the profession to be socially responsible with regard to their behaviour in research and promote public trust in scientific community's self-regulation. Some professional codes are armed with sanctions, linked to membership disbarment for misbehaviour, even if they lack coercive power. They are part of 'a web of prevention' to minimise the potential risk of dual-use.[6] The ethical guidelines and related codes of conduct, awareness-raising and educational activities pertaining to the responsible use of research and the prevention of misuse constitute an important pathway to implement the WHO global framework of governance for responsible science and dual use. They are not exhaustive at this moment and could be added or updated in light of best practices at all levels.

18.4 Recent Dual-Use Scenarios

In recent decades, advances in life sciences were stunning and outpaced most national policymakers' attempts to regulate them, which created massive regulatory gaps for ethical discussions about the broader social implications of dual-use issues arising from these new areas. The following three lines of research have clearly far-reaching potential for dual use that demands our attention and action.

18.4.1 Synthetic Biology

Built upon the knowledge of molecular, cellular and biology systems, synthetic biology seeks to design and construct new biological entities or redesign existing ones. It has apparent dual-use application in various aspects. The same tools used in synthetic biology to invent new medical countermeasures and diagnostics capabilities could be easily used to bioengineer viruses, such as making pathogens more harmful or recreating known pathogenic viruses. For example, Swiss scientists synthesised the coronavirus de novo from its genetic sequence posted online by Chinese researchers almost simultaneously with the outbreak of the COVID-19 pandemic in 2020. No doubt, this study might increase our understanding of SARS-CoV-2 infections and support preparedness and strategies to combat coronavirus infections; however, it also brought the potential risks that synthesising dangerous virus like SARS-CoV-2 could be misapplied for harmful purposes in other settings.

At present, national regulations on deadly pathogen research remain vague, inconsistent and insufficient. The current debate about such studies is as much about whether they should have been conducted at all, as about the scope of pathogenic research subject to special oversight, who should have access to the details of the research and how to globally collaborate to set the playing field. Before necessary governmental oversight mechanisms are in place, researchers are responsible to make good-faith efforts to harness the power of synthetic biology, prevent against misuse and take into account of appropriate ethics concerns before such studies are conducted in labs or open environments.

18.4.2 Gene Editing

Gene editing techniques, such as CRISPR and related techniques, have generated both excitement and dual-use concerns. They can provide considerable promise to create novel solutions to genetic diseases and enhancement, but such capacities can be used to induce harmful genetic modifications in biological entities (e.g. microbes, plants, animals and humans) or produce bioagents that pose risk to public safety because many of the bioagents modified or designed for therapeutics have dual-uses of concern. A recent example is the 'gene-edited babies' occurred in 2018.[7] It showed that, while the use of CRISPR for certain therapeutics was viewed as being appropriate, its use for reproduction purposes still were regarded as far less acceptable, particularly when there was an alternative solution that could be more mature and less risky since the full impact of human heritable genetic alterations remain unknown. After the incident, whether or under what condition alterations of DNA in human sperm, eggs or embryos are acceptable remains an issue to be hotly debated both at national and international levels.

Uses of gene editing techniques in contexts outside of human biology to modify plants, agricultural products, animals and insects also raise dual-use concerns that require regulatory oversight and governance. In addition, rapid commercialisation of these techniques and lowered barriers to use gene-editing techniques by public research and do-it-yourself communities could foster risk incurred by both inadvertent misuse or intentional development of products that threaten public safety. More engaged discourse to define the needs, values and ethics of use gene editing techniques is necessary amongst all stakeholders, in order to establish agreed-upon international norms and oversight mechanisms at all levels to ensure that all aspects of these biotechnological tools and methods not be used in ways that threaten public safety.

18.4.3 Neuroscience

Human brain research has brought new dual-use concerns about novel neuro-technologies such as brain–computer interfaces, warfighter enhancement, manipulation of emotional states, etc., that have profound social and ethical implications.[8] These neuro-technologies initially are developed to assist or improve the diagnosis and treatment for humans with mental and neurological diseases. They also can be used to manipulate individual brain functions for military, political or commercial purposes and even challenge the basic notions in humanity, including what it means to be human. They pose a wide range of risks to everyone in the near future when such neuro-technologies land in the wrong hands. Malevolent uses of neuroscience could be performed not only by individual actors but also by rogue States and organisations.

The risks associated with the advancements in novel neuro-technologies have raised global ethical and security concerns,[9] and urged national regulators, researchers and other private actors to take concrete actions to minimise their harmful societal implications through improving the relevant tools and governance mechanisms, including developing globally coordinated ethical and societal guidelines to promote responsible practices and awareness-enhancing initiatives at various levels, and ensure that the uses of neuro-technologies are ethical and responsible based on a foundation of equity and public good.

18.4.4 Digital Biological Data

The recent trend of open science and data sharing worldwide has resulted in a stunning amount of digital data becoming available online for reuse in the life sciences, including whole genome or individual DNA/RNA sequence data; transcriptome, proteome or metabolome data describing the biological activity of a given cell or organism and epidemiological data, amongst others. The benefits of these data are apparent, as they facilitate new research for a better understanding of the fundamentals of life, but they cannot, by default, be considered only beneficial. They have the dual-use potential and can be invaluable for hostile actors to cause harm. Current discussions on dual use are less concerned with the potential risks from digital biological data online. The ethical tension between the researcher's duty to data sharing in supporting open science and the responsibility of addressing the potential risk from the misuse of the raw data is challenging to tackle. Further discussion will be useful to examine how to reconcile the two issues so as not to place researchers in an untenable ethical position with regards to their data, and foster a biological digital data ecosystem that advances innovation while adhering to the principles of responsible conduct of research.[10]

Dual-use dilemma exists in various new areas of life-science research. We just describe a few amongst them, and new dual-use issues are emerging continuously. Ethics can play an important role in raising awareness, managing the potential risks and preventing against misuse of cross-cutting scientific advances if the potential dual-use implications are well discussed with all relevant stakeholders and appropriate monitoring measures are placed at institutional, local and international levels. Meanwhile, based upon the abovementioned principles, what counts as appropriate standards to responsible practices in specific situations must be open to continuous evaluation as part of an exchange amongst relevant stakeholders, in line with the rapid advances of S&T and their ethical, legal and societal implications.

18.5 Ethical Education for Future Dual Use

Adequate ethical education on biosecurity would help to equip scientists and other stakeholders to recognise and address dual-use concerns, particularly those in the position of developing policies and practices for the effective oversight of the dual use, to ensure that they will never lose sight of the significant, wide-ranging and far-reaching impact that their work has on the lives of individuals, groups and communities and on our shared humanity. The forms can be various, including bioethics seminars, case conferences, ethics consultations, Institutional Review Board (IRB) deliberations and other educational opportunities, particularly for early career researchers. Meanwhile, the cultural variations of dual-use understanding should be taken into account to make the dialogue more robust and effective globally.

S&T advances in the life sciences and converging technologies hold great promise for new and improved ways to support healthier populations worldwide while raising significant ethical, legal, societal, safety and security risks. States should foster ethical education and training at all levels to enhance awareness of these risks and minimise them. The efforts of ethical education and training are not only to promote awareness of ethical issues and knowledge of ethical concepts but also to influence attitudes and behaviour of relevant stakeholders with regard to their practices in daily work.

In summary, harnessing S&T advances in the life sciences and converging technologies to further social good and not cause harm is an on-going challenge to all stakeholders. There is no one-size-fits-all approach to the concerns about dual use. They need a comprehensive governance approach and engagement of all stakeholders to build a web of measures at all levels, from individual to community and from domestic to international, complementary and synergistic, to minimise the potential risk of dual use while not hampering the benefits of advancements of S&T. In the real world, States might differ in their tolerance for the potential risks of dual use, upon which they formulate their own regulations, policies and ethical requirements, and no society could or should operate with a true 'zero-risk' approach to dual use. The balance between freedom of scientific research and responsible stewardship of science is an art of governance, and ethics can play an important role in the complex process of decision-making, particularly in dealing with dual use, if deployed appropriately.

Author Biography

Dr. Leifan Wang is an associate professor at Tianjin University Law School and Center for Biosafety Research & Strategy. Her research focuses on public international law, international environment law, biosafety and biosecurity governance. She co-drafted the Tianjin Biosecurity Guidelines for Codes of Conduct for Scientists. She is the Council Member of Chinese Society of International Law. She obtained her Ph.D. in international law from Renmin University of China and admitted to the bar of New York State and P.R. China.

References

***1** Rappert, B. *et al* (2013) *On the Dual Uses of Science and Ethics*, ANU Press.

***2** World Health Organization. (2022) *Global Guidance Framework for the Responsible Use of the Life Sciences: Mitigating Biorisks and Governing Dual-Use Research*, World Health Organisation, Geneva.

3 Evans, N. G. *et al* (2022) Reconciling regulation with scientific autonomy in dual-use research. *The Journal of Medicine and Philosophy: A Forum for Bioethics and Philosophy of Medicine*, 47 (1), 72–94.

***4** UNESCO. (2005) *Universal Declaration on Bioethics and Human Rights*, UNESCO, Paris.

5 Wang, L. *et al* (2021) Bioethics in Chins biosecurity law: forms, effects, and unsettled issues. *Journal of Law and the Biosciences*, 8 (1), lsab019.

***6** McLeish, D. F. B. R. C. (2007) Introduction: A web of prevention? in Rappert, B. and McLeish, C. (Eds.) *A Web of Prevention*, Routledge.

***7** Greely, H. T. (2019) CRISPR'd babies: human germline genome editing in the 'he Jiankui affair'. *Journal of Law and the Biosciences*, 6 (1), 111–183.

8 Dando, M. (2020) *Neuroscience and the Problem of Dual Use: Neuroethics in the New Brain Research Projects*, Springer, Cham, Switzerland.

9 Goering, S. *et al* (2021) Recommendations for responsible development and application of *neurotechnologies. Neuroethics*, 14 (3), 365–386.

10 Bezuidenhout, L. (2013) Data sharing and dual-use issues. *Science and Engineering Ethics*, 19 (1), 83–92.

19

Where Is the Governance of Dual-Use Science Going?

Nancy Connell[1] *and Gigi Gronvall*[2]

[1] *Rutgers New Jersey Medical School, Newark, NJ, USA*
[2] *Center for Health Security, Johns Hopkins University Bloomberg School of Public Health, Baltimore, MD, USA*

Key Points

1) The term 'dual use' in the context of biological governance has lately been narrowed and even limited to representing only pathogen research. We support the view that all research is dual use, and individual projects lie along a continuum or spectrum of potential hazards. Placing a research effort on this continuum is difficult, as not all outcomes or consequences are necessarily predictable, but sophisticated risk analyses can aid in these estimates while maintaining the potential and excitement of scientific inquiry.
2) There is an inherent conflict between 'top-down' and 'bottom-up' governance structures. This tension can be reconciled by thinking about governance as hybrid and multi-layered. Some scientific fields have made progress in oversight (nuclear, chemical) while others are just beginning to understand the implications (artificial intelligence [AI]); we would argue that biological research lies in between and can learn from these other systems.
3) Governance is the mechanism to ensure accountability, providing oversight to ensure that risks are adequately prevented and mitigated. Thus, biosafety and biosecurity are central to the mechanism of governance. Safety standards in other industries (nuclear, maritime, airlines, electronic, food safety, construction, etc.) provide methodologies for carrying out operations and functions by agreeing on operating procedures. Biosafety and biosecurity standards will contribute to the setting of norms across borders.
4) Governance structures must be adapted to be relevant to changing environments, advances in technology and novel applications; in some cases, new structures and ways of limiting harm are urgently needed. Practitioners of science must be central

Essentials of Biological Security: A Global Perspective, First Edition. Edited by Lijun Shang, Weiwen Zhang, and Malcolm Dando.

to the development of these structures. Amongst the models under discussion are networked-based governance, transnational mixes of government, sub-government and stakeholders who work together to solve collective problems. Qualitative framework analysis can help evaluate risk by bringing interdisciplinary groups together and developing a common language.

Summary

The chapter begins with a brief discussion of rapidly emerging technologies such as genetic manipulation and artificial intelligence (AI). The inadequacy of the term 'dual use' to encompass the range of scientific development is discussed. The term 'dual use' forces analysis into a binary choice ('dual use' or not 'dual use'), and we observe that experimental risk lies along a continuum. Essential to discussion of risk are the experts, those designing and using these technologies and the chapter proceeds to survey the history of self-regulation. The collective process of creating an international governance mechanism for dual-use research requires standardisation of language and concepts. The chapter describes the value of frameworks – decision models – to help clarify and standardise the approach being used. International, national and local oversight mechanisms, including monitoring of advancing technologies, have begun to populate agencies and institutions whose remit covers dual-use governance. Other fields have begun to develop hybrid governance models, combining elements of both hierarchical (top-down) and network-based (collaborative) approaches to governance, operating at multiple levels. These approaches recognise the need for centralised authority and coordination while also valuing the benefits of collaboration, participation, and decentralised decision-making. It is likely that no single 'top-down' formula for governing research and ensuring responsible life science will be created. Rather, a layered, hybrid approach spanning self-governance, education as well as local, national and global oversight will continue to evolve in response to the moving target of biological research.

19.1 Background: Genetic Technologies and Their Applications

The ability to read, write and edit DNA with speed and precision has had massive impacts across science and society. Research into the microbiome, pathogens and molecular epidemiology has enhanced our ability to detect, identify, track and protect against bacteria and viruses. We have greater understanding of microbial evolution, population dynamics, function and diversity, all of which are critical for creating more effective therapeutics and understanding the role of different microbial species in the environment. These advances in genetics are critical for enabling the growth of the bioeconomy. Utilising genetic engineering in microbial species has allowed bio-based strategies for manufacturing that are considered more sustainable than previous methods. Foundational knowledge and gene editing tools, especially tools that work at scale, enable the growth of the bioeconomy, built on the groundwork of synthetic biology (synbio).

Synbio is a multidisciplinary field comprising the convergence of engineering with biotechnology and genetics. Synbio is a subset of the broader field of 'engineering biology', collectively projected to transform the entire world within the next two decades, with an estimated value of $4–30 trillion US dollars. The novelty of engineering biology derives from the application of engineering principles to the design of genetically engineered organisms. Synbio adopts the 'design, build, test' model of engineered design, introducing both precision and convenience in the design of new organisms. In addition to bioproduction, synbio has entered the field of biosensing, allowing organisms to perform detections in disease diagnosis, hazard detection, food/water safety, physiological state, etc. Biosensing is amongst the most extensively developed applications in the biotechnology arena and will be a key element in the advance to the 'Internet of Living Things' – the network of objects in which data are collected and used to carry out tasks in real time.

19.2 Dual-Use Science: Evolving Story of a Dualistic Term

During the Cold War, the concept of 'dual use' was applied to those technological and scientific advances with both military and civilian applications. Nuclear technologies led to such products, as nuclear power and detection and imaging technologies; aerospace research led to satellite technology, telecommunications, navigation and weather prediction, hypersonic and stealth technology, materials, alloys and advanced composites. The development and dissemination of many of these technologies have raised concerns about their potential misuse. In the case of nuclear developments, multiple regulatory agreements and international treaties promote the peaceful use of nuclear technology, aiming to minimise the risk of proliferation and misuse for military purposes and fostering the use of nuclear technology for peaceful purposes.

Artificial intelligence (AI) tools are rapidly developing, and biology has already benefited from advances in data science and computing. AI technologies are already contributing to healthcare and medicine. For example, the prediction of 3-dimensional protein structures from the 2-dimensional chain of amino acids sequence – an insoluble problem before AI – has accelerated drug design. However, AI technologies raise fundamental questions about power and transparency in the use of these technologies. When and by whom are AIs used? What datasets are used and how are they labelled? How are algorithms designed, structured and validated? Were the data obtained with consent? When operating at speed and scale, and with interoperable systems or immutable biometric data, such as DNA, these questions become even more urgent. The possibility of calls for governance from its own developers. The Center for AI Safety, a non-profit organisation, released a one-sentence statement: 'Mitigating the risk of extinction from AI should be a global priority alongside other societal-scale risks, such as pandemics and nuclear war',[1] reminiscent of the calls for control of nuclear technology by the Manhattan Project in 1945.[2] What can dual-use life sciences learn from these regulatory crises, and vice-versa?

In recent years, the term 'dual use' has been used with sharp focus on biological research, where scientific knowledge and discoveries can have both beneficial and potentially harmful applications, most notably about biological pathogens. This expansion of the term's use recognises that ethical considerations and responsible governance are important in navigating the dual-use nature of knowledge and technologies. Indeed, the concept of dual use has evolved to represent and reflect the complex interplay amongst technological advancements, societal implications and the need for responsible stewardship of knowledge and innovation in both civilian and military contexts. In fact, the term dual use in the context of biological governance has lately been further narrowed and even limited to represent all pathogen research. Application of the term 'dual-use' to describe the nature of a specific experimental approach forces a binary interpretation, i.e. an experiment is either dual use or not dual use. We support the view that *all* research is dual use and individual projects lie along a continuum or spectrum of potential hazards; some work may cause legitimate concern. Positioning a single research effort on this continuum can be challenging, as not all outcomes or consequences are predictable and risks can change over time, but sophisticated risk analyses may aid in these estimates while maintaining the potential and promise of scientific inquiry.

19.3 Begin with the Experts: Models of Self-regulation

19.3.1 From Asilomar to Napa

In 1975 at Asilomar State Beach in California, scientists gathered to debate the regulatory options for a powerful new technology: the ability to cut and splice DNA molecules, also known as recombinant DNA (rDNA) research. Consensus was reached on safety issues and formed the basis of the 1976 US National Institutes of Health's (NIH) *'Guidelines for Research Involving Recombinant DNA Molecules',*[3] a scientifically responsive document that has been modified multiple times, as scientific approaches change. The goals for the Asilomar conference were to define responsible conduct of research and develop a transparent biosafety governance framework that enabled advances in the life sciences while promoting the safety of researchers, the public and the environment. The success of this meeting demonstrated that scientists can come to consensus quickly and effectively – more than a political process would allow.

Forty years later, in January 2015, scientists gathered again, this time in Napa Valley, CA, to debate not the safety of technology but also the societal and ethical implications of using genome editing in the human germline. While consensus was not reached and guidelines were not created, a set of general recommendations for how to proceed were indicated, for example, continuing dialog should be carried out with full transparency and a moratorium on human germline manipulation should be discontinued until international ethical and safety guardrails have been created. These two cases illustrate how the scientific community may implement impactful moratoria on specific experiments and that self-regulation can result in the development of guidelines to ensure the safe handling of emerging biotechnology. It is essential for the scientific community to lead in this role, as emerging technologies are too rapidly evolving, with many uncertain risks. The expertise needed to understand potential risks and benefits requires specific, sub-domain expertise.[4]

19.3.2 Tools for Self-regulation: Risk and Benefit Analyses: Useful Frameworks

The collective process of creating an international governance mechanism for dual-use research requires standardisation of language and concepts. Frameworks – decision models – help clarify and standardise the approach being used. Frameworks can ensure that estimates include all the components that the collaborating decision makers believe are important, and secondly, they enable comparisons across multiple estimates. Developing shared frameworks increases the usefulness of estimates of risk and allows comparison across national and conceptual boundaries. Several frameworks or decision trees, have been elaborated for analysis of the risks and benefits of research with dual-use potential. Qualitative frameworks serve as a tool to evaluate risks and benefits and determine how to address them both. The process of using the frameworks to discuss scientific and technological capabilities organises information in ways that illuminate unstated assumptions, clarify areas of agreement and disagreement, bring forward questions and facilitate productive discussions. In this way, the frameworks enable potential security risks to be assessed in a systematic way to inform policymakers and support the goal of evidence-informed policy. A National Research Council (NRC) study tested the utility of framework analysis of dual-use scenarios in the context of the Biological Weapons Convention (BWC), showing that individuals representing multiple disciplines – expert and non-expert – can use specific frameworks to analyse risks and benefits of specific experiments.[5]

19.3.3 A Patchwork of Layered Oversight, from Global to Local

International, national and local oversight mechanisms, including monitoring of advancing technologies, have begun to populate agencies and institutions whose remit covers dual-use governance. Here, we discuss several more influential and mature mechanisms. For example, early oversight of dual-use biological sciences began with the 1972 BWC.[6] By default, this treaty has hosted for decades discussions of advances in the life sciences, as they relate to the development of biological and toxin weapons. This kind of work at the margins of the BWC was carried out by concerned scientists and other members of civil society, representing NGOs and academic institutions from around the world. Offensive programmes in biological weapons (BWs) led to lists of specific agents (bacteria, viruses, fungi and toxins) with weapons potential. The language of Article I of the Convention refers only to 'microbial or other biological agents or toxins whatever their origin or method of production, of types and in quantities that have no justification for prophylactic, protective or other peaceful purposes'.

A similar focus by civil society was found in The Pugwash Conferences on Science and World Affairs,[7] begun in 1958 and continuing to this day, with a mission 'to bring scientific insight and reason to bear on, namely, the catastrophic threat posed to humanity by nuclear and other weapons of mass destruction.' The International Atomic Energy Agency (IAEA)[8] was established in 1957 within the United Nation (UN), has its own treaty and reports to both the General Assembly and the Security Council of the UN. Nuclear materials are detectable, quantifiable threats and these characteristics perhaps influenced early deliberations of the BWC, whose framers nonetheless acknowledged the possibility of less proscribed threats of the future, beyond the production and stockpiling of specific, existing biological agents such as bacteria and viruses. While novel organisms were anticipated by the framers, the nature of biological threats has since moved well beyond existing pathogenic agents. Thus, early oversight mechanisms (e.g. verification or attribution discussions related to bioweapons; biosafety programmes) focused on three components: the organisms, the equipment required to produce them and the information, required either to create or to weaponise and disseminate. This focus on specific organisms and toxins remains a limitation in current biothreat governance models.

Today, the BWC remains a clearing house for these discussions. All but a handful of countries in the world are signatories, and while these uninvolved States Parties might be amongst those seriously lacking in life sciences oversight, the BWC stands to serve as an international body contributing to the effort. Unfortunately, the BWC lacks a formal scientific advisory body. Bringing together scientific and technical expertise in relevant fields such as biology, biotechnology, public health, infectious disease and biosecurity would assist representatives of the States Parties in conducting assessments of emerging scientific developments, potential dual-use research and advances in biotechnology. Such a body would foster international cooperation, support risk assessment and compliance and enhance the implementation of the treaty, ensuring that the BWC remains effective in preventing the development, production and use of BWs while addressing evolving challenges in the field of biotechnology.

Complementing the work of the BWC, the World Health Organisation published *The Global Guidance for the Responsible Use of the Life Sciences: Mitigating Biorisks and Governing Dual-Use Research* in 2022.[9] The report is the product of global contributions from multidisciplinary experts and provides a framework for countries to develop ethical, beneficial and safe application of life sciences research and technology. The document provides a comprehensive collection of multiple sources of dual-use identification and/or management tools. There has been a steady stream of work in this arena beginning in 2004 with the NRC's *Biotechnology in an Age of Terrorism* report, also known as the Fink report. The WHO guidance posits three core pillars of biorisk governance: biosafety, laboratory biosecurity and the oversight of dual-use research. Responsibility in life sciences research seeks to integrate formal and informal measures covering these concepts. Singling out dual-use science in this larger discussion, there is particular emphasis on the value of the following three areas in the oversight of dual-use research: (1) regulation, policies and legislation; (2) risk/benefit analysis; and (3) training, awareness-raising and codes of conduct. The first of these approaches, regulatory policies, are generally considered 'top-down' governance, while risk/benefit analysis and awareness-raising are 'bottom-up' activities.

A third example of an international body contributing to dual-use governance is *The International Genetic Engineering Machine Competition* (iGEM).[10] iGEM was started 20 years ago in the United States and now has participants from 66 countries and regions with 6000 students competing each year in 400+ teams. Multidisciplinary students work together in teams to identify problems to be solved, designing solutions using synbio and engaging in self-organisation to address societal problems safely and securely. Intrinsic to the iGEM process of project development is a dedicated biosafety and biosecurity programme. The iGEM Safety and Security Committee comprises a team of specialists from a range of fields with expertise in biosafety, biosecurity and risk assessment. Its members represent multiple sectors of industry, academia and government. The committee oversees iGEM's safety and security programmes, offering guidance on potential safety and security concerns and systematically reviewing each of the proposals. The competition has a number of checks for dual use along the life cycle of the projects: the parameters for evaluation also include human practices and rules for communication and transfers of microorganisms. Thus, students who compete in iGEM have early career exposure to the basics of responsible science. Over 40 000 students, instructors and judges have passed through the iGEM system since 2004.

These organisations – the BWC, WHO and iGEM – establish context and frameworks for evaluation of national-level oversight that will help to standardise governance mechanisms across boundaries.

19.3.4 Self-regulation: The Basis of Scientific Enterprise

Researchers naturally self-organise into collaborations from the 'bottom up'. The Asilomar conference in 1970 and iGEM are good examples of this process of embedded responsible science. To 'govern science' and control risk, these models illustrate the 'shaping' of a technology as it develops and forming the 'rules of the road' as norms for the technology are developed. In both cases, governance is 'intra-community'. Sundaram[11] describes the management of technology as 'shape and steer' rather than 'how it is regulated', a view promulgated by Voeneky: 'a multi-layer governance that consists of rules of international law, supranational and national law, private norm-setting and hybrid forms that combine elements of international or national law and private norm setting'.[12] Sundaram goes on to support the notion of framing social and political factors within the same boundaries as science and technology (S&T), arguing that governance mechanisms must recognise that these are 'sociotechnical systems'.[13] 'Steering' by the hybridised, layered approach of laws, norms and continuous education and reevaluation of the moving target that is dual use in the life sciences permits innovation and creativity while remaining aware of risk.

19.3.5 Oversight Along the Life Cycle of Research: Universities Are Sites of Layered Governance

A 2018 NASEM report, *Governance of Dual-Use Research in the Life Sciences: Advancing Global Consensus on Research Oversight: Proceedings of a Workshop (2018)*,[14] examined the kinds of oversight activities that are distributed across the research life cycle. Some activities cut across all stages, like national advisory boards, outreach conducted by governmental agencies to relevant research communities or systematic self-governance measure adopted by research communities. In early stages (conception, initial project design), institutional review committees and technical analysis or risk mitigation processes. At the funding stage, the funders themselves may require analysis and statements of intent from both researchers and institutional officials. During the conduct of the research, risk mitigation plans will be implemented as well as worker training in risk management; DNA synthesis requests may be screened; 'culture of responsibility' developed and maintained through education and awareness raising across university spheres. When results are disseminated, policies of the journals come into play and export control regimes will oversee dissemination of both material and information. Finally, during translation and product development, intellectual property, patent restrictions, licenses and material transfer agreements will be enforced. These examples demonstrate the involvement of multiple sectors and players beyond government regulators, who contribute to the layered system of approaches and points of intervention.

19.3.6 Toward an International Model: International Collaboration in Science and Technology (ICST)

The United States was the first to create a national advisory board to provide advice, guidance and recommendations to the US government (USG) on the biosecurity implications of the life sciences. Created in the wake of the anthrax attacks via the postal service in 2001, the National Scientific Advisory Board for Biosecurity (NSABB)[15] has been tasked to define dual research and address its oversight. Early guidance in the United States for oversight limited its focus to 15 specific organisms and reviewed seven specific kinds of experiments (see tables), inspired by the Fink Report. The board has recently (2023) released draft USG potential pandemic pathogen care and oversight (PC3O) and dual-use research of concern (DURC) policies. The report includes 13 recommendations which propose to increase transparency and clarify roles and responsibilities of institutions and investigators in evaluating risk, to promote an integrated approach to oversight of dual-use oversight. In contrast, the European Union's approach to dual-use oversight is more focused on export controls.[16] The Canadian DURC oversight programme has a well-integrated system of guidelines employing risk-assessment, expert committees, mitigation strategies and education to prevent the misuse of scientific knowledge and technologies.[17] The nongovernmental 'Pathogens project' is an effort to map Biosafety level-4 laboratories around the world.[18] Of the 27 countries tracked, 18 were deemed to have dual-use governance frameworks (legislation, oversight, awareness-raising, whistleblower protection or self-governance measures); of these, several had only whistleblower-protection and self-governance measures.

Advances in the life sciences are characterised by international efforts: international collaborations in science and technology (ICST) are rising rapidly. The incentives for ICST include increased visibility, sharing of costs, physical resources and complementary capabilities and exchange of ideas. International collaborations will require alignment of oversight mechanisms. Sovereign nations that implement legally binding regulations of life science research must be aware of the impact of their requirements on international collaborative structures. A vivid example is found in the critical response of a group of virologists to the recent NSABB recommendations to the USG: 'we provide these responses out of concern for the potential consequences of excessive oversight: policy that unduly restricts microbiology regardless of associated risk will limit our preparedness for and ability to respond to public health threats, constrain efforts to control infectious diseases of all kinds, diminish our international engagement in these critical global issues and damage US economic competitiveness'.[19]

19.3.7 International Standards for Biosafety and Biosecurity

Safety-critical sectors, like nuclear, railways and aerospace have lessons for the life sciences. Safety-critical systems require evidence-based safety cases for that system so that operations can be certified against technical standards. Controversy over the origin of the SARS-CoV-2 pandemic has led to calls for renewed interest in funding and support of applied biosafety programmes. Many industries – especially high-reliability organisations (HROs) such aerospace – have established risk analysis frameworks for safe operations, but in contrast to pharmaceutical development where there have long been stringent controls, the standards for laboratory research work are relatively recent.

There are several relevant International Organisation for Standardisation (ISO) standards related to biorisk management and biosafety, which provide guidance for organisations and laboratories working with biological materials. One such standard is ISO 35001:2019, which pertains to medical laboratories and addresses the requirements for quality management and competence. ISO 22468:2021, titled *Biotechnology – Requirements for laboratories performing research, development, or analysis utilizing recombinant DNA technology*, provides guidelines for organisations working with genetically modified organisms (GMOs) and recombinant DNA technology. Additionally, ISO 21387:2019, *Microbiology of the food chain – Requirements and guidelines for conducting challenge tests of food and feed products*, focuses on the assessment and management of risks associated with microbiological hazards in the food industry. ISO 35001:2019 *Biorisk management standard for laboratories (ISO35001)* could be adopted internationally with a third party responsible for certification and validation. Similarly, WHO's 194 member states should be urged to come to consensus on a minimal biosafety standard.

19.4 Alternative Governance Structures

As discussed above, S&T do not exist in vacuum, separated from politics and society. Recognition of the integration of science throughout society and the economy has led to alternative governance structures in several other adjacent scientific fields such as environmental science (i.e. climate change), public health, urban planning and ecological diversity.

19.4.1 Hybrid Governance Models

Hybrid governance models combine elements of both hierarchical (top-down) and network-based (collaborative) approaches to governance, combining the two approaches and operating at multiple levels. They recognise the need for centralised authority and coordination while also valuing the benefits of collaboration, participation and decentralised decision-making. They are often adaptive and context-specific, tailoring governance approaches to the specific needs and complexities of the issue at hand. They draw on various governance tools and mechanisms based on the unique characteristics of the problem or policy area. They encourage the inclusion of multiple perspectives, knowledge systems and expertise to enhance decision quality and legitimacy. These models also address the need for accountability and transparency by establishing mechanisms for monitoring, evaluation and feedback.

The UN High-level Political Forum on Sustainable Development (and the Sustainable Development Goals for 2030) offers a good example of hybrid governance mechanisms: non-State actors participate in decision-making and reflect a move towards including self-regulation or deregulation in seeking local solutions to environmental problems. Public–private partnerships and governance networks (see below) comprise hybrid governance mechanisms, combining different rationales and action steering models.

19.4.2 Network-Based Governance

Networked-based governance refers to a mode of governance that involves collaboration, coordination and decision-making amongst diverse actors through interconnected networks. It is an alternative approach to traditional hierarchical or centralised forms of governance. In network-based governance, power and authority are distributed amongst various actors who come together voluntarily to address complex issues or achieve common goals. Governance networks consist of multiple actors, such as government agencies, non-governmental organisations (NGOs), civil society groups, private sector entities and academic institutions. These actors form interconnected relationships based on shared interests, expertise and resources. Decision-making in network-based governance is decentralised, as multiple actors contribute to the process. They can respond more effectively to emerging issues and incorporate diverse perspectives and expertise. Responsibility and accountability are shared amongst network participants. Each actor brings their own resources, knowledge and capacities to contribute to the collective effort. Network-based governance operates at multiple levels, from local to global, and often involves actors from different sectors, such as government, civil society and private sector. A specific example is cancer network governance. Although on a small scale, multidisciplinary collaborative decision-making and patient involvement in cancer networks have yielded highly valued and satisfying results. This kind of collaborative approach to research oversight, based on a set of clearly defined principles – would be adaptable to local and national governance programmes.

19.4.3 Transnational Governance

Transnational mixes recognise that, as opposed to 'global' governance, many interactions are State-to-State and are not universally applicable. Looking specifically at the environmental arena, transnational governance refers to the collective actions and mechanisms undertaken by various actors, including governments, international organisations, NGOs and other stakeholders, to address environmental challenges that transcend national boundaries. It involves cooperation, coordination and decision-making processes that extend beyond the authority of individual states. Countries come together to negotiate and establish international agreements and protocols to address specific environmental issues. Examples include the Paris Agreement on climate change, the Convention on Biological Diversity and the Montreal Protocol on ozone depletion. International organisations, such as the United Nations Environment Program (UNEP) and the World Bank, facilitate global environmental governance by providing platforms for coordination, policy development and capacity-building amongst nations. NGOs, civil society groups, businesses and academic institutions play significant roles in transnational environmental governance. They contribute to research, advocacy and implementation of environmental initiatives, exerting pressure on governments and influencing policies. Transnational governance efforts often involve financial mechanisms to support environmental initiatives, including funds, grants and investment mechanisms. For example, the Green Climate Fund provides financial assistance to developing countries to address climate change. Finally, effective transnational governance relies on transparent information sharing, data exchange and knowledge dissemination amongst stakeholders.

19.5 Conclusion

Life science technologies all have the potential to be dual-use technologies. Each has the potential to either make substantial improvements or, if misused, lead to harm to human, animal, plant and environmental health. Applied research into biosafety and biosecurity practices can be used to standardise methodologies across regions as appropriate and progress our ability to appropriately manage risks. Existing and future governance structures will need to be adapted to be relevant to changing environments, advances in technology and novel applications, and in some cases, where existing governance structures are inadequate, new structures and ways of limiting harm will be needed. In their 2020, paper on biosecurity governance, Evans and colleagues suggest that new ways of biology governance are emerging as quickly as biology is advancing, yet these models are not evaluated, compared or assessed and may have implicit assumptions and warrant deeper analysis.[20] Thus, it is likely that no single 'top-down' formula for governing research and ensuring responsible life science will be created. Rather, a layered, hybrid approach spanning self-governance, education as well as local, national and global oversight will continue to evolve in response to this moving target.

Author Biography

Dr. Nancy Connell is Professor Emerita at Rutgers New Jersey Medical School where she directed biosafety-level three laboratory research in bacterial pathogenesis and drug discovery from 1992 to 2018. She is past Professor at Johns Hopkins Center for Health Security (2018–2021) and most recently Senior Scientist at the US-National Academies. With a Ph.D. in microbial genetics from Harvard University, Connell studies the impact of biotechnology advances on biosecurity, biosafety and biodefense, focusing on their impact on the implementation of BWC. She has long been involved in the development of regulatory policies associated with biocontainment work and DURC.

Dr. Gigi Gronvall is an Associate Professor in the Department of Environmental Health and Engineering at the Johns Hopkins Bloomberg School of Public Health and Senior Scholar at the Johns Hopkins Center for Health Security. She is an immunologist by training and leads work to advance indoor air quality and rapid diagnostic tests. Dr. Gronvall's work also focuses on minimising the technical and social risks of the life sciences while advancing biosecurity, biosafety, ethics and the bioeconomy. She is the author of Synthetic Biology: Safety, Security and Promise (2016) and Preparing for Bioterrorism: The Alfred P. Sloan Foundation's Leadership in Biosecurity (2012).

References

1 Center for AI Safety. (2023) *Statement on AI Risk.* https://www.safe.ai/statement-on-ai-risk.

2 US Department of Energy. (2023) *Manhattan Project Background Information and Preservation Work.* https://www.energy.gov/lm/manhattan-project-background-information-and-preservation-work.

3 US Department of Health and Human Services, N.I.o.H. (2019). *NIH Guidelines for Research Involving Recombiannt or Synthetic Nucleic Acid Molecules.* https://osp.od.nih.gov/wp-content/uploads/NIH_Guidelines.pdf.

4 Kwik, G. *et al* (2003) Biosecurity: responsible stewardship of bioscience in an age of catastrophic terrorism. *Biosecurity and Bioterrorism: Biodefense Strategy, Practice, and Science*, 1 (1), 27–35.

***5** Bowman, K. *et al* (2020) Assessing the risks and benefits of advances in science and technology: exploring the potential of qualitative frameworks. *Health Secure*, 18 (3), 186–194.

6 Office for Disarmament Affairs. (2023) *Biological Weapons Convention*, United Nations, New York.

7 Pugwash Council. (2023) *Pugwash Conferences on Science and World Affairs*, Pugwash, Rome.

8 Agency, I. A. E. (2023) *International Atomic Energy Agency*, IAEA, Vienna.

***9** World Health Organization. (2022) *The Global Guidance for the Responsible Use of the Life Sciences: Mitigating Biorisks and Governing Dual-Use Research*, WHO, Geneva.

10 iGEM Foundation. (2023) *The International Genetic Engineering Machine Competition.* https://IGEM.ORG.

11 Sundaram, L. S. (2021) Biosafety in DIY-bio laboratories: from hype to policy: discussions about regulating DIY biology tend to ignore the extent of self-regulation and oversight of DIY laboratories. *EMBO Reports*, 22 (4), e52506.

12 Voeneky, S. *et al* (2018) *Human rights and legitimate governance of existential and global catastrophic risks*, Human Rights, Democracy, and Legitimacy in a World of Disorder, Cambridge University Press, Cambridge, 139–162.

13 Jasanoff, S. (2015) *Dreamscapes of Modernity: Sociotechnical Imaginaries and the Fabrication of Power*, University of Chicago Press, Chicago.

***14** National Academies of Sciences, Engineering, and Medicine. (2018) *Governance of Dual Use Research in the Life Sciences: Advancing Global Consensus on Research Oversight: Proceedings of a Workshop*, The National Academies Press, Washington.

15 US Departmenrt of Health and Human Services. (2023) *National Scientific Advisory Board on Biosecurity*, Washington, D.C, National Institutes of Health (Office of Science Policy).

16 Himmel, M. and EU Non-proliferation and Disarmament Consortium. (2019) *Emerging Dual-Use Technologies in the Life Scinces: Challenges and Policy Recommendations on Expeort Control*, Stockholm International Peace Research Institute, Solna.

***17** Government of Canada, P.H.A.o.C. (2018) *Canadian Biosafety Guideline – Dual-Use in Life Science Research*, Public Health Agency of Canada, Ottawa.

18 Bipartisan Commission on Biodefense. (2023) *Pathogens Project: Creating the Framework for Tomorrow's Pathogen Research*, Bulletin of the Atomic Scientists, Geneva.

19 Lowen, A. C. *et al* (2023) Oversight of pathogen research must be carefully calibrated and clearly defined. *mSphere*, 8 (2), e0006623.

***20** Evans, S. W. *et al* (2020) Embrace experimentation in biosecurity governance. *Science*, 368 (6487), 138–140.

20

Towards an International Biosecurity Education Network (IBSEN)

Kathryn Millett[1] *and Lijun Shang*[2]

[1] *Biosecure Ltd, Cheltenham, UK*
[2] *Biological Security Research Centre, School of Human Sciences, London Metropolitan University, London, UK*

Key Points

1) A key component of effective management of biological risks is a sustained, flexible and well-supported approach to education of life scientists.
2) Biological security education is adaptable to particular circumstances and communities worldwide.
3) Efforts to date in the biological arena have been fragmented and with varying success.
4) There is an urgent and crucial need for a holistic and centralised approach under one institutional umbrella, i.e. a proposed International Biological Security Education Network (IBSEN).
5) Comparable approaches informed by the efforts in analogous threat fields, such as the International Nuclear Security Education Network (INSEN) under the International Atomic Energy Agency (IAEA) and the Chemical Weapons Convention's (CWC) Advisory Board on Education and Outreach (ABEO) can provide key lessons for an IBSEN.

Summary

The chapter begins by arguing the importance of a globally relevant and continually evolving International Biosecurity Education Network (IBSEN) as a crucial component of improving biosecurity and previous recognition of the need for systematic and sustained education for life scientists. The chapter next discusses how diverse actors and stakeholders have increasingly created opportunities to educate life scientists on biosecurity issues but emphasises that such efforts have so far been fragmented, with initiatives varying widely in focus, format, content and scope, and that overall biosecurity and dual-use awareness levels remain low amongst life and associated scientists.

Essentials of Biological Security: A Global Perspective, First Edition. Edited by Lijun Shang, Weiwen Zhang, and Malcolm Dando.

The chapter then specifically looks at the structure, approach and lessons learned from the International Nuclear Security Education Network (INSEN) and work of the Advisory Board on Education and Outreach (ABEO) with a view to drawing out parallels and differences that could be applied to the establishment of a comparative network for biosecurity education. Lastly, the chapter calls for a biosecurity programme under an IBSEN and framework for biosecurity education best practice, including implementation of the Tianjin Code of Conduct.

20.1 Introduction

The challenges associated with keeping biology safe have recently come to occupy global public attention following the COVID-19 pandemic and the ongoing debates over its origins. However, for decades prior, academics and policymakers have long sounded the alarm that human error and the deliberate misuse of the life sciences and technologies constitute serious threats to human, animal and plant life, and that dual-use research can create biosafety and biosecurity risks comparable in magnitude to global pandemics[1]:

> 'Scientific and technological advances in the life sciences and converging technologies can raise significant ethical, legal, societal, safety and security risks ... The same scientific information and technologies that can generate potential benefits for health and society could also accidentally or deliberately be misused and potentially cause harm to humans, nonhuman animals, plants and agriculture, and the environment ...'

Educating scientists and practitioners in the life sciences and fostering responsible research practices and scientific integrity are amongst the most effective strategies to anticipate and prevent misuse of life science research – not least because they are on the frontlines of driving innovation and new knowledge. However, life scientists often do not consciously consider that their work could be misused. To better address this and prevent misuse, cultivating a strong and global culture of responsible science through the promotion of codes of conduct and sustained educational programmes is essential.

This chapter outlines the need for sustained education and awareness-raising of dual-use research issues, discusses previous efforts in educating scientists, draws lessons from educational frameworks employed in the nuclear and chemical fields and lays out the case for establishing an International Biological Security Education Network (IBSEN) to provide structure, strategy and sustainability in efforts to cultivate a robust culture of responsibility amongst life scientists.

20.2 The Need for Biosecurity Education, Awareness-Raising and a Culture of Responsibility in the Life Sciences

Biologists stand at the vanguard of advances in the life sciences and, as such, are key to ensuring that biology is not misused to cause harm. Formal biosecurity educational programmes teach audiences in the life sciences how to recognise dual-use and ethical issues that might arise in their work, discuss why dual-use issues constitute a real risk and build competencies and knowledge on how to take appropriate steps to mitigate biorisks.

Security and disarmament civil-academic society has long recognised the central role played by life scientists in preventing the misuse of biology and the promise of education and codes of conduct in creating a culture of responsibility. For example, in 2007, the US National Science Advisory Board for Biosecurity (NSABB) stated:

> '... one of the best ways to address concerns regarding dual-use research is to raise awareness of dual-use research issues and strengthen the culture of responsibility within the scientific community'.

Equally, national governments at an international level have come to appreciate the role that education and creating a culture of responsibility can play. Article IV of the Biological and Toxin Weapons Convention (BTWC) requires its States Parties to 'take any necessary measures to prohibit and prevent the development, production, stockpiling, acquisition or retention of the agents, toxins, weapons, equipment and means of delivery '. This is further supported by the World Health Organisation (WHO) which has also emphasised the role education, training and awareness-raising amongst life scientists can play in managing biological risks in its *Global Guidance Framework* (2022) for responsible use of the life sciences (2022).

20.3 Past Efforts in Educating Life Scientists and Establishing a Culture of Responsibility

Since the early 2000s, a wide range of educational activities have been undertaken, and numerous codes of conduct have been developed by entities such as the American Society for Microbiology, the Royal Netherlands Academy of Arts and Sciences, the Organisation for Economic Cooperation and Development and the DIYBio community.

Education and awareness-raising efforts picked up pace in the mid-2000s pioneered by a series of biosecurity seminars conducted by Rappert and Dando under the University of Exeter and the University of Bradford, United Kingdom in which 90 seminars were initially held across 13 countries in 2005. Experts from the University of Bradford then developed the first online biosecurity course (Biosecurity Education Module Resource) in 2006 that provided background on biosecurity issues and the BTWC, a train-the-trainers course in 2011, and a national series of short biosecurity educational courses that were rolled out in the former Soviet States and the Middle East. This was followed in 2015 by the publication of two seminal biosecurity educational textbooks, *Preventing Biological Threats, What You can Do: A Guide to Biological Security Issues and How to Address Them* and an accompanying *Biosecurity Education Handbook*, designed for both undergraduate students and educators.

More recently, educational and outreach initiatives have become more innovative and diverse in both format and purpose (see Table 20.1) ranging from a series of podcasts, pop-up lectures and exhibits at museums and science festivals, free webinars, the publication of graphic novels and cartoons, development of paper-based escape games and mobile apps (see Figure 20.1).

Table 20.1 Selection of education and awareness-raising initiatives for life scientists.

Title	Year	Focus	Implementer	Activity type
Bioethics and responsible research	2012	Bachelor- and Master-level students	Bradford University, United Kingdom	Team-based learning seminars
Introduction into biosafety, biosecurity and dual-use concerns in biotechnology[a]	2014	Bachelor- and Master-level students	Bradford University, United Kingdom	Team-based learning seminars
Biosafety, biosecurity and dual-use issues in biotechnology	2014	Bachelor- and Master-level students and life science educators	Bradford University, United Kingdom	Team-based Learning seminars
Biohazardous threat agents and emerging infectious diseases	2019-ongoing	Masters-level students and professionals	Georgetown University, United States	Masters course and online certificate
IFBA global mentorship program	2019-ongoing	IFBA certified professionals	International Federation of Biosafety Associations, Canada	Mentorship programme
IFBA certification of Biorisk management and certification of biosecurity	2017-ongoing	Young professionals	International Federation of Biosafety Associations, Canada	Online professional certification
Stanford biosecurity	2017-ongoing	University students	Stanford University, United States	In-person seminars, student outreach campaigns, biosecurity certificate program/professional development certificate programme, internships
Next generation biosecurity: responding to 21st century Biorisks	2018-ongoing	Young scientists (but open to any learners)	Biosecure, University of Bath, United Kingdom	Online course

Global partnership initiated biosecurity academia for controlling health threats (GIBACHT)	2021-ongoing	Public health professionals in Africa, the Middle East, South Asia and Central Asia	Bernhard Nocht Institute for Tropical Medicine (BNITM), Robert Koch Institute (RKI), Swiss Tropical and Public Health Institute (Swiss TPH) and the African Field Epidemiology Network (AFENET) under the German Biosecurity Programme	Online course modules and year-long biosafety and biosecurity fellowship
iGEM Safety and security program responsible science, ethics and biosecurity	2014-ongoing	High school, university students graduate students, researchers and technicians in life science	iGEM Foundation	Online resources, workshops and Lecture series
Emerging leaders in biosecurity fellowship	2017-ongoing	Post-graduate students and early career professionals	Johns Hopkins Center for Health Security	Part-time 10 month fellowship, including workshops, symposium and networking events
Next generation for biosecurity competition	2017-ongoing	University students and early career professionals	Nuclear Threat Initiative (NTI), United States	Essay competition for winning publications and field visit to global health security conference
Biosecurity central	?	All	Talus Analytics and Georgetown University Center for Global Health Science and Security	Online resource library
Biosecurity in education	?	University students	Biosecurity Office, Netherlands	Guest lectures and seminars; e-learning modules
Engineering life: Synbio, bioethics and public policy	?	University students	Johns Hopkins University	Online course
Poisons and pestilence podcast	2022	All	Dr. Brett Edwards	Podcast series
BSc in biosafety and biosecurity	2023	Undergraduate students	Masinde Muliro University of Science and Technology MMUST	Undergraduate Bachelor of Science degree programme

[a] This series of seminars was part of a project by the EU CBRN Centres of Excellence Project 18 and included further series of seminars in different countries throughout 2014.

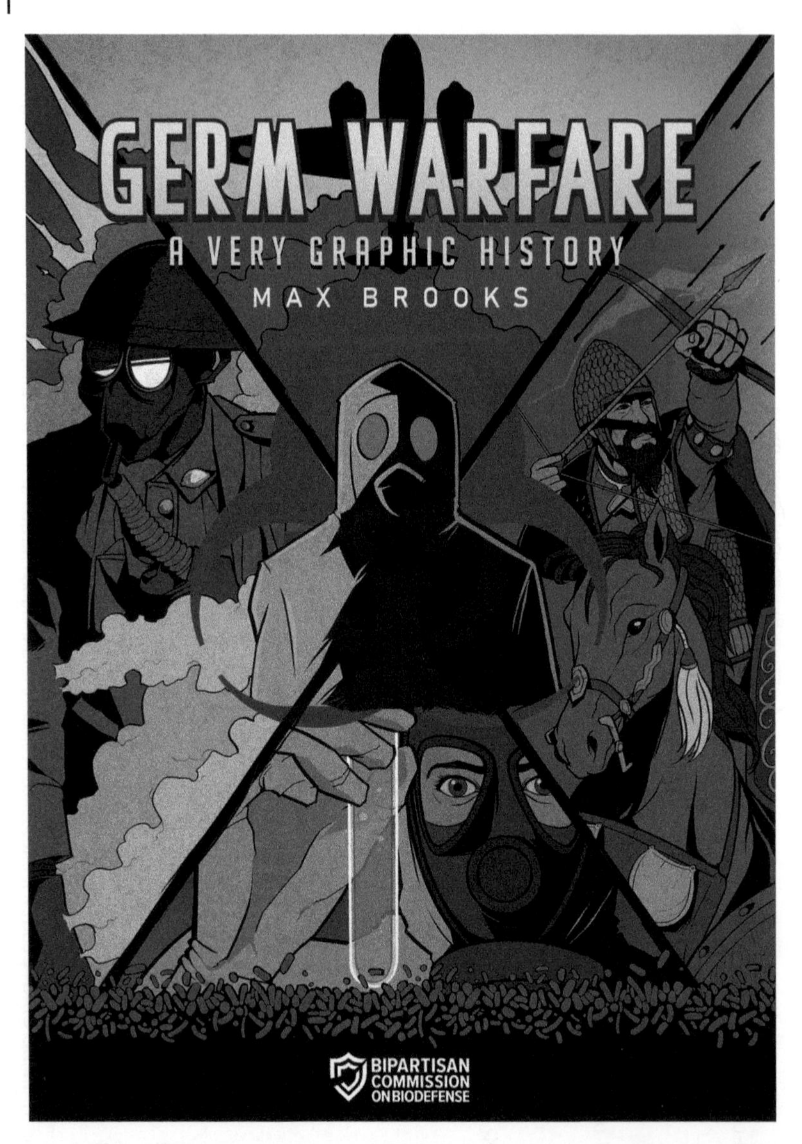

Figure 20.1 Example of biosecurity education graphic novel produced by the US Bipartisan Commission on Biodefense (2019). *Source:* Germ Warfare: A Very Graphic History by Max Brooks; 2017 © Bipartisan Commission on Biodefense.

More formally, leadership fellowship programmes have been established, such as the *Emerging Leaders in Biosecurity Fellowship* (ELBI) at the Johns Hopkins Center for Health Security and the *Fellowship for Ending Bioweapons* hosted by the Council on Strategic Risks, both of which provides instruction and networking for those in established biosecurity-related careers, while the *Youth for Biosecurity Initiative* hosted by the BTWC Implementation Support Unit provides an interactive training and awareness-raising programme for young scientists in the early stages of their careers. NTI's *Next Generation for Biosecurity Initiative* organises an annual essay competition on specific biosecurity issues that provides an opportunity for winners to publish their work and attend biosecurity and global health-related international meetings.

The International Federation of Biosafety Associations (IFBA) has developed a set of seven individual international professional certifications in areas of biorisk management, including a specific certification for biosecurity and cyberbiosecurity, predominantly aimed at early career professionals. The IFBA also runs a number of awareness-raising activities such as its *Biosafety Heroes* program, while its *Global Mentorship Program* is designed to encourage existing professionals to help others in the field and create new pathways for young scientists to forge careers in the biorisk management field. More recently, the IFBA has partnered with Masinde Muliro University of Science and Technology (MMUST) in Kenya in developing a pilot undergraduate university-level degree with a view to rolling out the degree course at further universities worldwide. In 2018, the first massive open online course (MOOC) on biosecurity and dual-use issues was published by the University of Bath and Biosecure Ltd. entitled *Next Generation Biosecurity: Responding to 21st Century Biorisks*. The course incorporates the teaching of the University of Bradford's *Preventing Biological Threats, What You Can Do: A Guide to Biological Security Issues and How to Address Them* as well as the IFBA biorisk management professional certification to educate learners to achieve a baseline of knowledge that could be translated into a professional certification. The course is regularly updated with new modules, case studies and educational materials to take into account developments in the biosecurity field.

Another significant programme is that of the annual International Genetically Engineered Machine (iGEM) competition which incorporates biosecurity themes into the competition itself. The iGEM competition is aimed at high school and university students, where multidisciplinary – and often cross-border – teams compete to design, build and test synthetic biology projects geared towards real-world issues. Integral to the competition is the stipulation that all teams work safely and responsibly. To that end, iGEM has instituted a comprehensive and wide-ranging *Responsibility Program* under which is housed a robust *Safety and Security Program* to ensure all teams avoid harming themselves or others during their work. It also actively promotes responsible science both within and without its community beyond the competition itself through its *iGEM Community* programme that brings young synthetic biologists into contact with the meetings of the BTWC.

The above initiatives provide a small snapshot of the biosecurity-related education and awareness-raising activities currently underway. Myriad more training and education is also undertaken by States, professional associations, international and regional organisations and others. The US Departments of Health and Human Services and Agriculture co-chair the International Working Group (IWG) on strengthening the culture of biosafety, biosecurity and responsible conduct in the life sciences which serves as a forum 'for collaboration and community of practice ... to develop guiding principles and educational/training resources to support and promote a culture of biosafety, biosecurity, ethical and responsible conduct in the life sciences' produces an annual *Guide to Training and Information Resources on the Culture of Biosafety, Biosecurity and Responsible Conduct in the Life Sciences*. This Guide lists 22 courses, credentialing and repositories of training and educational resources offered or produced by professional associations, governments, international organisations and non-governmental organisations (NGOs).

20.4 Challenges Faced by Biosecurity Education and Awareness-Raising

Despite the burgeoning number of educational, outreach and awareness-raising initiatives that currently exist, there remain a number of fundamental challenges that hamper their overall and lasting effectiveness and undermine efforts to promote a global culture of responsible conduct.

Firstly, there is a *continuing lack of awareness* amongst life scientists about the possibility of dual-use aspects emerging from their research and the potential for misuse of biology. Similarly to conclusions reported following a series of dual use and biosecurity seminars in 2005, a survey of current biosecurity education projects in 2022 also concluded that '... there has *not* been a significant improvement in the knowledge of the problems of dual use and biological security in general amongst the life science community' and that 'it has proven very difficult to shift the culture of the life science community'. Another survey on biosecurity education programmes that same year used iGEM as a case study and further concluded that overall, the 'educational machinery [has] so far failed to integrate teaching about dual-use research issues' with only 41% of respondents understanding the meaning of the term 'dual-use', and over half reporting that they are not taught dual-use issues at their university.[2] The WHO reiterated this challenge in its 2022 *Global Guidance framework*, stating:

> 'A chronic and fundamental challenge in biorisk management is a widespread lack of awareness that work in the area of the life sciences could be conducted or misused in ways that result in health and security risks to the public. The lack of awareness is unsurprising, given that biorisks are often overlooked or underemphasized in both educational curricula and on-the-job training. If they are unaware of the potential for misuse and potential malicious application, stakeholders cannot accurately weigh the risks and benefits of proposed research ...'

When considering the promulgation of codes of conduct in particular, the simple existence of creating codes of conduct cannot translate into a broader culture of responsibility without effective measures to educate target audiences about their existence and importance. A number of studies have identified continuing issues faced by educators in incorporating biosecurity and dual-use issues into curricula, including: an absence of space in existing curricula; the absence of time and resources available to develop new curricula; an absence of expertise and available literature on biosecurity education; and general doubt and scepticism about the need for biosecurity education from some educators and scientists. This lack of dual-use education has a knock-on effect in devaluing biosecurity and dual-use considerations as a fundamental component of life science education. Nevertheless, over half the students from the iGEM survey were concerned about the misuse of biology and were keen to learn more. Another challenge to creating a holistic culture of responsibility is that education and awareness-raising efforts so far have been *fragmented and can vary in quality and content.* Although Table 20.1 and Figure 20.1 demonstrate that there are many educational and outreach initiatives underway and that many of them are highly innovative and seek to make use of varied means of teaching core principles, it is difficult to know to what extent these materials and messaging are consistent. In addition, there are scattered localised, small-scale and short-term educational activities being undertaken worldwide for which it is difficult to assess the quality and impact of these efforts – or even be aware of them – as there is no central body that collates and shares this information, let alone to reflect lessons learned to be shared. Added to this, is that 'biosecurity' as a term means different things in different languages and different contexts. With no central guidance on what makes impactful biosecurity education, it is likely that each initiative has varying success and impact and teaches at different standards.

This leads us to a further common problem: *low government priorities and sustainable funding*. The fragmentation of efforts and their short-termism is a consequence of the relative low priority with which most national governments assign promoting a culture of responsibility. Without broader recognition and government buy-in, it is difficult to secure the sustained funds needed to ensure that continuous multi-generational efforts are implemented. Furthermore, while a number of countries have committed to improving biorisk management both domestically and internationally, there is a notable disparity between focus on biosafety competency-building and education and awareness-raising on dual-use issues. For example, biological security projects and programmes under the *Global Partnership Against the Spread of Weapons and Materials of Mass Destruction (GP)* predominantly prioritise biosafety training within professional laboratory settings versus broader education and awareness-raising activities relating to dual-use issues and the responsible use of the life sciences at the student and young professional level. In the period 2017–2022, spending on biological security-related activities totalled 311 distinct projects by 20 GP partners valued at over $1.6 billion (USD). Of these, the author identified over 80 projects that provided laboratory-level biosafety and biosecurity training compared with 30 projects that included elements of education and awareness-raising on dual-use issues and building a culture of responsibility in the life sciences. Of the latter, only seven specifically declared a main focus on educating students or young professionals totalling less than $4million USD combined. Ultimately, this all translates to the fact that we are simply not doing enough, and we cannot do more without sustained support and a more strategic approach that crosses the globe and enjoys the backing of a wide range of stakeholders.

20.5 Comparable Approaches Implemented in Analogous Frameworks in the Nuclear and Chemistry Fields

Building a culture of responsibility within the life sciences with education as its foundational base is a complex, but necessary undertaking, requiring sustained commitment and active participation across a broad array of stakeholders, including, but not limited to, national governments, scientific and technological communities, academia, research institutions, educators, professional scientific organisations and associations, funding bodies, industry and civil society. It is clear that developing appropriate and effective educational materials that can apply across many life sciences disciplines – and embedding them as much as possible within the multitude of educational institutions worldwide – is a daunting and unique task. However, as Novossiolova and Pearson noted in 2011, looking to analogous educational frameworks already successfully in place in similar fields is a useful starting point.[3]

In this regard, the work of the International Nuclear Security Education Network (INSEN) of the International Atomic Energy Agency (IAEA) and of the Advisory Board on Education and Outreach (ABEO) under the Organisation for the Prohibition of Chemical Weapons (OPCW) provide useful models and lessons for consideration in building a similar framework for biosecurity education.

20.5.1 IAEA Nuclear Security Culture and the International Nuclear Security Education Network (INSEN)

The IAEA has long recognised the importance of a robust security culture[4] in which education and training programmes play an integral part in creating an overall safety and security culture across the nuclear field:

> 'The fundamental principles of nuclear security include embedding a nuclear security culture throughout the organizations involved. By the coherent implementation of a nuclear security culture, staff remain vigilant of the need to maintain a high level of security'.

A nuclear security culture is defined as 'The assembly of characteristics, attitudes and behaviour of individuals, organizations and institutions which serves as a means to support and enhance nuclear security' and emphasises the fundamental role played by training and education in its establishment and sustainability:

> 'Such a regime can be strengthened through appropriate training and education at all levels, and in all organizations and facilities involved in nuclear security, by preparing the next generation of professionals with knowledge, expertise and understanding of the importance of nuclear security'.

Instilling the correct beliefs and attitudes throughout facilities lies at the core of a universal and sustained nuclear security culture – a key problem that still besets biosecurity efforts within the life sciences communities:

> 'Beliefs and attitudes that are formed in people's minds over time become casual factors in behaviour and affect how people respond to security issues and events .... For nuclear security, effectiveness depends upon the extent to which these beliefs and attitudes are commonly held and manifest themselves in appropriate behaviour and practices'.

This involves making sure that all relevant parties understand that both (i) a credible threat exists and (ii) that nuclear security is important:

> 'Where an effective nuclear security culture exists, [relevant personnel] ... hold a deep-rooted belief that there is a credible insider and outsider threat and that nuclear security is important.
>
> These beliefs form the foundation of nuclear security culture and are vitally important because they affect behaviour that ultimately influences the effectiveness

of nuclear security to achieve objectives relating, for example, to nuclear non-proliferation and counter-terrorism. Without a strong basis of beliefs and attitudes, an effective nuclear security culture will not exist. Nuclear security should be a concern of everyone working in the facility, related locations or organization – including to a certain extent the members of the public – and not of the organization's security specialists alone'.

Following its first articulation of the importance of a security culture in 2000, the IAEA instituted a continuously updated series of documents since 2006 entitled the *IAEA Nuclear Security Series* which provides 'international consensus guidance' on all aspects of nuclear security. Under this series, the Agency has published several documents related to developing educational courses to build nuclear security cultures. Having noted in 2008 that 'a systematic approach to training and qualification [is needed] for an effective nuclear security culture', the IAEA first published technical guidance in 2010 on a university-level model educational programme which has since been updated in 2021. The model curriculum was created in tandem with academic experts and aimed at master's degree level or an academic certificate programme for use by university curriculum developers and others in their educational institutions.

Strengthening education and awareness-raising as the prime factor in establishing a flourishing nuclear security culture was further reinforced in 2010 with the establishment of the INSEN. The INSEN is a partnership between the IAEA and universities, research institutions and other stakeholders 'to promote sustainable nuclear security education'. Established during an IAEA workshop on nuclear security education to discuss how better to assist States in this area, the network aims 'to enhance global nuclear security by developing, sharing and promoting excellence in nuclear security education' and in support of which, the network collaboratively develops and shares educational and professional development materials for students and faculty. To date, their activities have included, inter alia, the development of, and quality assurance on, peer-reviewed textbooks and computer-based teaching tools, the establishment of faculty and student exchange programmes, surveys on the effectiveness of nuclear security education and the development and implementation of specialised degree programmes and courses in addition to sponsorship of professional development courses.

In terms of structure, membership of the network is informal and open to any educational and research institution or competent authority interested in, involved in or planning, future nuclear security education and currently counts 198 members from 66 countries and international organisations.' The network currently comprises three working groups that meet annually and focus on:

- **Working Group I – Exchange of information and development of teaching materials for nuclear security education**: coordinate and assist in the creation and development of peer-reviewed textbooks and other instructional materials for nuclear security academic programmes in different languages, including updating the IAEA Nuclear Security Series 12 model curriculum as appropriate.
- **Working Group II – Faculty development and cooperation amongst universities**: enhance faculty development through activities including identification of educational institutes willing to host in-depth courses for teachers, assistance in developing and implementing tailored curricula and facilitating student and faculty exchanges.

- **Working Group III – Promotion of nuclear security education**: engage with nuclear security competent authorities and appropriate institutions to promote nuclear security education.

In addition to the working groups, Ad Hoc Groups are convened when necessary to address issues on nuclear security education that transcend the mandate of INSEN working groups. During annual meetings, members review the activities of the working groups, discuss and identify issues to be addressed and collectively task working groups to create action plans accordingly. The Chairperson of the meeting is elected for one year only.

The IAEA provides support for INSEN through convening the annual meetings, compiling subjects for discussion in consultation with INSEN members and reporting on the implementation of activities. The IAEA also hosts and maintains the INSEN online hub on its restricted *Nuclear Security Information Portal* (NUSEC) which provides INSEN with the infrastructure for promoting, managing, disseminating and preserving nuclear security; for communication and exchange of information; and for storing information and establishing databases with relevant materials. The IAEA also organises meetings for its working groups and provides a crucial role in promoting and disseminating the work of the network to its member states.

20.5.2 Chemical Weapons Convention and the Advisory Board on Education and Outreach (ABEO)

Almost since its inception, the Organisation for the Prohibition of Chemical Weapons (OPCW) has recognised and promoted the role of education and outreach (E&O) as essential for the future implementation of the Chemical Weapons Convention (CWC) and in preventing the re-emergence of chemical weapons. The OPCW's efforts began in 2001 with the launch of the *Ethics Project* that aimed to increase awareness of the OPCW and its objectives amongst relevant professions and in higher education. It has since expanded and evolved due to the establishment, firstly, of the temporary working group (TWG) on E&O under the Scientific Advisory Board (SAB) in 2011, and secondly, the ABEO in 2015.

In its Final Report in 2014, the SAB TWG emphasised that E&O are integral components of preventing the re-emergence of chemical weapons, and made two key recommendations:

- E&O with respect to the responsible use of science, particularly as it is relevant to the Convention, should remain a core activity of the OPCW, so as to achieve and maintain a world free of chemical weapons; and,
- An ongoing expert advisory group on E&O with respect to the responsible use of science, particularly as it is relevant to the Convention, should be established to help the OPCW fulfil its mandate for E&O, and to ensure that activities and practices are grounded in science education and communication research findings and effective practices.

Accordingly, in 2015, States Parties decided to establish the ABEO with a mandate to advise the Director-General or States Parties on matters of education, outreach and awareness-raising and public diplomacy concerning the CWC and its international and domestic implementation in relation to States Parties and key stakeholder communities. The Board comprises a gender and geographically balanced membership of 15 individuals drawn from CWC States Parties with appropriate expertise in subjects such as education, science communication, the chemical industry, dual/multiple-use issues related to chemistry and the life sciences and ethics and the CWC. Each member serves as an independent expert.

The Board operates at very low cost, meeting twice a year in person at the Hague with funding provided under the annual OPCW general budget for flights, hotels and per diems. Additional external funding for specific projects can also be sought. To facilitate communication between Board members between meetings, the OPCW hosts an electronic discussion platform which also allows for hosting subgroups in which ABEO Members, as well as observers, can develop ideas and discuss working papers on topics decided at preceding meetings.

In line with its strategic plan, the ABEO has focused on activities that meet its key E&O goals, namely:

a) Provide advice on E&O activities to the Director-General, and to States Parties and other stakeholders that is effective, sustainable, cost-effective and benefits from the latest advances in E&O theory or best practice.
b) Develop a portfolio of E&O activities and projects that benefits the broadest range of stakeholders.
c) Increase awareness of the work of the OPCW amongst key target audiences, particularly non-specialised audiences.
d) Improve the reach of the OPCW's activities, also through e-learning, both at the national and regional levels, including through the systematic translation of materials into the OPCW's official languages.

Amongst its first activities, the ABEO produced a number of recommendations on activities to be undertaken by the OPCW such as the publishing of a dedicated webpage for the CWC's 20th anniversary year along with a series of commemorative events, and youth outreach and engagement of civil society during sessions of the Conference of the States Parties. Working groups were established to (a) consider how to assist National Authorities with carrying out E&O activities; and (b) engage specific stakeholder communities such as scientific associations, industry and professional organisations; as well as ways to engage with other international organisations in promoting peace and disarmament education and youth outreach. Currently, the Board will also continue to focus on the following: e-learning, raising awareness of the OPCW's mission worldwide, providing assistance to National Authorities upon their request, and connecting with the chemical industry, academia and professional associations.

While mainly an advisory body and often called upon to provide advice on specific issues by the Director-General of the OPCW Technical Secretariat, such as developing a portfolio of specific E&O activities and projects that the organisation, States Parties and the ABEO and its individual members should pursue as a matter of priority, the ABEO has initiated and produced, or provided substantive input to shape a wide range of educational and outreach enterprises at the OPCW, including, inter alia, factsheets, educational videos, course modules and materials, OPCW display materials at its Visitor Centre and for events such as commemorating the centenary of the first large-scale use of chemical weapons at Ieper in Belgium.

20.5.3 Key Lessons from the INSEN and ABEO for an International Biosecurity Educational Network (IBSEN)

20.5.3.1 Comprehensive Understanding Learned from Related Initiatives

As pointed out by Perkins et al in 2019[5]:

> '... to improve the culture of biosafety, biosecurity, and responsible conduct, the life sciences will have to pay more attention to lessons learned in other fields and to adapt those tools and frameworks to the life sciences context'.

Experiences from the INSEN and ABEO demonstrate the utility of creating bodies that focus on the creation and rolling evolution of educational and awareness-raising materials, and especially the impact of international networks that collaborate and share materials in close collaboration with treaty regimes.

An examination of the activities and set up of INSEN and the ABEO provides a number of key lessons that should inform considerations on the establishment of an international biosecurity education network (IBSEN). It is worth noting that a number of these lessons overlap with the lessons from past biosecurity education and awareness-raising activities identified in the *WHO Global Guidance Framework* (2022).

20.5.3.2 A Biosecurity Education Network Must Be Underpinned by Firm and Sustained Commitment from States and the Future Network's Host Body (Such as the Biological and Toxin Weapons Convention)

A key component of the success of the INSEN is the IAEA and States' full appreciation that nuclear security is first and foremost a national responsibility and that human resource development and a robust nuclear security culture are key factors in preventing the misuse of nuclear and radiological materials and knowledge.

The IAEA has consistently incorporated support for nuclear security education within its strategic plans as one of its core activities.' Similarly, while slower to fully link education and awareness-raising efforts to the success of chemical security, the OPCW has recognised that 'public engagement, education and awareness-raising' must become 'an integral part of OPCW activities' and over time has developed a number of resources including e-learning modules (that are currently under revision with the input of the ABEO), the FIRES Documentary Video Project, and the Multiple Uses of Chemistry website of resources for students and educators.

20.5.3.3 Sustained Financial Support Observed and Maintained

Without sustainable financial support, E&O efforts cannot provide necessary up-to-date educational and training materials, hold valuable workshops and training events and achieve maximum reach and effectiveness.

The IAEA has integrated financial support for the INSEN into its general budget and provides in-kind support such as hosting and organising INSEN meetings. In 2023, the planned budget line for the subprogramme on 'education and training programmes for human resource development' which includes support for INSEN, as well as the funding for the development of materials, totalled €454,524.

The OPCW also financially supports the work of the ABEO, but its lack of direct financial support beyond funding in-person ABEO meetings has hindered the development of new materials and broader engagement with relevant stakeholders, such as industry and educational establishments. A report by the ABEO in 2021 stressed a significant weakness affecting the work of the Board was the lack of secure funding for activities which contributed to 'a sense of inability to perform serious long-term tasks that require human, financial and institutional expenditure' and a 'barrier to undertaking E&O activities'. Indeed, it was pointed out that certain costs related to ABEO work were borne by ABEO members themselves, which is somewhat shocking when taking into consideration that the OPCW's overall budget totalled €75,988,858 for 2023. Lacking a dedicated budget line for active projects, the ABEO has relied on external funding sources for some projects. For example, EU Council Decision 2019/538 is providing funding for 'a tailored programme for E&O on chemical safety and security management for youth/students in schools/universities in the context of the peaceful uses of chemistry' and support for ABEO advice on the design and execution of new e-learning modules.

20.5.3.4 Strategic Vision and Clear Pathways Are Required for Communication and Collaboration Between a Network and Treaty Bodies/States

The development and implementation of a comprehensive strategic plan to set clear goals and priorities, focus energy and resources and set responsibilities and pathways for communication between relevant stakeholders is essential to provide direction and put effective education, awareness-raising and outreach activities into practice.

The INSEN meets annually to set priorities, review, provide updates on activities and organise future work of its working groups, in light of the needs identified by network members and in support of the IAEA Nuclear Security Plan. In this way, INSEN is able to fluidly adjust and redirect its resources to where they are most needed and continue to ensure that all stakeholders are working towards a common goal.

The ABEO, however, as an advisory body, is more reactive in that it is able only to respond to ad hoc requests from the OPCW Director-General, Technical Secretariat and States Parties. This lack of a clear strategy and its detrimental effect has also been noted within the OPCW itself with regard to its education projects:

> '... the development of the Organisation's external e-learning offering had occurred in an ad hoc way, with relevant units required to independently identify needs and funding for new modules. This resulted in an external e-learning offering that lacked a certain overall coherence and was aimed at a limited range of external stakeholders. The fact that the Organisation had no internal expertise on e-learning compounded this issue'.

In addition, without clear, regular lines of communication, defined responsibilities and mechanisms for engagement between all the relevant stakeholders within networks and the treaty regime it supports, any initiatives undertaken risk being hampered.

20.5.3.5 Diversity of Memberships and Engagement with a Wide Range of Stakeholders

Part of the INSEN's success has been due to its extensive and broad membership. As an open network, the network has expanded year on year to its current membership of almost 200 institutions across all geographical regions, with more than 80% providing nuclear security education at their home institutions. The 2022 INSEN annual meeting featured presentations from over 20 INSEN members from different countries and its annual rotation of chairs and vice-chairs of the network and working groups ensures continued geographically diverse engagement. The appointment of regional group representatives helps highlight regional perspectives and priorities, while meetings of the regional groups provide opportunities for regional collaboration. The openness of the network to any relevant institution that implements or plans to implement nuclear security education ensures that the nuclear education field as a whole is kept apprised of efforts worldwide and encourages innovation and new approaches from across the world.

Equally important is the ability to engage with a wide variety of stakeholders from target audiences such as students, teaching faculty and National Authorities to liaison with States, media, research councils, relevant science and technology communities, industry, civil society, professional societies and associations and others. The flexibility and ability to liaise with stakeholders is especially important in relation to a biological security education network due to the breadth of disciplines within the life sciences and the dynamic pace of advances.

The INSEN meetings and ABEO members' presentations at OPCW meetings demonstrate that these meetings present an opportunity for educators to share knowledge and discuss best practices amongst each other as well as with international organisations and agencies. They discuss, compare, learn from each other and establish links and collaborations. They also serve to keep each other apprised of key issues in the nuclear and chemical security fields such as developments in science and technology that affect the security environments, and bring support to bear on important issues such as the promotion of gender and diversity in addition to enabling the exchange of information on curricula, teaching methodologies and identifying priority topics both internationally and regionally. Together, the IAEA and INSEN have spearheaded efforts to encourage gender parity through initiatives such as the *Women in Nuclear Security Initiative* (WINSI) and the *Marie Sklodowska Curie Fellowship Programme* since 2016.

20.6 Conclusion

Interest in the life sciences as a career prospect is booming following the COVID-19 pandemic and the continuing revolution in biotechnology. Investment in the life sciences continues to experience rapid growth and there has been a significant expansion in the number of new biological facilities under construction worldwide. A rapidly expanding global workforce in the biological sciences creates an imperative to ensure that those embarking on a career in the life sciences, wherever they are, are taught from the earliest stages how best to identify, prevent and mitigate issues relating to the potential misuse of the biological sciences.

Strengthening biosafety, biosecurity and the responsible conduct of the life sciences relies on cultivating and sustainably embedding a widespread culture of responsibility which is essential for ensuring that people follow safety and security procedures and that they act responsibly in new or unfamiliar scenarios. Appropriate education, training and the promulgation of codes of conduct are key to achieving this in the biological arena. However, so far educational and awareness-raising initiatives have been sporadic and fragmented, of variable quality and content, and the overall and lasting impact has been difficult to assess. As argued by Australia in 2011 'the frequent lack of awareness of aspects related to biosecurity and the obligation of the Convention amongst life scientists has to be addressed more urgently, strategically and comprehensively'.

Much more must be done to achieve consistency and cohesiveness in the quality and scope of biosecurity education and awareness-raising, ensure that efforts are sustained and sustainable, and develop, promote and embed codes of conduct (that ideally incorporate elements of the Tianjin Biosecurity Guidelines). The establishment of an IBSEN could play a significant role in this effort and could help overcome a number of the challenges identified by the WHO that have thus far beset efforts to educate biologists.

The decision at the 2022 Biological Weapons Convention Ninth Review Conference to establish a new 'Working Group on the strengthening of the Convention' with a mandate to address issues including 'Measures on national implementation of the Convention' presents a renewed opportunity to take decisive action to pioneer new biosecurity education and awareness-raising initiatives. Founding and sustainably funding an International Biosecurity Education Network – a concept already supported by a number of experts – would be a significant step forward in heightening biological security and ensuring that life scientists have the tools and knowledge to realise their obligations to prevent and mitigate the misuse of biology.[6] Last but not least, the continuous and creative civil society collective inputs would enhance the biological security education and eventually catch up with the rapid advancements in science and technology.

Author Biography

Ms. Kathryn Millett is the Director of Biosecure Ltd., a private company dedicated to safeguarding the bioeconomy and has been active in the field relating to CBRN risks for almost 25 years. Kathryn develops and conducts bespoke biosecurity training for students and professionals in the life sciences and is the author of the only free, open-access online course on biosecurity and managing biorisks called '*Next Generation Biosecurity: Responding to 21st century Biorisks*' which has attracted over 5000 learners. She has also created online and in-person teaching courses and resources for UN agencies, governments, private industry and the iGEM Foundation. Formerly the Director of the BioWeapons Prevention Project (BWPP), Kathryn was also Managing Editor of the *Mine Action Monitor*, *Cluster Munitions Monitor* and the *Bioweapons Monitor* and has run the safety and security screening for the International Genetic Engineered Machine Competitions since 2015.

Dr. Lijun Shang is the Professor of Biomedical Sciences at School of Human Sciences in London Metropolitan University (LMU). He is the founding director of Biological Security Research Centre at LMU. His research focuses mainly on ion channels in Health and Disease. Since 2015, he expanded his research interest into biochemical weapons and science convergence, as he wished to incorporate studies of the social impact of the advances in the life sciences. Since 2020, Professor Shang has been leading a series of projects in the effort to provide a civil society input into the broad BTWC.

References

*1 World Health Organization. (2022) *Global Guidance Framework for the Responsible Use of the Life Sciences: Mitigating Biorisks and Governing Dual-Use Research*, World Health Organization, Geneva. ISBN: 978-92-4-005610-7. (electronic version).

*2 Vinke, S. *et al* (2022) The dual-use education gap: awareness and education of life science researchers on nonpathogen-related dual-use research. *Health Security*, 20, 1.

3 Novossiolova, T. and Pearson, G. S. (2011) *Biosecurity Education for the Life Sciences: Nuclear Security Education Experience as a Model*. Bradford Briefing Paper No. 5, University of Bradford.

*4 International Atomic Energy Agency. (2008) *Nuclear Security Culture: Implementing Guide*, IAEA Nuclear Security Series No.7, International Atomic Energy Agency, Vienna. ISBN: 978-92-0-107808-7. (electronic version).

5 Perkins, D. *et al* (2019) The culture of biosafety, biosecurity, and responsible conduct in the life sciences: a comprehensive literature review. *Applied Biosafety*, 24, 1.

*6 Shang, L. *et al* (2022) Key issues in the implementation of the Tianjin biosecurity guidelines for codes of conduct for scientists: a survey of biosecurity education projects. *Biosafety and Health*, 4, 5.

Appendix A

The Tianjin Biosecurity Guidelines for Codes of Conduct for Scientists**

The **Tianjin Biosecurity Guidelines for Codes of Conduct for Scientists** focus on the prevention of intentional misuse of bioscience research, as per the articles and norms of the BWC, though the prevention of unintentional harm is equally important and closely intertwined. With the inclusion and implementation of elements from the Tianjin Biosecurity Guidelines for Codes of Conduct for Scientists, institutions, professional organizations, and all scientists can increase biosecurity and minimize risks of misuse and harm.

Advances in the biological sciences bring about well-being for humanity, but the same advances could be misused, particularly for the development and proliferation of biological weapons. To promote a culture of responsibility and guard against such misuse, all scientists, research institutions and governments are encouraged to incorporate elements from the **Tianjin Biosecurity Guidelines for Codes of Conduct for Scientists** in their national and institutional practices, protocols and regulations. The ultimate aim is to prevent misuse of bioscience research without hindering beneficial outcomes, in accordance with the articles and norms of the Biological and Toxin Weapons Convention (BWC) and in advancement of progress towards achieving the UN Sustainable Development Goals.

1) *Ethical Standards*
   Scientists* should respect human life and relevant social ethics. They have a special responsibility to use biosciences for peaceful purposes that benefit humankind, to promote a culture of responsible conduct in biosciences and to guard against the misuse of science for malicious purposes, including harm to the environment.
2) *Laws and Norms*
   Scientists should be aware of and observe applicable domestic laws and regulations, international legal instruments and norms relating to biological research, including those on the prohibition of biological weapons. Scientists and their professional bodies

*For purposes of this document, 'scientists' are practitioners engaged in work that includes biological science, including those involved in funding, education, and training; research and development (in the public and private sectors); project planning, management, dissemination, and oversight.
**From https://www.interacademies.org/sites/default/files/2021-07/Tianjin-Biosecurity-Guidelines-Codes-Conduct.pdf

Essentials of Biological Security: A Global Perspective, First Edition. Edited by Lijun Shang, Weiwen Zhang, and Malcolm Dando.

are encouraged to contribute to the establishment and further development and strengthening of relevant legislation.

3) *Responsible Conduct of Research*
   Scientists should promote scientific integrity and strive to prevent misconduct in research. They should be aware of the multiple applications of biological sciences, including their potential use for developing biological weapons. Measures should be taken to prevent the misuse and negative impacts of biological products, data, expertise or equipment.
4) *Respect for Research Participants*
   Scientists have a responsibility to protect the welfare of both human and non-human research participants and to apply the highest ethical standards in research conduct, with full respect for the subjects of research.
5) *Research Process Management*
   Scientists should identify and manage potential risks when they pursue the benefits of biological research and processes. They should consider potential biosecurity concerns at all stages of scientific research. Scientists and scientific institutions should put in place oversight mechanisms and operational rules to prevent, mitigate and respond to risks, and establish a culture of safety and security.
6) *Education and Training*
   Scientists, along with their professional associations in industry and academia, should work to maintain a well-educated, fully trained scientific community that is well versed in relevant laws, regulations, international obligations and norms. Education and training of staff at all levels should consider the input of experts from multiple fields, including social and human sciences, to provide a more robust understanding of the implications of biological research. Scientists should receive ethical training on a regular basis.
7) *Research Findings Dissemination*
   Scientists should be aware of potential biosecurity risks that might result from deliberate misuse of their research. Scientists and scientific journals should strike a balance when disseminating research findings between maximising benefits and minimising harm and communicate widely the beneficial aspects of research while minimising potential risks that could result from such publication.
8) *Public Engagement on Science and Technology*
   Scientists and scientific organisations should play an active role in encouraging public understanding and interest in biological science and technology, including its potential benefits and risks. They should communicate scientific facts and address concerns, uncertainties and misunderstandings to maintain public trust. Scientists should advocate for peaceful and ethical applications of the biosciences and work collectively to prevent the misuse of biological knowledge, tools and technologies.
9) *Role of Institutions*
   Scientific institutions, including research, funding and regulatory bodies, should be aware of the potential for misuse of bioscience research and ensure that expertise, equipment and facilities are not used for illegal, harmful or malicious purposes at any stage of bioscience work. They should establish appropriate mechanisms and processes to monitor, assess and mitigate potential vulnerabilities and risks in scientific activities and dissemination, and establish a training system for scientists.

10) *International Cooperation*
Scientists and scientific institutions are encouraged to cooperate internationally and to collaborate in the pursuit of peaceful innovations in and applications of the biosciences. They should promote learning and exchange opportunities to share best practices in biosecurity. They are encouraged to actively provide relevant expertise and assistance in response to potential biosecurity threats.

Index

Essentials of Biological Security: A Global Perspective, First Edition. Edited by Lijun Shang, Weiwen Zhang, and Malcolm Dando.

l

m

t